The Overburdened Planet

Paul Appleby

The Overburdened Planet

Making Sense of Sustainability

Springer

Paul Appleby
Norwich, UK

ISBN 978-3-032-01509-9 ISBN 978-3-032-01510-5 (eBook)
https://doi.org/10.1007/978-3-032-01510-5

This Springer imprint is published by the registered company Springer Nature Switzerland AG
The registered company address is: Gewerbestrasse 11, 6330 Cham, Switzerland

Preface

> *Climate change is the single greatest threat to a sustainable future but, at the same time, addressing the climate challenge presents a golden opportunity to promote prosperity, security and a brighter future for all.*
>
> *Ban Ki-Moon, Former Secretary-General of UN*

In the preface to their seminal book *Cents and Sustainability,* written in 2010, Michael Smith, Karlson Hargroves and Cheryl Desha wrote that 'the impact of today's civilization on its environment is growing to a scale that threatens to degrade it past the point at which it can continue to support life as we know it'. Well we're not there yet but 15 years on we are still heading in the same direction. There are still people in power today who dismiss climate change as just 'weather', or even worse 'a hoax'. In the background, there are organisations that are intent on exploiting natural resources to the limit and behind them are investors who only focus on the returns on investment, on dividend payments, on bonuses, on fuelling lifestyles augmented with multiple homes, expensive cars, yachts and cocktail parties with the rich and famous. Some of these people fund politicians and political parties, spend large amounts of money on lobbying, organising conferences, funding research that supports their narrative—the list goes on.

And yet every day it seems, in front of our very eyes, the news outlets report extreme weather events, temperature records being exceeded, species becoming extinct, glaciers receding, sea ice melting, oceans being degraded, communities being inundated, fires destroying forests and homes, droughts decimating lives and livelihoods. Again—the list goes on.

In her extraordinary book *This Changes Everything*, Naomi Klein assembled a charge sheet against the vested interests that have been responsible for pouring treacle into the pool through which action on climate change and planetary destruction is swimming. In her introduction, she asks the questions 'What is wrong with us? What is really preventing us from putting out the fire that is threating to burn down our collective houses?' In attempting to frame an answer to these questions, she refers to a 'conflict with deregulated capitalism' and 'an elite minority that has a stranglehold over our economy, our political process, and most of our major media outlets'.

The book you are (hopefully) about to read is not intended as a polemic. It is not going to rant and rage about the damage humans are doing to the planet, but it will describe the damage that is being done, some of the research behind this and what may happen in the long term, as projections are currently foreseeing. The politics around the environment and climate change can be pretty toxic and there have been numerous conflicts and frictions that have at their root cause disagreements over natural resources—and oil in particular.

And then there is the rate at which we are exploiting these natural resources. In their introduction to the aptly titled *The Burning Question*, Mike Berners-Lee and Duncan Clark claim that 'the money bound up in oil, coal and gas reserves has blocked political progress and clouded the analysis' and the fossil fuel sector is 'currently working hard not just to resist carbon cuts but to grow its reserves and markets as fast as possible'. This may have been written 12 years ago, but global oil production continues to rise year on year after a slight downturn during the COVID pandemic.[1] It could be said that this continual exploitation is akin to leaving the taps running whilst the bath is overflowing. Carbon Tracker is an independent financial think tank based in London which carries out authoritative research into the impact of the energy transition on capital markets and potential investment in the fossil fuel sector. In 2022, they produced an update to their seminal report *Unburnable Carbon* from 10 years earlier which provided a catalyst for the 'leave it in the ground' movement. Ten years on, the situation has worsened such that 'to limit warming to 1.5 °C 90% of fossil fuel reserves must remain in the ground as unburnable carbon'. These reserves are held by companies in a handful of financial centres and 'listed companies are exposed to significant (energy) transition risk'.[2] It is interesting to note that since this report was published, both Shell and BP have announced more limited ambitions for reducing the carbon intensity associated with their operations and have increased their focus on fossil fuel exploitation. Of course the implications of leaving 90% of reserves in the ground are enormous and create an existential threat to the industry, which is why the transition to renewables is so vital.

A parallel but interlinked slow motion disaster has also been unfolding to the natural world in the biosphere. A lot of this damage is linked to food production and forestry, aggravated by global warming and extreme weather. Inevitably, as with fossil fuel exploitation, money is at the heart of the problem. Large corporations, their supply chains and investors are frequently responsible for much of the damage. Large parts of the world have turned away from locally produced, sustainable food production to large-scale factory farming and production, with forests cleared for livestock rearing or crops and vast areas of monoculture created having little biodiversity and associated with the excessive use of chemicals for pest control and as fertilisers. This is happening not only on land but also in the oceans, which are suffering from overexploitation and unsustainable fishing techniques, such as bottom trawling.

Throw into this heady mix population growth, the transition in many countries from agrarian to industrial economies and associated demand for an increased

[1] https://yearbook.enerdata.net/crude-oil/world-production-statistics.html

[2] https://carbontracker.org/reports/unburnable-carbon-ten-years-on/

standard of living, consumer goods, meat-based and ultra-processed diets, etc., and the associated massive increases in infrastructure and energy required to support these, then we end up with a planet that cannot support us in the long term. The 'One Planet Living' movement is predicting that we're heading for a world where our cumulative impacts will be equivalent to two planets by 2030 and is already well above that in much of the developed world.

Of course it is easy when you look at all of the evidence set out in the following pages to say 'we're all doomed' and it is true that much of the damage that has already been done is irreversible. However, albeit very late in the day, there are signs of movement, particularly amongst millennials—people born since the year 2000—many of whom are showing signs of anger at what previous generations have been doing to the planet. Some countries, such as the UK, have established a 'net zero' narrative which, of course, politicians are arguing over, whilst legislation, such as the Building Regulations, is slowly catching up. In fact the UK is making pretty good progress in decarbonising the electricity grid, partly through wind and solar farms, battery storage and partly through new nuclear power stations, although the latter is very much a long-term plan.

The final chapter to this book explores the core issue of the relationship between the obsession with growth in gross domestic product (GDP) and damage to the planet. It is clear that in democracies with a short electoral cycle, there are few politicians who would risk committing to shrinking the economy. However, there are plenty of examples in the developed world of countries that have maintained economic growth whilst limiting the rise in greenhouse gas (GHG) emissions. For example, Sweden has reduced GHG emissions by 80% since 1990 whilst doubling GDP.

The future is in your hands, but it is important to understand the scale of the challenge. The aim of this book is to provide the reader with the context for this challenge and hopefully to make informed decisions on the way ahead.

Norwich, UK
June 2025

Paul Appleby

Acknowledgement

With thanks to John Pinnington for your invaluable comments on my early draughts and to Liz for your endless patience

Contents

1 **Introduction: Tales from a Planet Under Stress** 1
References 11

2 **Geopolitics and the Environment: Anthropogenesis Versus Nature** 13
2.1 The Debate 13
2.2 Agreements... 19
2.3 ...and Disagreements 26
2.3.1 Pipelines 29
2.3.2 Conflict Minerals 34
2.3.3 Conflict and Water 35
2.3.4 Environmental Impact 37
2.4 Extreme Weather 49
2.4.1 Storms and Flooding 52
2.4.2 Drought and Dust Storms 59
2.4.3 Inversions and Smog 61
2.4.4 Extreme Cold and Snow 62
2.5 The Politics of Food Production and Agriculture 64
2.6 The Politics of Water 75
2.7 The Politics of Waste 77
2.8 The Politics of Biodiversity and Ecosystem Resilience 86
References 90

3 **Climate Change and the Interconnectedness of All Things** 91
3.1 Introduction 91
3.2 Greenhouse Effect and Global Warming 92
3.2.1 Global Warming Projections 99
3.2.2 Temperature Change 101
3.3 Sea Level Rise, Snow and Ice Retreat 109
3.4 Ocean Acidification and Oxygen Content 119
3.5 Ocean Ecosystem Impacts 121
3.6 Permafrost Thaw 123

3.7 Freshwater Resources ... 125
3.8 Terrestrial Ecosystems and the Biosphere ... 126
3.9 Pollination and Pests ... 129
3.10 Impact on Animals ... 131
3.11 Extreme Weather ... 132
3.11.1 Social and Economic Impacts ... 136
3.11.2 Cyclones, Flood Risk and Extreme Rainfall ... 139
3.11.3 Heat Waves, Drought and Drylands ... 141
3.11.4 Fire Risk ... 145
3.11.5 Permafrost Thaw ... 146
3.11.6 Food and Agriculture ... 148
3.11.7 Fisheries and Aquaculture ... 149
3.11.8 Forests ... 152
3.11.9 Human Health ... 153
3.11.10 Surface Air Pollution ... 156
3.11.11 Poverty and Living Standards ... 158
3.11.12 Impacts on Cities, the Built Environment and Infrastructure ... 160
References ... 160

4 Greenhouse Gases: Emitters, Consumers and Sinks ... 163
4.1 Introduction ... 163
4.2 Energy, Fuel Reserves and Resources ... 165
4.2.1 Fossil Fuels ... 165
4.2.2 Nuclear Power ... 170
4.2.3 Renewables ... 171
4.3 Electricity Generation ... 172
4.4 Cogeneration ... 177
4.5 Community and District Energy ... 178
4.6 Carbon Capture and Storage ... 185
4.7 Buildings ... 189
4.8 Industry ... 191
4.9 Transport ... 191
4.10 Waste ... 192
4.11 Agriculture, Forests and Other Land Use (AFOLU) ... 193
4.12 Greenhouse Gas Sinks and Stores ... 196
References ... 196

5 Exploitation of Finite Resources: Reaching the Peak? ... 197
5.1 Introduction ... 197
5.2 Fossil Fuels ... 199
5.3 Minerals ... 203
5.4 Ancient Woodland and Peat ... 206
5.5 Water ... 209
References ... 212

6 Ecological and Carbon Footprinting: One Planet to Live On . . . 213
6.1 Introduction . . . 213
6.2 Ecological Footprinting . . . 214
6.3 Carbon Footprinting . . . 219
6.4 Water Footprinting . . . 221
6.5 Biodiversity Indicators . . . 225
References . . . 229

7 Buildings and Construction: New Build Versus Retrofit . . . 231
7.1 Introduction . . . 231
7.2 Energy and Greenhouse Gases . . . 234
7.2.1 Passive Design . . . 235
7.2.2 Energy Efficient Building Services . . . 236
7.2.3 Embodied Carbon and Other Environmental Impacts . . . 260
7.3 Noise and Air Quality . . . 264
7.4 Water Use and Drainage . . . 268
7.5 Waste Water Recycling and Rainwater Harvesting . . . 272
7.6 Landscaping, Biodiversity and Sustainable Drainage . . . 276
7.7 Socio-Economic Impacts: Sustainable Communities . . . 286
7.8 Sustainable Construction . . . 288
7.9 Retrofit Versus New Build . . . 291
7.10 Urban Regeneration Versus Green Field Development . . . 294
References . . . 296

8 Sustainable Transport: From Global to Local . . . 299
8.1 Introduction . . . 299
8.2 Demand: Commuting, Leisure, Business and Goods . . . 301
8.3 Greenhouse Gases . . . 304
8.4 Other Environmental Impacts . . . 305
8.4.1 Air Quality . . . 305
8.4.2 Noise . . . 308
8.4.3 Water Pollution and Ecological Impacts . . . 313
8.5 Alternatives to Fossil Fuel Engines . . . 317
8.5.1 Evolution of Battery Technology . . . 318
8.5.2 Charging Infrastructure . . . 321
8.5.3 Hydrogen and Fuel Cell Vehicles . . . 322
8.6 The Cycle Revolution . . . 324
References . . . 326

9 Products, Manufacture and Consumption . . . 329
9.1 Introduction . . . 329
9.2 Pollution from Industry and Farming . . . 330
9.2.1 Environmentally Persistent Waste . . . 330
9.2.2 Microplastics . . . 331

9.2.3 Other Industrial Pollutants ... 334
9.2.4 Non-energy Sources of Greenhouse Gases ... 336
9.3 Sustainability of Supply Chains ... 338
9.3.1 Food Miles and Carbon Footprint: Local Versus Imported ... 340
9.3.2 Forestry, Timber, Paper and Card Production ... 342
9.3.3 Fashion and Textiles ... 344
9.4 The Retail Sector: Physical Shops Versus Online Sales ... 346
9.5 Overproduction, Obsolescence and Waste ... 347
References ... 349

10 Conclusion: Towards a Different Future ... 351
10.1 Introduction ... 351
10.2 Green Growth or 'Degrowth'? ... 352
10.3 The Way Forward ... 360

Index ... 367

About the Author

Paul Appleby is a specialist in sustainability with a background in environmental engineering and low-carbon design of buildings. He is a Chartered Engineer with a First Class Honours Degree in Environmental Engineering and is an Honorary Fellow of CIBSE and a Fellow of the RSA. He advises design and masterplanning teams on the integrated sustainable design of buildings and communities. He has worked in the construction industry as a consultant, lecturer and researcher for some 55 years, including 25 years of international consultancy experience with URS Corporation (now AECOM), running practices specialising in building health and sustainability and working with some of the world's leading architects and developers on high-profile and award winning projects like Battersea Power Station, Scottish National Heritage HQ Inverness, Tate Modern extension, Greenwich Millennium Village and the Pinnacle tower Bishopsgate.

He has authored some 70 publications, including key guidance published by CIBSE and others. His book, Integrated Sustainable Design of Buildings, was listed among Cambridge University's 'Top 40 Sustainability Books of 2010'. Its follow-up, Sustainable Retrofit & Facilities Management, was published in January 2013. Both were published by EarthScan.

Paul has 10 years of design review experience, having participated in more than 100 panel meetings. He has been a Design Council Built Environment Expert since 2012, a panel member for Design South East since 2015, and was a member of the HS2 Independent Design Panel for 4 years. Paul has also been actively involved with the UK Green Building Council, participating in a number of Task Groups, including developing Retrofit Incentives and, in concert with the World GBC, exploring the relationships between productivity, health and well-being and sustainability in the office environment.

Chapter 1
Introduction: Tales from a Planet Under Stress

Human endeavour has unleashed unprecedented economic growth in recent decades in the pursuit of higher living standards. However, the sheer scale of economic and population growth has overwhelmed progress in curbing environmental degradation. Providing for a further 2 billion people by 2050 will challenge our ability to manage and restore the natural assets on which all life depends.

OECD Environmental Outlook to 2050—The consequences of inaction, 2012

Firstly, I would like to make a sincere apology for the subtitle of this book. It is quite obviously a cynical attempt to increase sales amongst dyslexic Jane Austen fans! On the other hand, I hope that the reader will forgive me the pun, since I am attempting to cut through the misinterpretations and confusion that frequently shroud the term 'sustainability'.

Sustainability will of course mean different things to different people, depending on the context. In this book, I will be referring to it in a number of contexts; for example: the ability of the planet to sustain life; the life expectancy of natural resources; creating 'development that meets the needs of the present without compromising the ability of future generations to meet their own needs'; as Brundtland stated so elegantly in her classic definition from 1987.[1]

In order to illustrate how this planet is currently under stress and is some distance from achieving Brundtland's definition of sustainable development it is worth referring to the example of China. Lester Brown [1], quoted in Smith et al. [2] *Cents and Sustainability* (p. 180), illustrates very succinctly China's problem:

[1]To download a copy of the Brundtland Report: Our Common Future visit: https://gat04-live-1517c8a4486c41609369c68f30c8-aa81074.divio-media.org/filer_public/6f/85/6f854236-56ab-4b42-810f-606d215c0499/cd_9127_extract_from_our_common_future_brundtland_report_1987_foreword_chpt_2.pdf (accessed 18 May 2025).

P. Appleby, *The Overburdened Planet*,
https://doi.org/10.1007/978-3-032-01510-5_1

> If, for example, each person in China consumes paper at the current American rate, then in 2030 China's 1.46 billion people will need twice as much paper as is produced worldwide today. If we assume that in 2030 there are three cars for every four people in China, as there now are in the United States, China will have 1.1 billion cars. The world currently has 860 million cars. To provide the needed roads, highways, and parking lots, China would have to pave an area comparable to what it now plants in rice. By 2030 China would need 98 million barrels of oil a day. The world is currently producing 85 million barrels a day and experts from the International Energy Agency have warned in 2007 that we may never produce much more than that.

As far back as 2005, key people in the Chinese leadership recognised the dangers in exponential economic growth. Interviewed in Spiegel Online, Pan Yue, then Deputy Minister of Environment, was remarkably frank in asserting that:

> This (Chinese economic) miracle will end soon because the environment can no longer keep pace. Acid rain is falling on one third of the Chinese territory, half of the water in our seven largest rivers is completely useless, while one fourth of our citizens do not have access to clean drinking water. One third of the urban population is breathing polluted air, and less than 20% of the trash in cities is treated and processed in an environmentally sustainable manner. Finally, five out of ten most polluted cities worldwide are in China.[2]

An example of the cumulative effects of China's growth strategy was starkly evident during the early part of 2013 when Beijing experienced air pollution at record levels, indicated by PM2.5 concentrations (respirable particles) reported as reaching 886 $\mu g/m^3$. This compares with the US National Ambient Air Quality Standard (NAAQS) of 35 $\mu g/m^3$, averaged over 24 h, and the WHO 1 year average standard of 25 $\mu g/m^3$. Beijing is not the only city in China experiencing this level of air pollution, which has been attributed not only to an increase in car ownership but dirty fuel, coal-fired power stations, industrial pollution and building sites.

According to the EC Communication *Rio + 20: towards the green economy and better governance* published in 2010, a green economy '… generates growth, creates jobs and eradicates poverty by investing in and preserving the natural capital offers upon which the long-term survival of our planet depends'.[3] Similarly the UK Government's 2011 Policy Paper *Enabling the Transition to a Green Economy* states that 'a green economy will maximise value and growth across the whole economy, whilst managing natural assets sustainably'.[4]

There is a fundamental dichotomy in these statements which has its birth in the link that Malthus made in 1798 between growth and catastrophe, later addressed by the Club of Rome in their 1972 report *The Limits to Growth* [3], also by BioRegional and the world wildlife fund for nature (WWF) in their One Planet Living concept[5] and even more recently in the UK's Sustainable Development Commission (SDC)

---

[2] Spiegel Online: interview 'The Chinese miracle will end soon'. 7 March 2005 can be found at: http://www.spiegel.de/international/spiegel/spiegel-interview-with-china-s-deputy-minister-of-the-environment-the-chinese-miracle-will-end-soon-a-345694.html.

[3] http://eur-lex.europa.eu/legal-content/EN/TXT/PDF/?uri=CELEX:52011DC0363&from=EN

[4] https://www.gov.uk/government/publications/enabling-the-transition-to-a-green-economy

[5] https://www.bioregional.com/one-planet-living

report *Prosperity Without Growth*.[6] This dichotomy concerns whether growth is compatible with sustainability. Of course before this question can be addressed, it is important to consider what it is that we do or do not want to grow. Malthus was particularly concerned about population growth outstripping agricultural production. To this the Club of Rome report added growth in industrialisation, resource consumption and pollution. However, Governments are primarily concerned with economic growth or per capita increases in Gross Domestic Product (GDP). This is generally considered to be an indicator of prosperity and standard of living.

However, it is the type of growth that matters. As Robert Kennedy warned way back in 1968:

> The Gross National Product[7] includes air pollution and advertising for cigarettes, and ambulances to clear our highways of carnage. It counts special locks for our doors, and jails for the people that break them. GNP includes the destruction of the redwoods and the death of Lake Superior. It grows with the production of napalm and missiles and nuclear warheads. And if GNP includes all this, there is much that it does not comprehend. It does not allow for the health of our families, the quality of their education, or the joy of their play. It is indifferent to the decency of our factories and the safety of our streets alike. It does not include the beauty of our poetry or the strength of our marriages, or the intelligence of our public debate or the integrity of our public officials … GNP measures neither our wit nor our courage, neither our wisdom nor our learning, neither our compassion nor our devotion to our country. It measures everything, in short, except that which makes life worthwhile; and it can tell us everything about America—except whether we are proud to be Americans.

The UK Sustainable Development Commission's (SDC) assertion that 'the material profligacy of consumer society is depleting natural resources and placing unsustainable burdens on the planet's ecosystems' raised similar alarms to that so eloquently sounded by Kennedy. Its report goes on to state that 'there is an urgent need to establish clear resource and environmental limits on economic activity and develop policies to achieve them'. Its argument was that this profligacy increases the depletion of non-renewable natural resources, pollution, energy demand, carbon emissions and associated climate change. Lord Stern in his 2007 review of *The Economics of Climate Change* takes a different perspective.[8] He considers that 'tackling climate change is the pro-growth strategy for the longer term, and it can be done in a way that does not cap the aspirations for growth of rich or poor countries'. However, this requires enormous investment, driven by Governments, in infrastructure to reduce both risk and impacts of climate change. This is what both President Obama and President Biden attempted to implement in the USA, against significant political opposition.

Similarly in 1997 Australian 'ecological economist' Steve Dodds [4] expressed the view that:

---

[6] http://www.sd-commission.org.uk/data/files/publications/prosperity_without_growth_report.pdf

[7] GNP is usually similar to GDP but excludes products and services that generate income outside countries borders.

[8] For an executive summary of the Stern Review refer to http://webarchive.nationalarchives.gov.uk/20130129110402/http://www.hm-treasury.gov.uk/d/Executive_Summary.pdf (accessed 18 December 2013).

> Discouraging economic growth would do little to encourage a more sustainable society. Indeed engineering a recession is likely to have perverse consequences, increasing unemployment, reducing the willingness of consumers and firms to adopt improved social and environmental practices and encouraging farmers and resource based industries to eat into natural capital to maintain cash flows and living standards.

As can be seen from the Chinese example '…the current balance in many economies around the world makes it cheaper to degrade nature than to care for it' [2, p. 221].

However, there does not have to be a conflict between the SDC and Stern/Dodds strategies. It is important to define the type of growth that is compatible with a sustainable future. Kennedy was arguing for a reorientation of the types of products and services that constitute GDP. The SDC and Stern both foresee an economy that is geared up to support sustainable development and climate change mitigation, what the UN Environment Programme (UNEP) calls the 'Green New Deal' in its 2009 report *Rethinking the Economic Recovery: A Global Green New Deal*—borrowing from the Roosevelt New Deal of the 1930s and not to be confused with the UK Coalition's now defunct Green Deal for energy efficient retrofit.[9]

This has led to the concept of 'green growth', with a number of initiatives focussing on the developing and emerging economies in particular. In 2009, the Organisation for Economic Co-operation and Development (OECD) countries signed up to a Declaration on Green Growth on the basis that:

> A number of well targeted policy instruments can be used to encourage green investment in order to simultaneously contribute to economic recovery in the short-term, and help to build the environmentally friendly infrastructure required for a green economy in the long-term, noting that public investment should be consistent with a long-term framework for generating sustainable growth.[10]

The Global Green Growth Institute (GGGI) evolved from an initiative launched in 2010 by South Korea, with a vision that focuses on 'a low-carbon, resilient world of strong, inclusive, and sustainable growth'.[11] It has evolved into an international organisation that has instigated a number of research projects to support its Green Growth Planning and Implementation programmes.

In the UK, it is the construction industry that has been setting the green growth agenda, primarily via the UK Green Building Council and Building Magazine in a coordinated 'Green for Growth' campaign aimed at Government and the Treasury in particular. The theme of this campaign has been investment in sustainable buildings and infrastructure under five headings, namely:

- Long-term incentives for the Green Deal and subsequent incentive schemes;
- Ring-fenced spending for public sector retrofit programmes;
- Commitment to zero-carbon homes and non-domestic buildings;
- A roll-out of Display Energy Certificates; and:
- Long-term investment in low-carbon energy infrastructure.

---

[9] https://www.cbd.int/development/doc/unep-global-green-new-deal.pdf

[10] https://legalinstruments.oecd.org/en/instruments/OECD-LEGAL-0374

[11] http://gggi.org/about-gggi/background/organizational-overview/

Smith et al. refer to a 'decoupling (of) economic growth from environmental pressures '. One of the most dramatic examples of this occurred in the USA following the oil shocks in the 1970s. Between 1979 and 1986, the US economy grew by 27% whilst oil consumption fell by 17%. Whilst over a much longer term a number of countries, including Sweden, Denmark and the UK, have achieved decoupling of GDP from CO_2 emissions (see Fig. 1.1).

The overall growth in GDP shown in these statistics disguises some local disasters of course. For instance, the industrial bases in both the USA and the UK have suffered considerably from a number of factors such as competition from the 'tiger' economies, incompetence at national and local levels and over-specialisation.

Detroit is frequently given as an example of a once thriving city gone wrong:

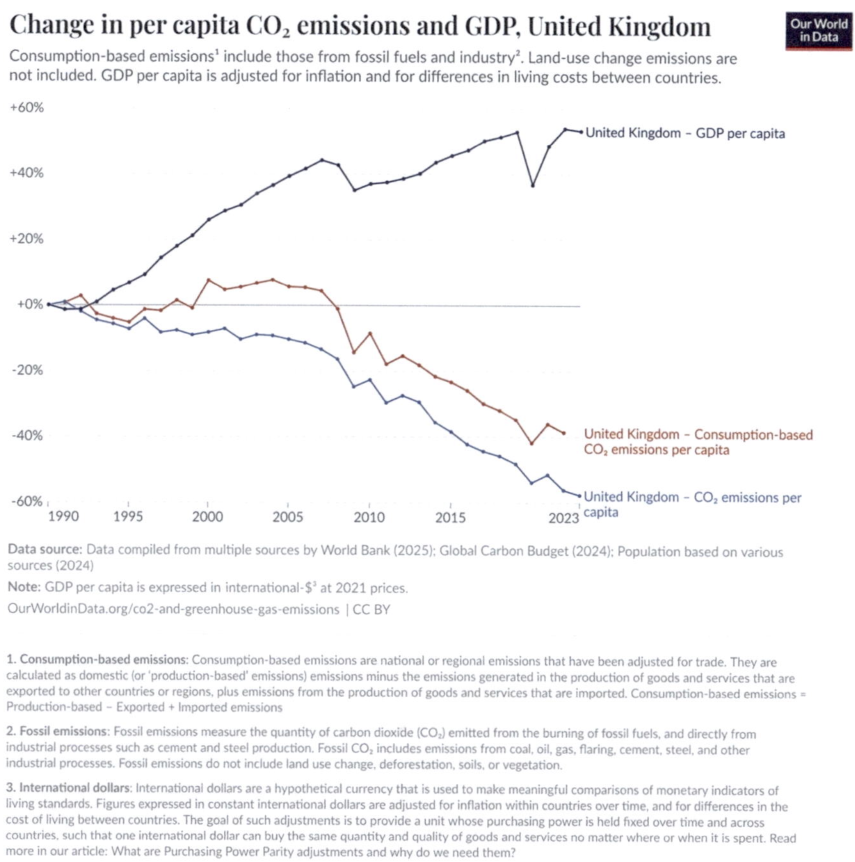

Fig. 1.1 Example of decoupling of CO_2 emissions from Gross Domestic Product (https://ourworldindata.org/co2-gdp-decoupling. From: Hannah Ritchie (2021)—'Many countries have decoupled economic growth from CO_2 emissions, even if we take offshored production into account' Published online at OurWorldinData.org)

> Detroit's population neared 300,000 at the turn of the century, just as Ransom Olds, David Buick, the Packard brothers and Henry Ford began their battle for automobile supremacy. By 1914, Detroit was making half the country's cars. By 1929 it was the fourth largest city in the country, with a thriving economy, a population of 1.6 million (nearly 160,000 working in the auto industry) and impressive new roads and skyscrapers.
>
> The 1973 oil embargo, the subsequent race to increase fuel efficiency and the rise of Japanese automakers like Honda proved the straw that broke the city's back. Within 2 years, U.S. auto production had fallen nearly 30 per cent, and Detroit was well on its way to becoming a byword for urban decay.[12]

By 2012, the population of Detroit had fallen to just over 700,000 and it was possible to buy a house for the price of a family car; unemployment and poverty levels were amongst the worst in the USA, whilst crime as indicated by per capita homicide rates was the highest in the USA and greater than Cape Town in South Africa and Cucuta in Columbia.

Where Detroit represents an example of the unintended consequences of a blinkered and monolithic approach to planning and economic growth, history is littered with examples of poor judgement and corporate greed having disastrous consequences. Rachel Carson in her landmark book from 1962—*Silent Spring*—provides an elegant quotation from Albert Schweitzer in her dedications, which laments that 'Man has lost the capacity to foresee and to forestall. He will end by destroying the earth'. *Silent Spring* is credited with providing a catalyst for an emergent environmental movement in the USA in the 1960s by providing a damning critique of the widespread application and devastating impacts of DDT as a pesticide:

> The most alarming of all man's assaults upon the environment is the contamination of air, earth, rivers and sea with dangerous and even lethal materials.

From Carson, R [5] *Silent Spring*.

Although the agricultural use of DDT has largely ceased, there are many other pollutants which are insidiously impacting our environment and our ecosystems. Take the humble plastic bag for example. It has been estimated that during the first decade of the twenty-first century up to an unimaginable one trillion plastic bags were being used every year.[13] In 2008, these contributed to some 3.5 million tonnes of non-recyclable plastic wrapping being discarded. Although the majority of plastic ends up in landfill, it has been estimated that between 1% and 3% of the plastic bags ever produced has been discarded as litter, much of which finds its way into our drains, our rivers and our oceans. Once it has escaped, a plastic bag poses a threat to both humans and wildlife. Many end up in the stomachs of living creatures: perhaps mistaken by a turtle for a jelly fish for example, causing a slow death through starvation. In 2002, a minke whale that washed up on a beach in Normandy was found to have 800 grams of plastic and other packaging in its stomach. According to the

[12] https://www.bloomberg.com/news/articles/2012-03-12/the-historical-roots-of-detroit-s-ruin

[13] https://www.businesswaste.co.uk/your-waste/plastic-recycling/plastic-waste-facts-and-statistics/#:~:text=One%20trillion%20plastic%20bags%20are,to%20break%20down%20in%20landfill

World Wildlife Fund tens of thousands of albatross chicks die every year by ingesting plastic from the waters around the remote Midway Atoll, part of the Hawaiian chain lying in the north Pacific, half way between North America and Japan (see Fig. 1.2). Plastic bags have periodically blocked drains in Bangladesh. This contributed to widespread flooding in 1998 resulting in plastic bags being banned throughout the country from 2002.

Once in the oceans, those bags and other plastics such as fishing nets and bottles that are not ingested whole by sea creatures are broken down by abrasion and photodegradation resulting in vast areas of fine floating particles becoming trapped in the oceanic vortices such as the North Pacific Gyre (see Fig. 1.3). These degraded particles, many of which are invisible to the naked eye, are ingested by the aquatic organisms that reside near the surface and that are known as neuston. Variously and colourfully known as the Eastern Garbage Patch or the Pacific Trash Vortex, this flotilla of waste also absorbs Persistent Organic Pollutants (POPs) such as PCBs and PAHs,[14] much of which find their way into the food chain. The Hawaiian chain of islands including the Midway Atoll, or Pehemanu, cuts right across this Gyre.

Fig. 1.2 A Midway Atoll albatross and the contents of its stomach. Source: Forest & Kim Starr (Creative commons licence) (From: https://commons.wikimedia.org/wiki/File:Starr-150331-0865-Cynodon_dactylon-dead_Laysan_Albatross_with_plastic-Harbor_Sand_Island-Midway_Atoll_(25153331792).jpg)

[14] PCBs are polychlorinated biphenyls which have historically been used in the manufacture of transformers and electric motors (although banned globally in 2001). PAHs are polycyclic aromatic hydrocarbons.

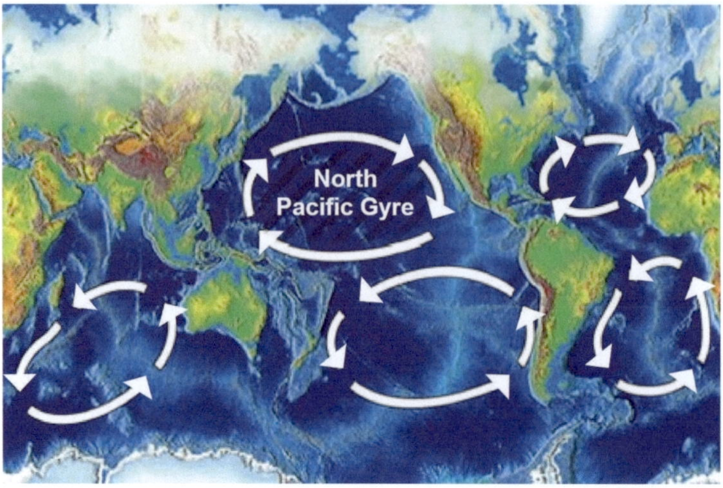

Fig. 1.3 The North Pacific Gyre (http://en.wikipedia.org/wiki/File:North_Pacific_Gyre_World_Map.png)

Of course the ubiquity of plastic packaging has closely followed the exploitation of oil as both a fuel source and a raw material for manufacture. As human populations have grown along with standards of living the demand for energy generation and natural resources has grown also. The exploitation of these natural resources and their impact once processed has irreversibly altered the planet, its atmosphere and the lives of its inhabitants.

Not every country has had access to sufficient cheap oil, gas or coal reserves to fuel its economic growth. Some have built nuclear power stations: France, for example, generates 75% of its electricity from nuclear power; whilst others have constructed large hydroelectric schemes. Norway, for example, meets around 99% of its electricity demand from hydro power.

So let us return to China, which, along with India, has been involved in a massive construction programme of coal-fired power stations. At the time of writing, there are approximately 2400 coal-fired power stations worldwide of which 1118 are in China and 173 in India. As of 2023, there are still dozens of new coal-fired power stations under construction or in the planning process, most of which are in China and India. According to the Centre for Research in Energy and Clean Air with two new starts per week, 'The coal power capacity starting construction in China (in 2022) was six times as large as that in all of the rest of the world combined'.[15]

Remarkably China also has the greatest hydroelectric capacity in the world. Where fossil fuel-firing generates around 80% of China's electricity, hydroelectric turbines provide 17% or so. Hundreds of dams have been constructed resulting in

[15] https://energyandcleanair.org/publication/china-permits-two-new-coal-power-plants-per-week-in-2022/

thousands of square miles of land being flooded with sometimes devastating social and environmental impacts. Between 1949 and 1999, some 17.5 million people were displaced by hydroelectric schemes, with many going uncompensated.

For example, the Three Gorges Dam scheme (Fig. 1.4), started in the 1990s and completed in 2012, created a 600 km long reservoir that flooded 14 cities, 140 towns, 1352 villages and approximately 100,000 acres of some of the most fertile land in China, which had previously produced 40% of China's grain and 70% of its rice. According to an account written in 2007:

> Of the 1.2 million relocated so far by the Three Gorges Dam, 500,000 are peasant farmers, many of whom earn just 1/3 of the World Bank poverty level of $1 per day. The resettlers were not offered substantial compensation to cover the cost of new housing. Most were moved into urban centres near Chongqing, or into the new cities built specifically for relocation, on steep unstable slopes that have now been determined unfit for development or farming.[16]

It has been estimated that in the long run some six million people will have to be relocated because of the Three Gorges Dam and associated unforeseen consequences, including major landslides, although the exact numbers are difficult to predict because of China's ongoing urbanisation project.

The impact on flora and fauna has been disastrous in some instances, with 50% of forest destroyed, 57% of plant species endangered and freshwater fish badly

Fig. 1.4 The Three Gorges Dam on the River Yangtze, China (https://commons.wikimedia.org/wiki/File:Three_Gorges_Dam_(2872397048).jpg)

[16] Link no longer available but similar article can be found at: https://tjmcdonaldthree.weebly.com/impacts-on.html

impacted by changes in water temperatures and flow regime, whilst many fish die attempting to negotiate the turbines. The project is also alleged to have accelerated the extinction of the Chinese river dolphin and drastically reduced numbers of the Yangtze sturgeon, which was declared extinct in the wild in 2022.

On the other hand if the Three Gorges Dam had not been commissioned some 50 million tonnes of coal would have to be burnt to generate the same amount of electricity each year, emitting around 100 million tons of CO_2, 2 million tons of sulphur dioxide and vast volumes of fine particulates. Prior to it being flooded deliberately, the Yangtze river valley was notorious for flooding, with an estimated 145,000 people having drowned in 1931 alone, and periodic floods in later years causing millions of dollars-worth of economic damage and numerous fatalities. It was claimed that the capacity of the reservoir would be sufficient to prevent future flooding downstream, although the Three Gorges Dam was breached in 2010.

And herein lie the horns of the dilemma on which national governments and decision-makers worldwide are caught: there is always a down side. There is no truly benign way to harness natural resources to generate energy to meet the needs of this planet's ever increasing and demanding population. We will see in Chaps. 3 and 4 that the most recent predictions of global warming indicate that in order to hold average global temperature rise to two degrees kelvin (2C) or lower only 20% of the reserves of coal, oil and gas held by the world's largest companies can be burnt, thus wiping billions of dollars off the value of these companies. For example, 'HSBC analysts have recently concluded that oil giants such as BP—beloved of UK pension funds—could have their value cut in half if the world decides to tackle climate change' (from Guardian article by Duncan Clark, 2013).[17] However, this same article reveals that 'pension fund managers and other investors are allowing listed fossil fuel companies to spend the best part of $1 trillion a year (comparable to the US defence budget, or more than $100 per person on the planet) to find and develop yet more reserves.' President Obama has said that although 'American emissions are falling due to rising auto efficiency standards and gas displacing dirtier coal in the energy mix … the US is extracting carbon and flowing it into the global energy system faster than ever before … (through exporting) more of the coal to countries such as China' which in turn exports goods to the US markets. In other words, once the fuel is in the market place a use will be found for it.[18]

The following chapters will provide a more detailed analysis of the issues raised in the examples described above. In the next chapter, an overview of the human impact on the planet compared with natural forces will be provided, including an historical perspective on the impacts of politics, conflicts and human endeavour on social, economic and environmental sustainability. This will lead into a discussion on 'climate change and the interconnectedness of all things' (Chap. 3) and

---

[17] Article published in online Guardian dated 17.04.13: *Why can't we quite fossil fuels* by Duncan Clark. At: https://www.theguardian.com/environment/2013/apr/17/why-cant-we-give-up-fossil-fuels

[18] https://blogs.worldbank.org/climatechange/sweden-decoupling-gdp-growth-co2-emissions-possible

'greenhouse gas emissions: sources, consumers and sinks' (Chap. 4) demonstrating links between human activities such as construction, manufacture, transport and the like, and operational energy use, embodied energy, water consumption, waste generation, pollution, carbon emissions, and the resulting global temperature rise, extreme weather events, and impacts on health, well-being and ecology.

Chapter 5 then goes on to examine the exploitation of finite resources by reviewing how humans have extracted non-renewable natural resources with consequential social, economic and environmental impacts, the implications as extraction becomes more difficult and costly and the potential for running out of coal, oil, gas, minerals, et al., including destruction of primary forests and the massive increase in the demand for rare earth elements required for mobile phones, computers and other consumer electronics.

Chapter 6 will demonstrate how human activities can be expressed in terms of an ecological footprint and used to determine how many 'planet-equivalents' humans require to maintain living standards and growth. It will also explore how carbon and water footprinting as well as other benchmarks can be used as a basis for understanding choices from the global to the personal.

Chapter 7 will take a closer look at buildings and construction and the cradle to grave environmental impact of buildings and structures, setting out the pros and cons of retrofitting compared with demolishing existing buildings and replacing with new. How this can be done sustainably will be explored, allowing for future population growth and demographic changes. Also the role of urban regeneration will be compared with that of garden cities, whilst assessing whether green belt development is always undesirable: all within the context of developing sustainable communities.

Chapter 8 will explore the different modes of transport and what drives demand, their greenhouse gas emissions and other environmental impacts, alternatives to fossil fuel engines and likely transitions and how to get people off the roads.

Chapter 9 will investigate the dichotomy between economic growth and environmental impact, and the sustainability of supply chains, physical shops versus online shopping and the potential death of the high street and the challenges of local food production compared with the international marketplace.

Finally, Chap. 10 will conclude with a review of the options for a different way forward. It will investigate whether decoupling of economic growth from environmental degradation can work as a long-term strategy and discuss the merits of 'green growth' compared with post-growth or degrowth strategies.

References

1. Brown, L. (2008). *Plan B 3.0: Mobilizing to save civilization* (p. 14). W.W. Norton.
2. Smith, M. H., Hargroves, K., & Desha, C. (2010). *Cents and sustainability*. Earthscan.
3. Meadows, D. H., Meadows, D. L., Randers, J., & Behrens, W. W. (1972). *The limits to growth*. Pan Books.

4. Dodds, S. (1997). Economic growth and human well-being. In C. Hamilton & M. Dilsendorf (Eds.), *Human ecology: Ideas for an ecologically sustainable future* (pp. 99–124). Allen & Unwin.
5. Carson, R. (1962). *Silent spring*. Houghton Mifflin.

Chapter 2
Geopolitics and the Environment: Anthropogenesis Versus Nature

> *The history of life on earth has been a history of interaction between living things and their surroundings. To a large extent, the physical form and habits of the earth's vegetation and its animal life have been moulded by the environment. Considering the whole span of earthly time, the opposite effect, in which life actually modifies its surroundings, has been relatively slight. Only within the moment of time represented by the present century has one species—man—acquired significant power to alter the nature of his world.*
>
> *Silent Spring—Rachel Carson [1]*

2.1 The Debate

In recent years, there has been an increasingly quasi-religious tone to the debate about the impacts that humans are having on the planet compared with those that result from natural changes. Although these have focussed primarily on climate change, there have been interrelated debates across a wide range of geopolitical, social, economic and environmental issues, ranging from population growth to the impacts of different food production strategies.

One of the conclusions drawn from the Club of Rome's seminal first report published in 1972 on its 'project on the predicament of mankind'—*The Limits to Growth*—was that:

> If the present growth trends in world population, industrialisation, food production, and resource depletion continue unchanged, the limits to growth on this planet will be reached sometime within the next hundred years. The most probable result will be a rather sudden and uncontrollable decline in both population and industrial capacity (Meadows et al., [2]).

In his 1988 book *The Collapse of Complex Societies* archaeologist Joseph Tainter [3] refers to the phenomenon he defines as 'declining marginal returns in complex societies'. Using examples from history such as the Roman Empire and Mayan civilisation, Tainter explains how historically, as resources have become more difficult to extract they have become more expensive to obtain until 'as the number and

P. Appleby, *The Overburdened Planet*,
https://doi.org/10.1007/978-3-032-01510-5_2

costliness of organisational investments increase, the proportion of a society's budget available for investment in future economic growth must decline'. In his concluding chapter, Tainter suggests that because nations no longer operate in an economic vacuum, they are unlikely to reach the point of collapse individually. However, he does not exclude the potential for global collapse, concluding that:

> Even if the point of diminishing returns to our present form of industrialism has not yet been reached, that point will inevitably arrive. Recent history seems to indicate that we have at least reached declining returns for our reliance on fossil fuels, and possibly for some raw materials. A new energy subsidy is necessary if a declining standard of living and a future global collapse are to be averted. A more abundant form of energy might not reverse the declining marginal return on investment in complexity, but it would make it more possible to finance that investment.

An attempt was made to evaluate the relationship between the decline, referred to by both Meadows and Tainter, and climate change in *The Stern Review* published in 2007. Examining the long-term impacts of climate change Lord Stern concluded in his Executive Summary that:

> The evidence shows that ignoring climate change will eventually damage economic growth. Our actions over the coming few decades could create risks of major disruption to economic and social activity, later in this century and in the next, on a scale similar to those associated with the great wars and the economic depression of the first half of the 20th century. And it will be difficult or impossible to reverse these changes.[1]

Stern predicts that the 'costs of extreme weather alone could reach 0.5–1% of world GDP per annum by the middle of the century, and will keep rising if the world continues to warm'. However, he also estimates the cumulative impact of climate change under various scenarios, concluding that 'with 5–6°C warming—which is a real possibility for the next century—existing models that include the risk of abrupt and large-scale climate change estimate an average 5–10% loss in global GDP, with poor countries suffering costs in excess of 10% of GDP'.

According to the Intergovernmental Panel on Climate Change (IPCC) Fifth Assessment Report by its Working Group II published on 31st March 2014:

> By 2100, due to climate change and development patterns and without adaptation, hundreds of millions of people will be affected by coastal flooding and displaced due to land loss (high confidence). The majority affected will be in East, Southeast, and South Asia. The relative costs of adaptation vary strongly among and within regions and countries for the 21st century (high confidence). Some low-lying developing countries and small island states are expected to face very high impacts and associated annual damage and adaptation costs of several percentage points of GDP[2].

The question of where these people will settle is not clear, although it is likely that many will relocate within national borders, whilst the pressure on population density and agriculture is likely to be catastrophic in some instances.

---

[1] For an executive summary of the Stern Review refer to http://webarchive.nationalarchives.gov.uk/20130129110402/http://www.hm-treasury.gov.uk/d/Executive_Summary.pdf (accessed 18th December 2013).

[2] https://www.ipcc.ch/site/assets/uploads/2018/02/WGIIAR5-Chap5_FINAL.pdf.

We will be examining the world of climate change impact and prediction in more detail in the next chapter, whilst the potential for mitigation and adaptation will be discussed in Chap. 10.

In her preface to the landmark report for the UN World Commission on Environment and Development: *Our Common Future* Dr. Gro Brundtland expresses a similar view to the Club of Rome that 'Many critical survival issues are related to uneven development, poverty and population growth. They all place unprecedented pressures on the planet's lands, waters, forests and other natural resources, not least in the developing countries'.

In the early decades of the twenty-first century, very little progress has been made in addressing many of these issues. Advances in mitigation have tended to be offset by increases in population, industrialisation, urbanisation, living standards or some combination of these. In 2010 progress in securing Brundtland's legacy was reviewed in a comprehensive publication marking the 20th anniversary of *Our Common Future* [4].[3] The key issues that the authors conclude need to be addressed in the twenty-first century include, inevitably, climate change mitigation, as well as population growth, food production, deforestation, habitat loss and depletion of non-renewable resources.

In their introduction Smith et al. ask the stark question: 'are we destroying the world we are creating?' and state unequivocally that 'if climate change is not mitigated such changes (increased intensity of cyclones, hailstorms, flooding and bushfires) risk being complemented by sea-level rises, shifting rainfall patterns, reduced water availability and agricultural production, the further spread of vector-borne diseases, and loss of species, to name a few of the negative changes forecast this century'.

To this list can be added 'increased food production (which has) often relied on over-extraction of groundwater and freshwater that cannot be sustained'. Also the pollution impacts associated with pesticides and fertiliser which, although significantly diminished from the 1960s when they were so eloquently highlighted by Rachel Carson in *Silent Spring*, remain of concern—particularly in the developing world. For example, the World Health Organization (WHO) reports that, although there is a dearth of reliable data, it is likely that millions of people are affected by pesticide poisoning annually, including thousands of fatalities.[4]

Smith et al. also highlight the plight of forests in that 'the scale of deforestation, loss of habitat and species loss is reducing the earth's ecosystems' resilience and capacity to assimilate pollution and other environmental pressures—a service that we depend on'. Forests across the planet have shrunk in size over the centuries, normally correlating with population growth and associated urbanisation and

---

[3] To download a copy of Our Common Future visit https://gat04-live-1517c8a4486c41609369c68f30c8-aa81074.divio-media.org/filer_public/6f/85/6f854236-56ab-4b42-810f-606d215c0499/cd_9127_extract_from_our_common_future_brundtland_report_1987_foreword_chpt_2.pdf (accessed 18 May 2025).

[4] https://www.pan-uk.org/dramatic-rise-in-global-pesticide-poisonings-revealed/

industrialisation. The rate of depletion has varied with both time and geographical location: with the greatest tree loss for the developing world occurring more recently.

According to a Fact Sheet produced by the UN Framework Convention on Climate Change (UNFCCC):

> Deforestation, mainly conversion of forests for agriculture activities, has been estimated at an alarming rate of 13 million hectares per year (in the period 1990–2005). Deforestation results in immediate release of the carbon stored in trees as CO_2 emissions. It is estimated that deforestation contributed approximately 5.8 $GtCO_2$/yr to global greenhouse gas emissions in the 1990s. According to the IPCC in its Fourth Assessment Report, reducing and/or preventing deforestation is the mitigation option with the largest and most immediate carbon stock impact in the short term.[5]

The loss of ancient woodland and slow-growing tropical forests can be compared with the irreversible extraction of non-renewable natural resources, such as coal, oil and minerals. In many instances, these forests and associated habitats have been replaced with crops or urban sprawl, typically changing forever the landscape, wiping out ecosystems and altering water courses.

At the UN Climate Summit held in New York in September 2014, the New York Declaration on Forests was announced. This was a collective commitment by a diverse array of actors, including national and subnational governments, companies, indigenous groups and NGOs, to halve the rate of natural forest loss globally by 2020 and eliminate it completely by 2030. According to the original Declaration 'the conversion of forests for the production of commodities—such as soy, palm oil, beef and paper—accounts for roughly half of global deforestation. Infrastructure, urban expansion, energy, mining and fuel wood collection also contribute in varying degrees'. In its most recent form 'its ten goals include halting natural forest loss by 2030, restoring 350 million hectares of degraded landscapes and forestlands, improving governance, increasing forest finance, and reducing emissions from deforestation and forest degradation as part of a post-2020 global climate agreement'.[6]

If all of the original commitments had been met by the signatories, the UN estimated that the equivalent of between 4.5 and 8 Gt of CO_2 emissions per year could have been removed by 2030. However, as of early 2023, the 2020 deadline had been missed and '201 (40%) of the companies and financial institutions with the most exposure to, and influence on tropical deforestation still haven't set a single policy on deforestation'.[7]

The demand for these non-renewable resources continues to rise 'resulting in many nations and companies …. competing to shore up access to the remaining reserves, which some warn could lead to ongoing resource wars' [4]. Continual

[5] https://unfccc.int/files/press/backgrounders/application/pdf/fact_sheet_reducing_emissions_from_deforestation.pdf

[6] https://forestdeclaration.org/about/new-york-declaration-on-forests/

[7] https://forest500.org/publications/2023-watershed-year-action-deforestation

exploitation of these resources creates dual dilemmas. Firstly if the currently available fossil fuel reserves are burnt then, according to the Carbon Tracker Initiative,[8] approximately five times the volume of carbon dioxide required to produce two degrees of global temperature rise will be emitted. Secondly (and the clue is in the term 'non-renewable'), these resources will ultimately be exhausted. As resources become more difficult to extract, environmental, social and economic impacts will rise. This is already evident, for example, from the processing of coal tar sands in Canada, the hydraulic fracturing (fracking) of gas and oil in the USA and deep water oil exploration such as in the Gulf of Mexico and the Arctic. We will be examining the risks and limits of non-renewable resource extraction further in Chap. 4.

The Club of Rome's thesis on the 'limits to growth' was elaborated further by the World Wildlife Fund (WWF) International and BioRegional with their development of the concept of 'One Planet Living'. They estimated, in broad brush terms, that 'if everyone lived like the average American we would need five planets to live on'.[9] Their campaign is based around ten 'principles of sustainability' in an attempt to persuade policy makers, developers and the public at large to make the necessary changes to achieve One Planet Living. These address carbon emissions, waste, transport, materials, food, water, land use and wildlife, culture and community, equity and local economy, and health and happiness. At a more strategic level 'Planetary Boundaries' were first mooted by an international team led by Johan Rockstrom of the Stockholm Resilience Centre[10] in a paper for Ecology and Society in 2009. They suggested nine numerical thresholds which represent tipping points for key environmental criteria beyond which regional or even global ecological systems could be destabilised:

> Transgressing one or more planetary boundaries may be deleterious or even catastrophic due to the risk of crossing thresholds that will trigger non-linear, abrupt environmental change within continental- to planetary-scale systems.[11]

The nine boundaries proposed fall under the headings of:

- Stratospheric ozone layer.
- Biodiversity.
- Chemical dispersion.
- Climate change.
- Ocean acidification.
- Freshwater consumption and global hydrological cycles.
- Land system change.
- Nitrogen and phosphorous inputs to biosphere and oceans.
- Atmospheric aerosol loading.

---

[8] The Carbon Tracker Initiative report *Unburnable Carbon – Are the world's financial markets carrying a carbon bubble?* can be downloaded from https://carbontracker.org/reports/carbon-bubble/

[9] https://www.bioregional.com/one-planet-living

[10] https://www.stockholmresilience.org/research/planetary-boundaries.html

[11] http://www.ecologyandsociety.org/vol14/iss2/art32/

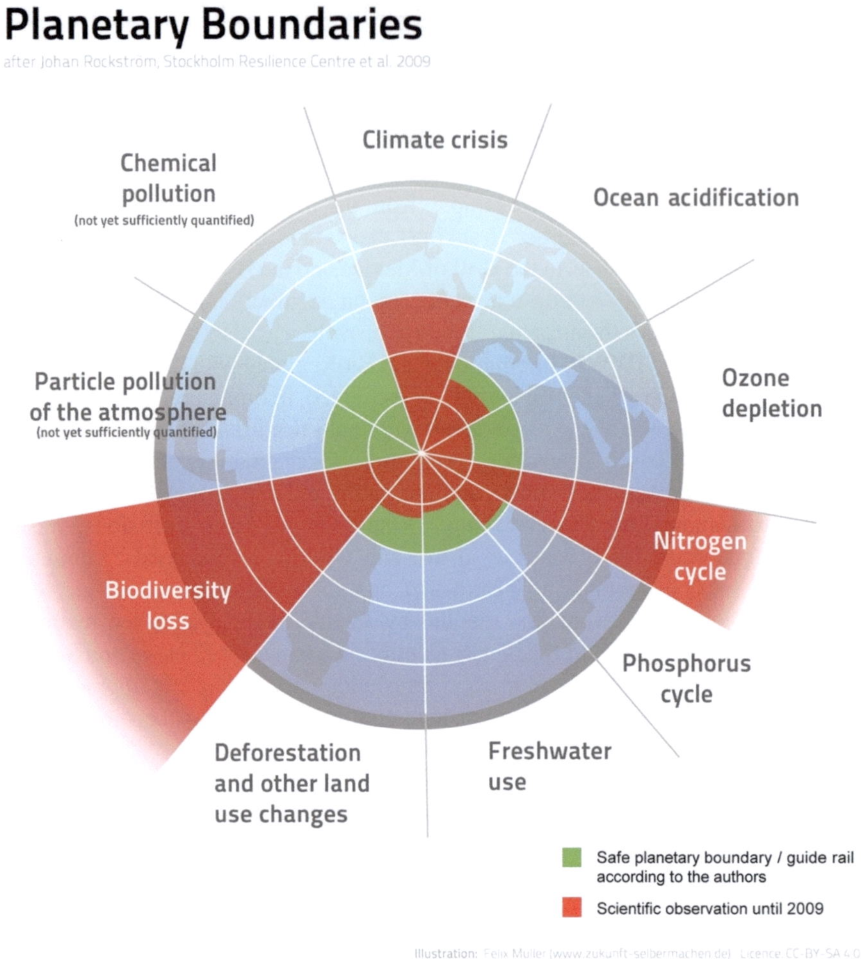

Fig. 2.1 Planetary boundaries (From: Felix Mueller at: https://commons.wikimedia.org/wiki/Category:Planetary_boundaries#/media/File:Planetary_Boundaries.png)

Rockstrom's team set out some preliminary thresholds for seven of the above based on the global criteria. They were not able to propose thresholds for chemical dispersion or atmospheric aerosol loading based on the knowledge available at the time. In their abstract, the authors state that:

'We estimate that humanity has already transgressed three planetary boundaries: for climate change, rate of biodiversity loss, and changes to the global nitrogen cycle'. This can be seen from the red areas highlighted in Fig. 2.1. The green area represents the 'safe operating space'.

Hence, it can be seen that Rockstrom's team estimated that in 2009 the threshold for climate change had been exceeded by some 50%, for the 'nitrogen cycle' by a factor of four and for biodiversity loss by more than ten times. We will be referring to the issues thrown up by this paper in later chapters.

2.2 Agreements...

Despite the best efforts of the United Nations and others, obtaining an international agreement on an effective strategy to manage global resources and mitigate against climate change has proven intractable. The global political machinations concerning the environment and climate change have been convoluted and frustrating. The internationalisation of environmentalism followed shortly after the founding of the United Nations in 1945, which turned its attention to the 'Protection of Nature' at a conference at Lake Success near New York in 1949. A number of UN bodies have since been established to coordinate activities relating to environmental protection. The UN Environment Programme (UNEP) was launched in 1972 at the UN Conference on the Human Environment in Stockholm. In collaboration with the World Meteorological Organization, this led to the First World Climate Conference in Geneva in 1979 and the founding of the Intergovernmental Panel on Climate Change (IPCC) in 1988.

Following the IPCC's first report some 2 years later the UN general Assembly created the UN Framework Convention on Climate Change (UNFCCC)[12] which was opened for signature at the Rio Earth Summit of 1992. This Summit also established the UN Commission on Sustainable Development (CSD),[13] the UN Convention on Biological Diversity (UNCBD)[14] and the UN Convention to Combat Desertification (UNCCD).[15] Each of these initiatives has resulted in a programme of international conferences that continue to this day. Since the first Conference of the Parties to the UNFCCC (CoP 1) took place in Berlin in 1995, landmarks have included CoP 3 in 1997, which produced the far reaching Kyoto Protocol, establishing targets for carbon emission reduction and forming the basis for subsequent CoP negotiations.

Similarly UNCSD has met regularly since 1992, with the 20th and final meeting—CSD 20 taking place in New York in September 2013. CSD has since been re-badged as the High-level Political Forum on Sustainable Development. Anniversaries of the 1992 Rio Earth Summit have been marked with further summits including the World Summit on Sustainable Development in Johannesburg in 2002 and, more recently, the Rio +20 Conference on Sustainable Development in 2012. Where the UNFCCC conferences focus on climate change, since 2004 the

[12] https://unfccc.int

[13] http://sustainabledevelopment.un.org/csd.html

[14] http://www.cbd.int/

[15] https://www.unccd.int

CSD meetings have been divided into seven 2-year implementation cycles (see Box 2.1) each considering a thematic cluster divided into a menu of cross-cutting issues.

Box 2.1 UN Commission on Sustainable Development Cycles, Clusters and Issues

Cycle	Thematic cluster
2004–2005 (CSD-12 & 13)	Water, sanitation, human settlements
2006–2007 (CSD-14 & 15)	Energy for sustainable development, industrial development, air pollution/atmosphere, climate change
2008–2009 (CSD-16 & 17)	Agriculture, rural development, land, drought, desertification, Africa
2010–2011 (CSD-18 & 19)	Transport, chemicals, waste management, mining, initiation of 10 year framework of programmes on sustainable consumption and production patterns
2012–2013 (CSD-20 & 21)	Forests, biodiversity, biotechnology, tourism, mountains
2014–2015 (CSD-22 & 23)	Oceans and seas, maritime resources, small island developing states (SIDS), disaster management and vulnerability
2016–2017 (CSD24 & 25)	Overall appraisal of implementation of Agenda 21,[16] the Programme of Further Implementation of Agenda 21 and the Johannesburg Plan of Implementation[17]

Cross-cutting issues considered under each cluster heading:

Poverty eradication, Changing unsustainable patterns of consumption and production, Protecting and managing the natural resource base of economic and social development, Sustainable development in a globalizing world, Health and sustainable development, Sustainable development of SIDS, Sustainable development for Africa, Other regional initiatives, Means of implementation, Institutional framework for sustainable development, Gender equality, and Education.

It can be seen from Box 2.1 that the scope of the UNCSD was very broad indeed and clearly overlapped with the remit of other UN bodies such as the UNFCCC, UNCBD and UNCCD (dealing with climate change, biodiversity and desertification, respectively). The UN Environment Programme (UNEP)[18] also has a

[16] Agenda 21 set out a wide range of environmental issues requiring global cooperation and committed to by signatories to the 1992 Rio Declaration. For more details refer to: http://sustainabledevelopment.un.org/content/documents/Agenda21.pdf

[17] The 2002 Johannesburg Plan of Implementation can be found at http://www.un.org/esa/sustdev/documents/WSSD_POI_PD/English/WSSD_PlanImpl.pdf

[18] https://www.unep.org

cross-cutting remit that covers much of the same ground, although the primary focus is on the management of environmental impacts and pollution, such as adaptation to climate change, ecosystem management, resource efficiency, environmental governance, disaster and conflict risk reduction and recovery. It also provides a home for a number of important Scientific Advisory Groups including the IPCC, the Ecosystem Conservation Group and teams of scientists providing support on matters relating to the marine environment and atomic radiation.

Towards the end of the 1990s, the UNEP, UN Development Programme (UNDP), World Bank and World Resources Institute (WRI) collaborated in the development of a Millennium Ecosystem Assessment (MEA), launched by UN Secretary General Kofi Annan in April 2000 and published by the WRI in 2005. In the Preface to the summary document titled *'Ecosystems and Human Well-being: Synthesis',* the purpose of the Assessment was set out as follows:

> The Millennium Ecosystem Assessment was carried out between 2001 and 2005 to assess the consequences of ecosystem change for human well-being and to establish the scientific basis for actions needed to enhance the conservation and sustainable use of ecosystems and their contributions to human well-being.[19]

The main findings of the Assessment included the alarming conclusion that 'over the past 50 years, humans have changed ecosystems more rapidly and extensively than in any comparable period of time in human history, largely to meet rapidly growing demands for food, fresh water, timber, fibre and fuel. This has resulted in a substantial and largely irreversible loss in the diversity of life on Earth'. Although exploitation of the natural environment has contributed to enhancements in well-being and economic development for some, this has been balanced by increased poverty elsewhere, and in the process, creating barriers to meeting the UN's Millennium Development Goals and subsequent Sustainable Development Goals (see below).

At about the same time as the MEA was launched, 29 UN bodies formed a partnership to establish and monitor a set of ambitious goals designed to tackle the challenges posed by a growing population with depleting resources globally (see section on Social and economic impacts in Chap. 3). They were established from a Resolution of the UN General Assembly in September 2000 known as the United Nations Millennium Declaration.[20] The headline Millennium Development Goals (MDG) were to:

1. Eradicate extreme poverty and hunger;
2. Achieve universal primary education;
3. Promote gender equality and empower women;
4. Reduce child mortality;
5. Improve maternal health;
6. Combat HIV/AIDS, malaria and other diseases;

[19] http://www.millenniumassessment.org/documents/document.356.aspx.pdf

[20] http://www.un.org/millenniumgoals/

7. Ensure environmental sustainability.
8. Develop global partnership for development.

These included targets to be achieved by 2015, some of which were met ahead of time and some that were not achieved by that year. They were replaced in September 2015 with 17 'Sustainable Development Goals' with revised targets for some of the original MDGs, whilst extending the scope to cover more detail, and with target dates extended to 2020, 2025 or 2030.[21] These include a greater emphasis on sustainability, including goals relating to sustainable energy generation, economic growth, consumption, production, agriculture, forest management and development, whilst combatting climate change and its impacts, such as desertification. Reference is also made to resilient infrastructure, inclusive, safe and resilient cities, conservation of oceans and marine resources, reversing land degradation and biodiversity loss, reducing inequality and promoting just, peaceful and inclusive societies.

The UNEP and the World Bank are, along with the likes of the World Resources Institute (WRI) and the World Wildlife Fund (WWF), members of the Natural Capital Coalition (NCC), formerly The Economics of Ecosystems and Biodiversity (TEEB) study team which was launched at a meeting of the G8 + 5 in 2007. NCC defines 'Natural Capital as the stock of natural ecosystems on Earth including air, land, soil, biodiversity and geological resources. This stock underpins our economy and society by producing value for people, both directly and indirectly'.[22]

WWF also partnered with the Nature Conservancy, University of Minnesota and Stanford University in the Natural Capital Project which was based around InVEST (Integrated Valuation of Ecosystem Services and Tradeoffs): 'a software tool that models and maps the delivery, distribution, and economic value of ecosystem services and biodiversity.[23] It was designed to support 'decision makers in visualising the impacts of decisions and identify trade-offs and compatibilities between environmental, economic, and social benefits'.

United Nations bodies have been involved with a number of related initiatives including the Natural Capital Declaration, the Wealth Accounting and Valuation of Ecosystem Services (WAVES) and the System for Environmental Economic Accounting (SEEA).

The Natural Capital Declaration (NCD) is a joint initiative of the UNEP Financial Initiative and Global Canopy Programme, the latter being a tropical forest think tank based in the UK. The NCD was launched at the Rio +20 conference in 2012 as 'a finance sector initiative, endorsed at CEO-level, to integrate natural capital considerations into loans, equity, fixed income and insurance products, as well as in accounting, disclosure and reporting frameworks'.[24]

---

[21] https://sustainabledevelopment.un.org/?menu=1300

[22] https://www.theoep.org.uk/natural-capital-committee#:~:text=The%20NCC%20defined%20natural%20capital,land%2C%20minerals%20and%20oceans"

[23] http://www.worldwildlife.org/projects/the-natural-capital-project

[24] https://www.unepfi.org/publications/natural-capital-declaration/

In parallel with this the World Bank launched WAVES in 2010 to ensure natural capital accounting (NCA) was factored into decision-making at a national level: 'working with central banks and ministries of planning and finance across the world to integrate natural resources into development planning through NCA, we hope to enable more informed decision-making that can ensure genuine green growth and long-term advances in wealth and human well-being'.[25] WAVES has been targeted at low- and middle-income developing countries. Since 2012, countries having some of the greatest conflicts between natural capital and human activity, such as Madagascar, Indonesia and the Philippines, have used WAVES to introduce NCA into their national accounting processes.

The UN Statistics Division developed a 'System of Environmental Economic Accounting' (SEEA) as far back as 1993, applying accounting techniques to a range of environmental decisions under the headings of Energy, Water, Land & Ecosystems, and Agriculture, Forestry & Fisheries. The protocols have been revised approximately every 10 years since their inception and the 2014 publication on *Experimental Ecosystem Accounting* sets out to 'define a measurement framework for integrating biophysical data, tracking changes in ecosystems and linking those changes to economic and other human activity'.[26]

Overlying much of this is the UN Convention on Biological Diversity, which is another initiative that was launched at the Rio Earth Summit in 1992, leading to the Strategic Plan for Biodiversity and the Aichi Biodiversity Targets being agreed by the Conference of the Parties (CoP) in 2010, along with the Intergovernmental Science-Policy Platform on Biodiversity and Ecosystem Services (IPBES) which resulted in the 2014 Gangwon Declaration of Biodiversity for Sustainable Development.

The three main objectives of the Convention on Biological Diversity are: 'The conservation of biological diversity, the sustainable use of its components and the fair and equitable sharing of benefits arising out of the utilisation of genetic resources, and their crucial importance to sustainable development'.[27] These objectives were translated into a 'Strategic Plan for Biodiversity for 2011 to 2020', the stated vision for which actually looks further into the future with the ambition that 'by 2050, biodiversity is valued, conserved, restored and wisely used, maintaining ecosystem services, sustaining a healthy planet and delivering benefits essential for all people'.

> The mission of the Strategic Plan is to take effective and urgent action to halt the loss of biodiversity in order to ensure that by 2020 ecosystems are resilient and continue to provide essential services, thereby securing the planet's variety of life, and contributing to human well-being, and poverty eradication.[28]

---

[25] https://www.wavespartnership.org/en/about-us

[26] http://unstats.un.org/unsd/envaccounting/seeaRev/eea_final_en.pdf

[27] https://www.cbd.int/hls-cop/gangwon-declaration-hls-cop12-en.pdf

[28] https://www.cbd.int/decision/cop/?id=12268

The Aichi Targets were a comprehensive set of requirements for the Parties (Nations) that have signed up to the Convention to be achieved by 2020.[29] Although there were 20 'targets' only three of them had a quantitative performance requirement that should have been achieved by 2020. Target six required Nations to at least halve the rate of habitat loss by 2020, whilst also achieving a significant reduction in degradation and fragmentation of habitats. Target 15 required restoration of 15% of degraded ecosystems, whilst the 11th Target was looking for an increase in the area of land assigned protected status: by 17% for terrestrial and inland water areas and by 10% for marine and coastal areas.

Other targets dealt with the integration of biodiversity strategies into development and poverty reduction planning, including such measures as the NCA techniques referred to above. Targets were also given for increasing awareness; incentivisation; sustainable production and consumption; sustainable management of marine resources, agriculture and forestry; pollution reduction; elimination of invasive alien species; protection of vulnerable ecosystems such as coral reefs; maintaining genetic diversity; safeguarding essential ecosystem services; respecting traditional knowledge; increasing research and allocation of resources.

Progress towards achieving these targets is monitored by the UNEP and reported in the Global Biodiversity Outlook,[30] the 2020 edition of which indicated that none of the Aichi Targets had been met and indeed that 'Biodiversity is declining at an unprecedented rate, and the pressures driving this decline are intensifying. None of the Aichi Biodiversity Targets will be fully met, in turn threatening the achievement of the Sustainable Development Goals and undermining efforts to address climate change'.

Following this perhaps pessimistic progress report the Kunming-Montreal Global Biodiversity Framework was launched in 2022, responding also to the 2019 Global Assessment Report on Biodiversity and Ecosystem Services produced by the IPBES which stated that:

> An average of around 25 per cent of species in assessed animal and plant groups are threatened, suggesting that around 1 million species already face extinction, many within decades, unless action is taken to reduce the intensity of drivers of biodiversity loss. Without such action, there will be a further acceleration in the global rate of species extinction, which is already at least tens to hundreds of times higher than it has averaged over the past 10 million years.
>
> The biosphere, upon which humanity as a whole depends, is being altered to an unparalleled degree across all spatial scales. Biodiversity—the diversity within species, between species and of ecosystems—is declining faster than at any time in human history.[31]

The World Meteorological Organization (WMO) is another member of the UN family which, through its sponsorship of the World Climate Research Programme, provides much of the scientific support behind the output from the UNFCCC and

[29] https://www.cbd.int/doc/strategic-plan/targets/compilation-quick-guide-en.pdf

[30] https://www.cbd.int/gbo/gbo5/publication/gbo-5-en.pdf

[31] https://www.cbd.int/doc/c/e6d3/cd1d/daf663719a03902a9b116c34/cop-15-l-25-en.pdf

IPCC and other climate-related programmes. Its Global Framework for Climate Services (GFCS), which was created following the 2009 World Climate Conference in Geneva, is 'to enable better management of the risks of climate variability and change and adaptation to climate change, through the development and incorporation of science-based climate information and prediction into planning, policy and practice on the global, regional and national scale'.[32]

The UNFCCC also has a scientific support body called the Subsidiary Body for Scientific Support and Advice (SBSTA) which provides an interface with the IPCC as well as advising on the 'impacts, vulnerability and adaptation to climate change' emissions from deforestation, development, transfer of environmental-friendly technologies and greenhouse gas inventories.[33]

The outcomes from the annual climate change Conferences of the Parties (CoPs) have involved intense negotiations as individual countries and lobby groups attempt to protect their interests and present their perspectives. However, since the original Convention in 1992 and the Kyoto Protocol in 1997 the language in the outcome reports has become more urgent. Where Kyoto set the strategy and targets, the Bali Road Map and Action Plan of 2005 established a shared vision and way forward, whilst the Cancun Agreements of 2010 committed all to 'establish clear goals and a timely schedule for reducing human-generated greenhouse gas emissions over time to keep the global average temperature rise below two degrees'. At CoP 18 in Doha in 2012, the sense of urgency became palpable, resulting in a commitment to 'speedily work towards a universal climate change agreement covering all countries by 2020, to be adopted by 2015'. With this in mind the Kyoto Protocol and associated mechanisms were extended for another 8 years until 2021. The sense of urgency continued and by 2016 research was indicating that the two degree goal was inadequate, resulting in the Paris Agreement being agreed at CoP 21 with the 'overarching goal … to hold the increase in the global average temperature to well below 2 °C above pre-industrial levels' and pursue efforts "to limit the temperature increase to 1.5 °C above pre-industrial levels'.[34]

During CoP 19, which took place in Warsaw at the end of 2013 there was an emphasis on funding of both mitigation and dealing with the impacts of climate change. The developed nations committed to a $100 billion fund to be established by 2020 'to support developing countries in their climate change actions', whilst also instigating the Warsaw International Mechanism for Loss and Damage.[35] This was brought sharply into focus by the ongoing impacts of Super Typhoon Haiyan that cut a swathe through the Philippines just prior to the conference. This

---

[32] https://gfcs.wmo.int/site/global-framework-climate-services-gfcs/about-gfcs

[33] For a summary of the key steps to the UNFCCC process refer to http://unfccc.int/key_steps/warsaw_outcomes/items/8006.php

[34] https://unfccc.int/process-and-meetings/the-paris-agreement

[35] For a summary of the key steps to the UNFCCC process refer to http://unfccc.int/key_steps/warsaw_outcomes/items/8006.php

mechanism was ultimately agreed in 2022 at CoP 27 at Sharm el-Sheikh in Egypt and referred to as the Sharm el-Sheikh Implementation Plan.[36]

2.3 …and Disagreements

Conflict between nations over access to energy resources evolved from a growing internationalisation of the oil trade in the latter part of the nineteenth century. As oil and gas began to replace coal to fuel the industrial revolution, many oil producing countries found their economic power increasing out of proportion to their political maturity. Furthermore, tensions remaining from the gradual dismantling of colonial rule, and the inherently artificial boundaries created, resulted in a fundamental instability that remains to this day.

In a 2008 issue of the Foreign Affairs journal (written before the so-called Arab spring uprising and the ongoing civil war in Syria), Michael L Ross wrote in his article *Blood Barrels: Why oil wealth fuels conflict*:

> The main reason (for an increase in oil-related conflicts) is that oil wealth often wreaks havoc on a country's economy and politics, makes it easier for insurgents to fund their rebellions, and aggravates ethnic grievances. Today, with violence falling in general, oil-producing states make up a growing fraction of the world's conflict-ridden countries. They now host about a third of the world's civil wars, both large and small, up from one-fifth in 1992. According to some, the U.S.-led invasion of Iraq shows that oil breeds conflict between countries, but the more widespread problem is that it breeds conflict within them.[37]

As the demand for fossil fuels grew during the twentieth century, oil and gas resources globally were drawn into a complex marketplace controlled by a combination of large international energy companies and oil producing countries. Before long, securing supplies of oil and gas became intertwined with national security interests, resulting, for example, in 1928 in the 'Red Line Agreement' being introduced to restrict the supply of oil and ensure that the major oil companies could control oil prices and world markets. Following a period of sustained instability in the Middle East in the 1970s, culminating in the Russian invasion of Afghanistan in 1979, President Jimmy Carter declared in his 1980 State of the Union Address that:

> An attempt by any outside force to gain control of the Persian Gulf region will be regarded as an assault on the vital interests of the United States of America, and such assault will be repelled by any means necessary, including military force.[38]

This subsequently became known as the 'Carter doctrine' which some believe 'remains the dominant factor in American thinking in the Middle East (and beyond)

---

[36] https://unfccc.int/news/cop27-reaches-breakthrough-agreement-on-new-loss-and-damage-fund-for-vulnerable-countries

[37] http://www.foreignaffairs.com/articles/63396/michael-l-ross/blood-barrels

[38] http://www.businessinsider.com/how-oil-has-driven-global-conflict-for-the-past-100-years-presentation-2012-6?op=1

today, and it makes it very clear (that) … the flow of oil must be preserved at all costs'.[39] This was behind American thinking in the lead up to the first Iraq war following the invasion of Kuwait in 1990 and was one of a number of factors that persuaded George W Bush to topple Saddam Hussein in 2003. Following this, much of the military effort in Iraq was spent in protecting refineries, oil exporting facilities and pipelines, such as that linking Kirkuk in the north to the Turkish border.

Some might argue that the Carter doctrine has shaped numerous US interventions in the Middle East and elsewhere over the last 30 years, including perhaps creating the conditions that stimulated the growth of Al Qaeda and increasingly ambitious terrorist attacks that included the destruction of the twin towers in New York on September 11th 2001:

> Retired Navy Admiral Dennis Blair, Obama's former director of national intelligence, has stated that "the modern U.S. military involvement tied to global oil started with the first Gulf War under President George H.W. Bush (and that) if Kuwait had been a country in central Africa that was invaded by its neighbors, we would not have rallied an international coalition and kicked out the neighbor who invaded it," and from that war "you can draw kind of a straight line" to Al Qaeda's resentment toward the U.S., the Sept. 11 attacks, the invasions of Afghanistan and Iraq "and several trillion dollars and 7,000 lives that we lost in that period".[40]

Even without pressure from external interests, many countries have imploded from the unintended consequences of discovering oil and gas. For example, what has become known popularly as the 'oil curse' has affected around half of OPEC (Organization of the Petroleum Exporting Countries) members, which resulted in them being poorer in 2005 than during the oil boom some 30 years earlier. In the Michael L Ross comment piece from 2008 referred to above referring to this curse, he suggests that:

> The affliction hits when a country becomes a significant producer and exporter of natural resources. Rising resource exports push up the value of the country's currency, which makes its other exports, such as manufactured and agricultural goods, less competitive abroad. Export figures for those products then decline, depriving the country of the benefits of dynamic manufacturing and agricultural bases and leaving it dependent on its resource sector and so at the mercy of often volatile international markets. In Nigeria, for example, the oil boom of the early 1970s caused agricultural exports to drop from 11.2 percent of GDP in 1968 to 2.8 percent of GDP in 1972; the country has yet to recover.
>
> Oil wealth also has political downsides, and those are often worse than the economic ones. Oil revenues tend to increase corruption, strengthen the hands of dictators, and weaken new democracies. The more money the governments of Iran, Russia, and Venezuela have received from oil and gas exports, the less accountable they have become to their own citizens—and the easier it has been for them to shut up or buy off their opponents. A major boom in oil prices, such as the one that took the price of a barrel from less than $10 in February 1999 to over $100 in March 2008, only heightens the danger.[41]

---

[39] http://www.businessinsider.com/how-oil-has-driven-global-conflict-for-the-past-100-years-presentation-2012-6?op=1

[40] Quote from energy policy website in 2013 (web link no longer works)

[41] http://www.foreignaffairs.com/articles/63396/michael-l-ross/blood-barrels

At the heart of many of these conflicts and tensions is the desire for each nation to achieve a security of energy supply, whilst benefitting from the economic and political resilience that energy security can provide. Achieving energy security is a delicate balance between reliable and safe energy generation; balancing the economics of domestic fuel production and supply with imports; robust distribution networks; the availability of domestic renewable sources; and demand reduction through energy efficiency improvements.

UK Government policy on maintaining UK energy security has evolved from making sure consumers can access the energy they need at prices that are not excessively volatile to combining energy security with a 'Net Zero Strategy' (see Sect. 7.2).[42] Having extracted most of the economically accessible oil and gas from the North Sea and having both nuclear and coal-fired power stations that are reaching the end of their lives, Britain is particularly exposed to fluctuations in international gas prices, whilst embarking on a major wind farm installation programme. This was vividly demonstrated by the impacts of the Russian invasion of Ukraine in 2022.

In the USA, coal and gas are the dominant fuels used for electricity generation, with the trend being towards gas catching up with coal firing as shale gas exploitation has grown (see Fig. 2.2).

The USA is a net exporter of coal and exports natural gas via pipelines to Canada and Mexico. Oil products are almost entirely used for transportation rather than electricity generation.

In the developing world, energy supply and security can be a little more chaotic. Pakistan, for example, has suffered from a combination of under-investment, natural disaster, corruption and theft that has resulted in some consumers experiencing power outages for up to 20 h a day. Important industries have been badly impacted by unreliable power supplies, which has had a detrimental effect on the economy as a whole.

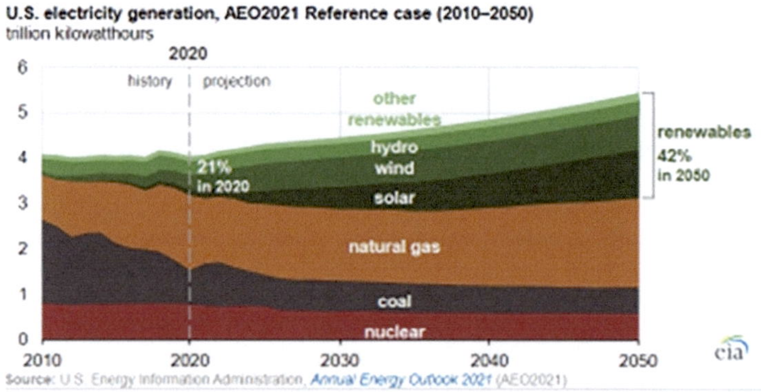

Fig. 2.2 Energy source for electricity generation in the USA, historical and future projection (https://upload.wikimedia.org/wikipedia/commons/6/63/US_Electricity_Generation.png)

[42] https://www.gov.uk/government/publications/powering-up-britain/powering-up-britain-energy-security-plan#:~:text=The%20policies%20and%20ambitions%20we,support%20up%20to%20480%2C000%20jobs

2.3.1 Pipelines

Russia has for many years been a dominant producer of gas and oil. Via the state controlled company Gazprom it has controlled around 20% of the world's gas reserves, including historically around 30% of that used by Europe. A complex lattice work of oil and gas pipelines links the main fields in Russia and beyond to Europe (see Fig. 2.3). There are political tensions between a number of the countries on this map. For example, Ukraine gained its independence in 1991 following the break-up of the Soviet Union, however, the Soyuz and Bratstvo (Brotherhood) pipelines conveyed something like 80% of the gas that Russia supplied to Europe, although this fell to 50% in 2011 as new pipelines came on line that bypass Ukraine.

Fig. 2.3 Proposed and operational gas pipelines linking Russia, Europe and beyond as of 2013 (https://upload.wikimedia.org/wikipedia/commons/d/d7/Major_russian_gas_pipelines_to_europe.png)

Historically more than 50% of Ukraine's demand has been supplied from Russia and for many years was heavily subsidised. But in the winter of 2005 to 2006, as Ukraine was increasing its ties to the European Union and Nato, Russia suddenly raised the price of gas from \$50 to \$230 for 1000 cubic metres, on the basis that the latter figure was closer to the market price at the time. Ukraine refused to pay the elevated price and so Russia shut off the supply in January 2006, when demand was high. Elsewhere in Europe gas pressure fell by around 30%, resulting in significant alarm and protest. A very similar pattern was repeated in January 2009 which impacted badly, not only on Ukraine, but also on supplies to Bulgaria and Slovakia, where a national emergency was declared. Western Europe was less badly affected than in 2006 because gas storage volumes had increased, although there was a major impact on wholesale prices.

The invasions of Ukraine by Russia in 2014 and 2022 had an even greater impact on gas supplies and the trickledown effect has impacted on many aspects of the global economy. Just prior to Russia's successful invasion and annexation of Crimea in February 2014, the Ukrainian leadership abandoned its EU Association Agreement in order to turn its face towards Russia, resulting in internal protest and division within Ukraine. This in turn led to violent suppression of protest, the Prime Minister being deposed and military intervention by Russia. Again Russia shut off the gas supply to Ukraine, although because of a mild winter Ukraine had significant volumes of gas in storage to reduce the impact in the short term.

As can be seen from Fig. 2.3 there were a number of pipelines under construction, proposed or being increased in capacity before the Russia/Ukraine conflicts. The Nord Stream comprises two 'lines' running side by side between Russia and Germany under the Baltic Sea, the first of which was brought on line in November 2011, whilst the second followed a year later, both of which were shut down by the Russians in 2022. At the time of writing the South Stream pipeline project, which should have been comprised of four lines run under the Black Sea from Russia, through Hungary and Bulgaria into Austria, has been cancelled. All of the pipelines running from Russia into Europe have been shut down.

In order to reduce dependency on Russia for gas in Europe, a number of new pipelines have been proposed to connect gas fields in Azerbaijan and Central Asia to Europe via Turkey and the Adriatic. After some 11 years of trying, the Nabucco pipeline project, which was to transport gas from Iraq and Azerbaijan across Turkey to Austria and beyond, was finally abandoned in 2013, whilst the Trans Anatolian Pipeline (TANAP) and the Trans Adriatic Pipeline (TAP) were completed in 2018. Between them they will supply between 26 and 36 billion cubic metres of gas per year from gas fields in Azerbaijan, which represents only 4–7% of European demand. The disruptions to gas supply caused by the disputes between Russia and Ukraine have led to those European countries most impacted to explore alternative supplies, including domestic shale gas exploration and the import of liquefied petroleum gas (LPG) from the USA.

There is considerable disparity between sources of data on the total lengths of oil and gas pipelines globally, but the greatest consensus seems to be around the

Fig. 2.4 The Trans-Alaska Pipeline System (TAPS) laid across the tundra (https://commons.wikimedia.org/wiki/File:Alaska_Pipeline.jpg)

2.2 million km mark, or enough to circumnavigate the planet more than 55 times.[43] Although it is unclear what proportion of pipelines run above ground, it seems likely that a significant proportion is either buried or run along the sea bed. In the USA, for example, many are buried alongside transport corridors, although in remote regions, pipelines tend to be laid above ground, partly due to cost and partly because of the risk of tundra melt and associated methane release, although the visual impact is significant (see Fig. 2.4).

As was demonstrated by the San Bernardino disaster in 1989, burying pipelines alongside railway lines carries its own risk. In this incident a goods train carrying trona—a mineral used in the manufacture of sodium carbonate—was derailed when passing through this Californian city, destroying a number of houses and killing four people. During the subsequent clean-up operation, a petroleum pipeline buried alongside the track was damaged and subsequently burst, releasing vapour that was ignited, creating a conflagration that destroyed more property and killed another two people.

1989 was a bad year for railway disasters attributed to leaking pipelines. In June of that year between 675 and 780 people were killed (reports vary) and hundreds injured when a liquefied gas pipeline leaked between the cities of Ufa and Asha in the Urals region of Russia. Gas accumulated in a hollow a few hundred metres from the pipeline and was ignited by sparks from two trains carrying some 1300 passengers travelling on the trans-Siberian railway that passed through this hollow. The resulting explosion destroyed 37 carriages and two locomotives. Many of those killed were children bound for holiday camps on the Black Sea coast.

[43] https://www.businesswire.com/news/home/20220218005210/en/Global-Oil-and-Gas-Pipelines-Market-Outlook-to-2025%2D%2D-Identify-Growth-Segments-and-Opportunities%2D%2D-ResearchAndMarkets.com

Fig. 2.5 Kalamazoo river: aerial view of oil sheen emitting from contaminated vegetation at the Ceresco dam area between the spill site and Battle Creek, Michigan (Photo: US Environmental Protection Agency: https://commons.wikimedia.org/wiki/File:Oil_Siphon_(4885250078).jpg)

Although this remains the worst pipeline disaster of all time in Russia, the largest death tolls have probably occurred in Nigeria, where leaks occur periodically in mostly elevated pipelines due to either attempted theft of petroleum or poor maintenance. In one incident in the village of Jesse, the cause of which remains disputed, a leak resulted in an explosion that is reported to have killed up to 1200 people, with a fire raging unabated for some 6 days until American firefighters reached the village, which lies some 340 km from Lagos.

Pipeline leaks can also produce major pollution incidents. One such incident occurred in the US state of Michigan in July 2010 when a 30 inch (760 mm) diameter pipeline known as Enbridge Line 6B ruptured spilling diluted bitumen from Canada's tar sands into the Kalamazoo river (see Fig. 2.5). Estimates vary but the US Environmental Protection Agency put the amount spilt over a 17 h period at one

Fig. 2.6 Oil leaking from the Trans-Alaska pipeline resulting from a shot fired by a local resident (https://www.fbi.gov/news/stories/north-to-alaska-part-4)

million gallons, contaminating some 56 km of river before being contained, damaging wildlife and ecosystems in the process. Like many of the pipelines criss-crossing the planet this one was more than 40 years old. Remedial work continued for many years afterwards.[44]

Pipelines are also vulnerable to terrorist attack or vandalism, particularly those above ground. The 1300 km Trans-Alaska Pipeline System (TAPS) shown in Fig. 2.6 has been subject to a number of random attacks, the most notorious occurring in October 2001 when a drunken resident from the tiny community of Livengood shot a hole in the pipeline with a high powered rifle, resulting in the loss of some 1.3 million L of crude oil and a clean-up taking many months and costing $13 million. At the time the pipeline handled around 20% of the US crude oil supply. In 2006 a jihadist website called for attacks against the pipeline, although no action followed.

Between 2011 and 2014, there were 26 attacks on the section of the pipeline carrying gas from Egypt to Israel, Jordan and Syria, crossing the Sinai Peninsula, which for a time was effectively under the control of militant Bedouins following the uprising in early 2011 that unseated President Mubarak of Egypt. Since 2014 attacks have continued on various sections of the pipeline, including one in 2020 on a section north of Damascus which resulted in major power outages in Syria.

[44] https://www.greatlakesnow.org/2020/12/10-years-after-oil-spill-kalamazoo-river-michigan/

2.3.2 *Conflict Minerals*

Some countries have experienced a massive increase in revenues over a short period due to the discovery of high value natural resources which, unless prudently invested, in some cases has been squandered on wasteful projects, lined the pockets of the rich and powerful, widened the divide been rich and poor and fomented unrest. Of course these problems have not only been associated with the exploitation of oil and gas but other valuable resources such as gold, diamonds, uranium, tantalum, tungsten and other minerals. Frequently there have been multiple factors causing unrest, such as poor working and living conditions, inadequate pay, localised pollution and environmental damage. To these can be added a lack of effective policing, territorial disputes and tribal tensions.

In the Democratic Republic of Congo (DRC) around 50% of the world's supply of tantalum is mined. Tantalum is a chemically inert metal that is critical to the manufacture of capacitors for use in mobile phones and other electronic devices worldwide. The DRC also has significant quantities of tin ore, gold and tungsten. The 'majority of miners work for militia groups that still control most of the country's mines'.[45] The conditions under which these miners have to work have been reported as being appalling. Profits from the mines were reportedly used to finance the Second Congo War, which officially ended in 2003 after millions of fatalities, many of which were incurred through fighting over the control of the mines. Minerals that are extracted from the Eastern Congo typically pass through a variety of intermediaries before being purchased by multinational electronics companies.

In an attempt to introduce due diligence to the purchase of minerals by the electronics industry both the USA and European Union have introduced legislation to the so-called conflict minerals market. In the USA, the misleadingly titled *Dodd-Frank Wall Street Reform and Consumer Protection Act* mandated the US Securities and Exchange Commission to create rules for addressing whether conflict minerals were benefitting armed groups in the DRC, and introduced a 'due diligence certification scheme obliging companies to source-check materials'.[46] The European Union introduced its Conflict Mineral Regulation in 2017 which 'requires EU importers of tantalum, tin, tungsten, and gold to source from responsible and conflict-free sources only'.[47] This is clearly quite a limited list and it is unclear why diamonds, for example, are not mentioned, particularly following their role in funding the war in Sierra Leone at the turn of the Millennium.

Conflict over energy resources has not been restricted to the oil and gas sector. The ability of certain countries to generate electricity from radioactive fuels has created considerable concern when their uranium enrichment programme has been

[45] http://www.mining.com/this-is-how-the-congo-supplies-conflict-minerals-to-the-it-world-36110/

[46] https://www.sec.gov/spotlight/dodd-frank/speccorpdisclosure.shtml

[47] https://europeanpartnership-responsibleminerals.eu/cms/view/73f644c4-da24-4134-b381-97e541e46a15/faqs-how-to-comply-with-the-eu-conflict-minerals-regulation

shrouded in secrecy and there has been the potential for the development of nuclear weaponry. Iran is of course the best known example of a country that has for decades invested in a nuclear power programme whilst ignoring UN Security Council and International Atomic Energy Authority (IAEA) requirements designed to both prevent nuclear weapons proliferation and ensure international safety measures are enforced.

Although the Soviet Union supported North Korea in the development of light water reactors in the late 1980s on condition that it signed the Nuclear Non-proliferation Treaty, this fizzled out after the break-up of the Soviet Union. Unlike Iran, North Korea has made little attempt to disguise its 'covert' nuclear weapons programme which is an ongoing threat. The combination of the loss of Soviet support and subsidised oil and the channelling of scarce resources into a grotesquely ambitious weapons programme led to the economic collapse of North Korea and the death by starvation of millions of its population in the 1990s.

One North Korean economy expert told a UN Commission of Inquiry on Human Rights in North Korea in 2013 that the North Korean famine '… "was a man-made, preventable tragedy. These people died needlessly, and the government is culpable". Impoverished North Korea has been spending millions of dollars developing nuclear weapons and missiles and has come under heavy international sanctions for years'.[48]

2.3.3 *Conflict and Water*

According to the data assembled by the UN and WHO in 2019/20:

- 2.2 billion people lack access to safely managed drinking water services.
- Almost two billion people depend on health care facilities without basic water services.
- Over half of the global population or 4.2 billion people lack safely managed sanitation services.
- 297,000 children under five die every year from diarrhoeal diseases due to poor sanitation, poor hygiene, or unsafe drinking water.
- Two billion people live in countries experiencing high water stress..[49]

In the 'Main Messages' from the 2012 World Water Development (WWD) Report: *Managing Water under Uncertainty and Risk* there is a warning that:

> Water shortages can cause conflicts of varying intensity and scale. Although conflicts may appear localized, they present challenges to the broader context of peace and security. Conflicts over water resources can also turn into or fuel ethnic conflicts—as ethnic conflict is most commonly fuelled by collective fears for the future, one can see how water scarcity could play into such fears.[50]

---

[48] http://www.rfa.org/english/news/korea/food-10312013185353.html

[49] https://www.un.org/en/global-issues/water

[50] https://sustainabledevelopment.un.org/content/documents/404water.pdf

For example, in the Middle East and North Africa where 'water stress' is highly cyclical, conflict has been endemic in certain countries. Wars have over the years destroyed water infrastructure in Beirut, Kuwait and Lebanon, for example. Some two-thirds of available surface water originates from outside of the region. At times, this has led to conflict with upstream countries.

According to UNESCO, 148 countries worldwide share river basins with other nations.[51] Take, for example, the Nile river basin, which runs through 11 countries namely: Burundi, Democratic Republic of Congo, Egypt, Eritrea, Ethiopia, Kenya, Rwanda, Sudan, South Sudan, Tanzania and Uganda. These countries are sometimes referred to as the 'eleven riparian countries' due to their proximity to the Nile river basin. At 6650 km, the river is the longest in Africa and an ever growing population lives in its proximity: estimates vary but in 2020 it was thought to be about 600 million, which represents half of the total population of the African continent at that time. Some 250 million people in these 11 countries have relied on the Nile for their water supply and any change in water flows has a major impact on the availability of water for these communities.

There has long been disagreement over who should benefit from the waters of the Nile and, despite 80% of the water originating from northern Ethiopia a post-colonial accord in 1959 awarded 65% to Egypt and 35% to Sudan.

Since 1999 there has been an ongoing process known as the Nile Basin Initiative that was originally intended to 'develop the river in a cooperative manner, share substantial socioeconomic benefits, and promote regional peace and security'.[52] However, the upstream riparian nations that were not party to the original accord grew impatient with the process and have developed their own 'Cooperative Framework Agreement', to which neither Egypt nor Sudan will commit.

Ethiopia has taken a major unilateral step in driving the process forward by constructing a dam in the Blue Nile Gorge. The Ethiopian Grand Renaissance Dam[53] lies close to the border with Sudan. This massive civil engineering project allows Ethiopia to provide irrigation water to large areas of a country which has experienced a great deal of water stress, whilst generating some 6000 MW of electricity. Unsurprisingly this has soured relationships between Ethiopia and Egypt, which at the same time has been going through major political upheaval.

Water stress occurs when demand exceeds supply for a defined period either due to over-exploitation or to water quality problems. According to data assembled by Earth.org in 2022 the greatest extent of water stress as measured by its 'Water Stress Index' (WSI) occurs in the Middle East and North Africa.[54] Where WSI is calculated from the ratio of domestic, industrial and agricultural consumption to renewable supplies of water from precipitation, rivers and groundwater.

---

[51] https://www.un.org/en/global-issues/water

[52] https://theowp.org/crisis_index/nile-river-conflicts/

[53] https://theowp.org/crisis_index/nile-river-conflicts/

[54] https://earth.org/data_visualization/eo-indexes-water-stress/

Note that numerous water stress indices have been developed since the late 1980s, a review of which can be found in a 2011 paper by Brown and Matlock of the Sustainability Consortium at the University of Arkansas.[55]

The WWD report referred to above includes a prediction that by '2030, 47% of world population will be living in areas of high water stress. Most population growth will occur in developing countries, mainly in regions that are already experiencing water stress and in areas with limited access to safe drinking water and adequate sanitation facilities'.

There are also examples where political change has occurred within countries as a result of disputes over water resources. For example, according to the WWD report referred to above:

> In Australia, longstanding conflicts between environmentalists and farmers in the Murray–Darling Basin form the historical backdrop to the Landcare movement and to multi-stakeholder forums for managing water in its basin context. Over an even longer period, competing demands of states have served as the basis for institutional development of the Murray–Darling Basin water management framework.[56]

We will be discussing the history of this conflict in more detail later in this chapter under the heading of 'The politics of water'.

2.3.4 *Environmental Impact*

Most of the issues discussed in this chapter are of geopolitical importance because of some combination of social, economic and environmental impact. We have seen how these factors can lead to conflict, but there are many other lower order impacts that have political implications.

Most human activity has an impact to some degree. Like the hypothetical beating of a butterfly's wings that causes a hurricane, a seemingly trivial action can sometimes have a significant impact. For example, the assassination in Sarajevo of the Archduke Franz Ferdinand of Austria on 28 June 1914 that set off a chain of events resulting in the outbreak of the First World War. Of course rarely does a single action result in an impact as significant as a war or disaster. Most activities are multi-factorial or cumulative and the discipline of environmental impact assessment (EIA) has evolved in order to predict the outcomes from a group of known activities, such as those involved in energy generation or industrial processes for example.[57]

Looked at from a geopolitical standpoint, it tends to be large scale and potentially polluting activities that create tension or conflict within or between nations.

[55] https://sustainabilityconsortium.org/download/a-review-of-water-scarcity-indices-and-methodologies/

[56] http://unesdoc.unesco.org/images/0022/002257/225741E.pdf

[57] http://www.eia.gov/todayinenergy/detail.cfm?id=13731

Impacts can be from normal operating conditions or due to an accident or criminal act. For example, the ongoing emissions of radioactive waste into the Irish Sea from the Sellafield nuclear site on the west coast of Cumbria in England have been of concern to the Irish and others since a series of major pollution incidents in the 1970s. Perhaps more dramatically the meltdown and subsequent explosions at Chernobyl nuclear power station in the Ukraine in 1986 produced airborne radioactive emissions that not only created extreme pollution from fallout locally in the Ukraine but also were distributed by fluctuating winds across much of the northern hemisphere, with those countries experiencing precipitation in the months after the disaster receiving the greatest deposition of radioactive 'rainout' [5].

Whereas many individual nations have the requirement for environmental impact assessment (EIA) for certain types of activity enshrined in law, transboundary impacts are subject to a voluntary agreement under the 1991 Espoo Convention. This agreement, otherwise known as the *UN Convention on Environmental Impact Assessment in a Transboundary Context*, comes under the auspices of the UN Economic Commission for Europe (UNECE), although not restricted to European countries. Signatories include the USA, Canada and the Russian Federation, as well as all European countries and former members of the Soviet Union. The Convention requires consultation between neighbouring countries where there are 'major projects under consideration that are likely to have a significant adverse environmental impact across boundaries'.[58] The Convention was followed by meetings in Kazakhstan and Kiev, resulting ultimately in the *Protocol on Strategic Environmental Assessment* being adopted in 2003, although not by the USA, Canada or the Russian Federation.

The impact of human activity on local communities can also have a political dimension. Potentially controversial projects such as nuclear power stations, wind farms, incinerators, oil and gas exploration, roads and railways, for example, can create a great deal of concern amongst those who may be impacted locally. Indeed some projects have resulted in considerable ramifications for local politicians.

Take, for example, the King's Lynn incinerator which became mired in controversy and threatened to blow a large hole in Norfolk County Council budgets. Although badged somewhat emotively as an 'incinerator' the proposed installation should more accurately be referred to as an energy from waste facility (see Box 2.2). The Environment Agency had confirmed that it met the R1 Energy Efficiency criteria under the Waste (England & Wales) Regulations 2011 and was therefore described as an 'energy recovery facility'. The local town council fought the scheme, with the County Council being caught between a rock and a hard place: either build the facility against the wishes of the local community or incur a massive cancellation penalty charge from the developers, Cory Environmental and Wheelabrator.

An EIA had been submitted to the Environment Agency as part of an Environmental Permit Application which includes air quality and noise and vibration assessments which concluded that the risk of harm to the nearest neighbours or

[58] http://www.unece.org/env/eia/eia.html

Box 2.2 King's Lynn Energy from Waste Facility
The proposed facility, at the Willows Business Park, Saddlebow Lane south of King's Lynn in North Norfolk, England, was intended to divert some 270,000 tonnes of mixed household, commercial and industrial waste from landfill per year and generate an estimated 24.2 MW of electricity, with the potential for using some of the steam generated to supply local businesses. The facility had been designed with multiple stages of flue gas cleaning comprising: oxides of nitrogen (NOx) control by ammonia or urea injection; acid gas abatement in a fluid bed reactor with injection of lime and water; and activated carbon injection for removal of vapour-phase heavy metals, dioxins, furans and mercury; as well as fabric filtration for removal of solids.

nature conservation sites would be low, even when taking account of cumulative impacts of road traffic and other nearby existing and permitted developments, such as an additional turbine for the neighbouring gas fired power station. Despite this a significant proportion of the King's Lynn community rose up against the proposed scheme, resulting in a local poll accumulating 65,516 votes (92.68% of respondents) it. This represented some 50% of the total population of King's Lynn and surroundings. The scheme was both abandoned by Norfolk County Council and 'called in' by the Secretary of State at the Department of Communities and Local Government (CLG).

Although the EIA indicated that the facility would create no discernible harm to human health or local ecosystems the local community in general chose to believe the opposite. This may be due to a perception that incinerators in general are polluting, noisy and smelly. One of the main concerns is the potential emission of dioxins (polychlorinated dibenzodioxins—PCDDs) and furans (polychlorinated dibenzofurans—PCDFs), which are formed during the combustion of chlorinated hydrocarbons, such as polyvinyl chloride (PVC), and are harmful to all living organisms. Polychlorinated biphenyls (PCBs), which were widely used as dielectric and coolant fluids in the manufacture of transformers, capacitors and electric motors until banned by the UN *Stockholm Convention on Persistent Organic Pollutants* in 2001, are also associated with furan emissions during combustion. These substances can be inhaled or ingested, and are potentially carcinogenic at extremely low concentrations, as well as impacting on the immune system and foetus development.

Although there is strong evidence that, under certain conditions, PVC produces the highest dioxin emissions of any material that might be incinerated, there are other materials that can produce dioxins and furans. Indeed 'any source of organic materials in the presence of chlorine or other halogens will generate dioxins and furans during combustion'.[59] The list includes waste wood, paper and other materials that may have been treated or contaminated with chlorinated pesticides or preservatives, for example.

[59] https://www.who.int/docs/default-source/food-safety/dioxins/who-ced-phe-epe-19-4-4-eng.pdf?sfvrsn=15246d22_2#:~:text=Any%20source%20of%20organic%20materials,dioxins%20and%20furans%20during%20combustion

PVC is used for many building products, including window frames, gutters and pipework, cables, and roofing membranes, as well as car interior finishes, packaging, fabric coatings, and so on. In an AEA Technology study for the European Commission (EC) in 2000 recycling rates across the 15 European Union and six Accession countries at the time were given as 3% of the 3.6 million tonnes of post-consumer waste arisings, with 82% going to landfill and 15% incinerated.[60] This report anticipated that by 2020 the proportion of PVC being disposed of by incineration could increase to 45%, with up to 22% being recycled.

It is evident that this analysis anticipated a major incinerator construction programme across Europe, hence the pollution control measures installed in these incinerators has to be excellent, particular with regard to prevention of dioxin emissions. The EC *Directive on the incineration of waste* from the same year[61] sets out the operating conditions required to minimise the emission of dioxins and furans (see Box 2.3). It is interesting to note that dioxins and furans are seen as pollutants that cross international boundaries since the legally binding emission limits were established in a Protocol on Persistent Organic Pollutants (POPs) to the 1979 UNECE *Convention on Long-range Transboundary Air Pollution* at 0.1 ng/m^3—that is a ten thousand millionth of a gram in one cubic metre of combustion gas emitted from a chimney.[62]

Box 2.3 Extract from Article 6 of 'EC Directive 2000/76/EC on the Incineration of Waste' Setting Out the Required Operating Conditions for Incinerating Municipal Waste

Incineration plants shall be designed, equipped, built and operated in such a way that the gas resulting from the process is raised, after the last injection of combustion air, in a controlled and homogeneous fashion and even under the most unfavourable conditions, to a temperature of 850 °C, as measured near the inner wall or at another representative point of the combustion chamber as authorised by the competent authority, for two seconds. If hazardous wastes with a content of more than 1 % of halogenated organic substances, expressed as chlorine, are incinerated, the temperature has to be raised to 1,100 °C for at least two seconds.

It is interesting to compare and contrast the designs for the proposed incinerator at King's Lynn and for the biomass power station in Teeside in Fig. 2.7. Apart from disguising the utilitarian nature of a power station in a stunning envelope, if it had come to fruition the Teeside scheme would have been set in a community park and incorporate viewing decks and other facilities celebrating engineering, open to the public. In contrast, the King's Lynn proposal was for an unapologetically industrial complex surrounded by a security fence. Despite the potential air quality impacts

[60] https://ec.europa.eu/environment/pdf/waste/studies/pvc/economic_eval.pdf

[61] https://eur-lex.europa.eu/eli/dir/2000/76/oj/eng

[62] https://unece.org/sites/default/files/2021-05/1979%20CLRTAP.e.pdf

Fig. 2.7 Proposed biomass power station on teeside by heatherwick studios (Image: Heatherwick Studios)

associated with biomass firing the Heatherwick Studio Teeside scheme gained almost universal support from the community in which it would have been placed.[63]

Like incineration and energy generation, oil and gas exploration and extraction are processes that occur on an industrial scale can be polluting and may be located close to human occupation. As readily accessible existing reserves of oil and gas have been exhausted the hunt has been on for alternative sources. This has led to the exploitation of the so-called shale gas (and oil) which, as can be seen in Fig. 2.8, is found extensively in sedimentary basins located worldwide.

As well as shale gas and oil the industry often refers to separate categories of gas and oil resources called 'tight gas' and 'tight oil'. Although frequently lumped in with shale formations and exploited in much the same way, tight gas and oil are found in silt or fine sand having a porosity perhaps two orders of magnitude higher than the muddy sediment in which shale gas and oil are found.

'In shale gas reservoirs, the natural gas is more closely bound to the rock, and sits in a fine scale array of relatively isolated and small pores and cracks'.[64] Hence shale gas (or oil) cannot be extracted conventionally and requires fracturing through the injection of water at high pressure (fracking) containing a quartz-rich sand, known as a proppant, which is used to 'prop open' the fractures. Between 15% and 70% of this water, also containing acid to help initiate the fracture as well as other chemicals to prevent corrosion and scale formation, flows back to the surface, along with whatever gases and other substances are released during the fracking process. This

[63] http://www.heatherwick.com/teesside-power-station/

[64] Link no longer available

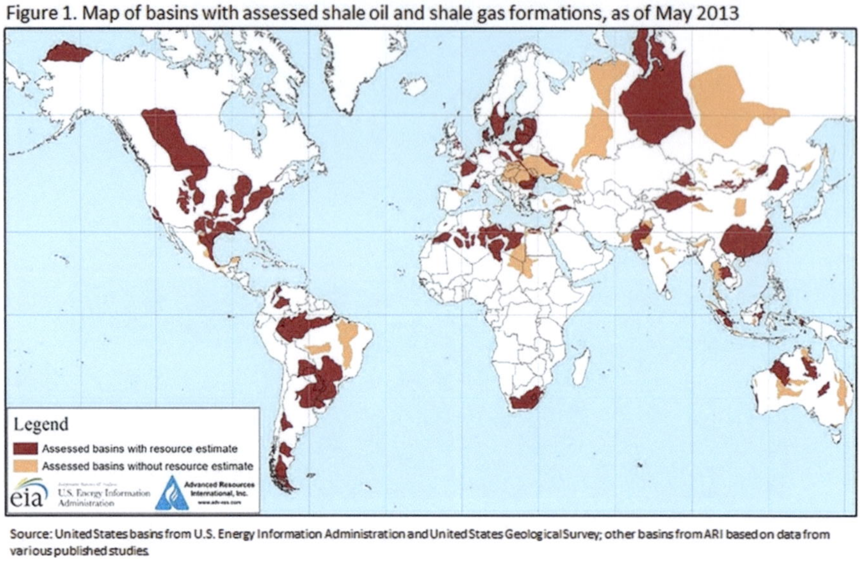

Fig. 2.8 Estimated distribution of shale gas and oil formations globally as assembled by the US Energy Information Association (http://www.eia.gov/analysis/studies/worldshalegas/; image: US Department of Energy; From: https://upload.wikimedia.org/wikipedia/commons/f/f6/EIA_World_Shale_Gas_Map.jpg)

flowback water is either stored for treatment and re-use, carried away in tankers for off-site treatment, recycled without treatment, injected into deep underground injection control (UIC) wells or allowed to evaporate. Some of the gases escape at the well head resulting in fugitive emissions of methane and other gases into the atmosphere. These have been variously estimated at between 1% and 9% of the total output from the well.[65] The problem is that methane has a global warming potential (GWP) some 25 times higher than carbon dioxide over a 100 year period or around 80 times higher over a 20-year period. This means that firing a power station with fracked gas could potentially result in greater global warming than for a typical coal-fired power station.

Methane emissions remain a problem across the US despite the fact that the US Environmental Protection Agency (EPA) launched the voluntary Natural Gas STAR Program in 1993 and chairs the Steering Group of the Global Methane Initiative, formed in 2004. Through the STAR Program 'EPA collaborates with oil and natural gas operators and industry stakeholders to encourage the identification and implementation of technologies and practices to reduce methane emissions from the oil and gas sector'.[66] The technology recommended by EPA for capturing up to 90% of

[65] https://www.carbonbrief.org/explained-fugitive-methane-emissions-from-natural-gas-production/

[66] https://www.epa.gov/natural-gas-star-program/natural-gas-star-program-and-methane-mitigation

methane from backflow is relatively inexpensive, with an estimated payback period of between 3 and 6 months.

There have been competing and contradictory studies of methane emissions in the USA. For example, a survey of '150 natural gas production sites with wells using hydraulic fracturing across the Gulf Coast, Mid-Continent, Rocky Mountain and Appalachian regions' published in September 2013 found that two-thirds had equipment installed to reduce fugitive methane emissions that resulted in typically a reduction of 99% in emissions.[67] However, the sites were identified by the fracking companies and are likely to represent best practice for the industry, typically meeting the EPA's 'green completions' or 'reduced emissions' criteria.[68] A study carried out by a team of scientists led by Scot Miller of the Department of Earth and Planetary Sciences at Harvard University estimated 'the spatial distribution of anthropogenic methane sources in the USA by combining comprehensive atmospheric methane observations, extensive spatial datasets, and a high-resolution atmospheric transport model'. Although a proportion of the methane measured is from ruminants, the amount of methane measured in the atmosphere above the south-central states of the USA indicated that fossil fuel extraction is the dominant source of atmospheric methane in that region.

A typical fracking site comprises a central 'pad' from which up to ten wells are drilled vertically into the gas-bearing rock, which can be kilometres below the surface, from where they spread out horizontally like the spokes of a wheel for a distance of up to 10 km (see Figs. 2.9 and 2.10). A similar technique is used for oil exploitation. A group of wells, which may constitute hundreds or thousands of wells, is usually referred to as a 'play'.

Analysis of production data by the US Energy Information Administration published in June 2013 concluded that at that time 11% of 'technically recoverable' global oil resources were comprised of tight oil, whilst 32% of gas resources were from shale gas. The data indicated that the exploitation of these resources had increased total resources globally by 11% for oil and 47% for gas.[69]

Globally much of the experience with fracking has been gained in the USA, although analysis by the US Energy Information Agency dating from 2013 indicates that technically recoverable shale oil resources in Russia exceeded those in the USA, whilst for shale gas, the resources estimated for China also exceeded those for the USA. Since the first exploratory wells using hydraulic fracturing were drilled in the USA in the 1930s, there have been thousands of wells drilled using this technology, although horizontal drilling did not become common until the late 1980s.

The potential environmental impact of fracking has been widely reported in the media, based on reports of water contamination, earthquakes, gas seepage, air pollution, noise disturbance and industrial scale movements of large vehicles.

According to the US Geological Survey 'Reports of hydraulic fracturing causing felt earthquakes are extremely rare. However, wastewater produced by wells that

[67] http://www.edf.org/climate/methane-studies/faq#3

[68] https://19january2017snapshot.epa.gov/sites/production/files/2016-06/documents/reduced_emissions_completions.pdf

[69] http://www.eia.gov/todayinenergy/detail.cfm?id=11611

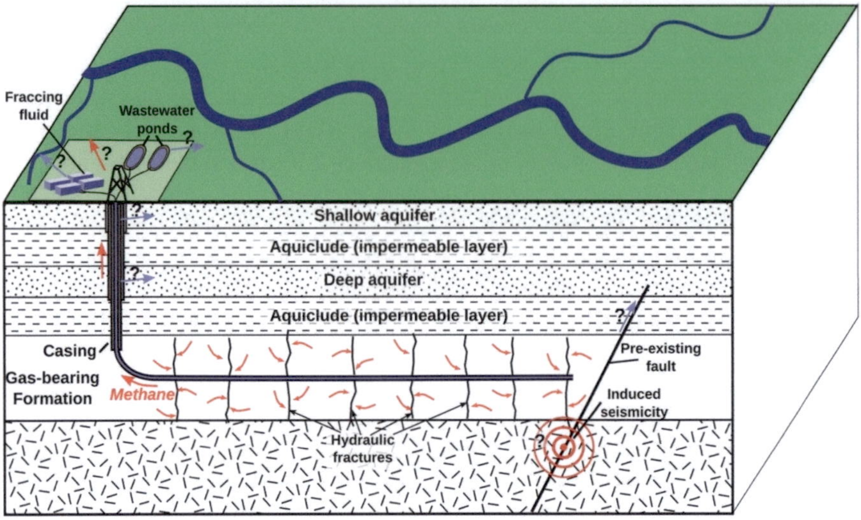

Fig. 2.9 Diagram showing typical section through an hydraulic fracturing well (https://commons.wikimedia.org/wiki/File:HydroFrac2.svg)

Fig. 2.10 Typical fracking site in the USA (https://upload.wikimedia.org/wikipedia/commons/9/90/Frac_job_in_process.JPG)

were hydraulic fractured can cause "induced" earthquakes when it is injected into deep wastewater wells'.[70] However, there have been instances where significant tremors have been experienced, including a shale gas hydraulic fracturing site in the North American state of Ohio for which it has been reported that 'over 109 small earthquakes … were detected during January 2011 to February 2012 in the Youngstown, Ohio area, where there were no known earthquakes in the past'. These shocks were close to a deep fluid injection (fracking) well. 'The 14-month period of seismicity included six felt earthquakes and culminated with (the greatest) shock on 31 December 2011 (measured at 3.9 MW on the moment magnitude scale*)'. The author concluded that the earthquakes were induced by the fluid injection at the well due to 'increased pore pressure along the preexisting subsurface faults located close to the wellbore' and at a depth of between 3.5 and 4.0 km in the 'Precambrian basement'.[71]

According to reports in the local press in Texas, the incidence of earthquakes having a magnitude of greater than 3.0 in the central and eastern USA has increased from an average of 21 per year between 1967 and 2000 to 100 per year since 2010. This has coincided with a fracking boom in these States. Some of these earthquakes have caused structural damage, such as the one having a magnitude of 5.6 that hit Prague, Oklahoma on 5th November 2011 damaging at least five homes, some towers and a road and resulting in injuries to two people. Experts have not been able to agree on the cause of this latter event however.[72] Across the USA, many operators have been disposing of waste water by injection into wells and it has been suggested that this process has been responsible for some of the higher magnitude earthquakes.

There have been numerous seismic events associated with fracking sites reported beyond the shores of North America, including relatively low magnitude earthquakes in the north west of England that have had a significant impact on the debate concerning the future use of fracking in the UK.

> On 1st April and 27th May 2011 two earthquakes with magnitudes of 2.3 ML and 1.5 ML[73] were detected in the Blackpool area. These earthquakes were immediately suspected to be linked to hydraulic fracture injections at the Preese Hall well (PH1), operated by Cuadrilla Resources Ltd. This well was hydraulically fractured during exploration of a shale gas reservoir in the Bowland basin. As a result of the earthquakes, operations were suspended at PH1 and Cuadrilla Resources commissioned a number of studies into the relationship between the earthquakes and their operations.[74]

Another key impact of concern relates to the potential contamination of drinking water. The gas or oil bearing sediments are typically many thousands of metres below the region where aquifers are found, hence the injection wells will be drilled through the aquifer in order to reach the sedimentary basin. Potential routes of contamination include:

---

[70] https://www.usgs.gov/faqs/how-hydraulic-fracturing-related-earthquakes-and-tremors

[71] http://onlinelibrary.wiley.com/doi/10.1002/jgrb.50247/abstract

[72] https://geowiki.ucsd.edu/sio15/case-studies/oklahoma-earthquake.html

[73] ML refers to the Richter scale which tends to be used for earthquakes below 3.5 ML but is roughly equivalent to the more recently devised moment magnitude scale (MMS).

[74] http://earthquakes.bgs.ac.uk/research/earthquake_hazard_shale_gas.html

- The leakage and spillage of water treatment chemicals being mixed at the well head;
- Leakage from the injection well;
- Leakage and spillage from wastewater ponds, tanks and injection wells.

The US Environmental Protection Agency (EPA) has carried out a major study into the potential impact on drinking water resources of fracking reporting on all of these issues.[75] The report concluded that the 'combinations of activities and factors (listed in Box 2.4) highlight, in particular, the vulnerability of ground water resources to activities in the hydraulic fracturing water cycle:'

Box 2.4 Factors That Potentially Contribute to Ground Water Pollution from Oil or Shale Gas Exploration

- Water withdrawals for hydraulic fracturing in times or areas of low water availability, particularly in areas with limited or declining groundwater resources;
- Spills during the management of hydraulic fracturing fluids and chemicals or produced water that result in large volumes or high concentrations of chemicals reaching groundwater resources;
- Injection of hydraulic fracturing fluids into wells with inadequate mechanical integrity, allowing gases or liquids to move to groundwater resources;
- Injection of hydraulic fracturing fluids directly into groundwater resources;
- Discharge of inadequately treated hydraulic fracturing wastewater to surface water resources;
- Disposal or storage of hydraulic fracturing waste water in unlined pits, resulting in contamination of groundwater resources.

Most reports to date refer to the presence of methane in drinking water and a study carried out by a team from Duke University, North Carolina, and published in the Proceedings of the National Academy of Sciences in 2013, analyzed 141 private drinking water wells located in northeastern Pennsylvania above the Marcellus Shale. Methane was detected in 82% of the wells, with those located within 1 km of a fracking well having concentrations of methane six times higher than those further away, with ethane concentrations 23 times higher. The authors conclude that there is a high probability that this contamination is from gases extracted from the Marcellus Shale.[76]

[75] https://cfpub.epa.gov/ncea/hfstudy/recordisplay.cfm?deid=332990

[76] http://www.pnas.org/content/early/2013/06/19/1221635110.abstract

It is interesting to note that in the USA, there are currently thousands of products that are available for use as additives to the proppant used for fracking. A significant proportion of these contain possible human carcinogens, yet the fracking process is exempt from the Safe Drinking Water Act (SDWA).[77] However, the US EPA Underground Injection Control (UIC) Program is responsible for regulating the construction, operation, permitting and closure of injection wells that place fluids underground for storage or disposal. This sets out 'minimum Federal requirements for the injection practices that protect public health by preventing injection wells from contaminating underground sources of drinking water'.[78]

Another concern is the presence of radon in wastewater from fracking wells. A report from Duke University published in October 2013 gives results from monitoring of local streams that take discharge water from a facility in western Pennsylvania treating wastewater from a shale gas fracking well. As well as elevated levels of chloride and bromide, 'radium levels in stream sediments at the point of discharge were (approximately) 200 times greater than upstream and background sediments and above radioactive waste disposal threshold regulations, posing potential environmental risks of radium bioaccumulation in localised areas of shale gas wastewater disposal'.[79]

Shale oil and gas extraction by hydraulic fracturing consumes a considerable amount of water, much of which is drawn from local rivers and streams. Although the amount of recycling is on the increase in the USA, as recently as 2011 only 8% of that drawn from natural supplies was recycled. A national study carried out by the Ceres Investor Network examined data available from the FracFocus.org data bank and the World Resources Institute on water consumption and water scarcity associated with fracking 'plays' between first January 2011 and 31st May 2013.[80] For the 39,294 wells fracked in that time some 97 billion gallons (437 million cubic metres) of water was used, compared with approximately 1700 billion cubic metres used in total by consumers during the same period. Overall some 55% of the plays were in areas experiencing drought, whilst in Colorado and California nearly all plays were in areas exposed to high or extremely high water stress. We saw earlier that high stress can be defined as when the ratio of withdrawal to supply of water is between 40% and 80% and 'extremely high stress' is when the demand exceeds 80% of supply.

The Ceres report refers to the massive Eagle Ford play in south Texas which has consumed 19.2 billion gallons during the time frame of the study, some 90% of which has come from groundwater. In one area the local aquifer levels have declined by between 30 and 100 m in recent years. Across the play, production was forecast to double during a 5-year time frame.

---

[77] http://water.epa.gov/lawsregs/rulesregs/sdwa/

[78] http://water.epa.gov/type/groundwater/uic/basicinformation.cfm#why_does

[79] For an abstract of the paper 'Impacts of Shale Gas Wastewater Disposal on Water Quality in Western Pennsylvania' by Warner, R et al go to: http://pubs.acs.org/doi/abs/10.1021/es402165b

[80] https://www.wri.org/insights/nearly-half-us-fracking-sites-overlap-water-stressed-regions

Perhaps it is not surprising that there is a growing resistance to fracking amongst both communities and politicians worldwide. In the USA, the position on fracking on Federal land depends on who is in the White House, whilst similarly politics plays a large part in the policies in individual States. For example, at one time fracking plays were either banned or put on hold in Vermont, New York State and a number of cities in Colorado. However, the invasion of Ukraine by Russia in 2022 created major stresses on supplies of oil and gas globally, opening up opportunities for fracking.

In Europe, where there has been a blanket ban on fracking in France, Germany and the Netherlands, for example, all countries which are in the front line of Russian restrictions, consideration has been given to resurrecting the fracking industry. In the UK, there was a moratorium from May 2011, following the seismic activity during exploration using fracking near Blackpool referred to above. This was lifted in December 2012, following which there have been a number of exploration licenses granted triggering a significant level of public protest and a subsequent ban in 2019 which remains in place at the time of writing.

The European Commission responded to concerns about fracking by publishing recommendations in 2014 'on minimum principles for the exploration and production of hydrocarbons (such as shale gas) using high-volume hydraulic fracturing'. This requires Member States to follow minimum principles when applying or adapting their legislation applicable to hydrocarbons exploration or production using high-volume hydraulic fracturing..[81]

Elsewhere there has been both frenzied activity from oil and gas companies fighting bans and moratoria on many fronts, along with a groundswell of anti-fracking campaigning and protest, much of which has been local and focussed on specific exploratory activities.

In the hunt for alternatives to fossil fuels, there has been a focus on short-rotation crops such as palm oil that can be converted relatively easily into a replacement for diesel and fuel oils. Although the majority of palm oil is used in food production, an increasing proportion is being dedicated to biofuel.

According to a report on the subsidy of palm oil for use as biofuels in the European Union:

> To a significant extent, palm oil production has been increasing at the expense of the global carbon stocks, such as tropical forests and peatland. For instance, in Kalimantan (Indonesia), from 1990 to 2010, 90 per cent of lands converted to oil palm were forested (47 per cent intact, 22 per cent logged, 21 per cent agroforests). If rainforest is cleared (or peatlands are drained) to convert the land for oil palm plantations, the carbon is released into the atmosphere, contributing to global climate change. The loss of tropical forests also results in loss of biodiversity, endangering some of the world's iconic species such as orang-utans.[82]

[81] https://eur-lex.europa.eu/LexUriServ/LexUriServ.do?uri=OJ:L:2014:039:0072:0078:EN:PDF

[82] From report published by Global Subsidies Initiative (GSI) and International Institute for Sustainable Development (IISD) which can be downloaded at http://www.iisd.org/gsi/sites/default/files/bf_eupalmoil.pdf see report for references.

This report states that the use of palm oil by EU countries increased from 0.4 million tonnes in 2006 to 1.9 million in 2012, whilst global production doubled between 2002 and 2012 to some 56 million tonnes and reached 77 million by 2022, at which time around 40% of this production was supplying the biodiesel market, with around 53% in the European Union.

Indonesian forests are of particular concern since, not only are the irreplaceable ancient trees being destroyed, but also the peat is being scrubbed up releasing methane and carbon dioxide resulting in significantly higher global warming potential than would be associated with normal deforestation.

The socio-economic impacts are also of concern. In an article on the Voice of America website from November 2013 Norman Jiwan, executive director of Transformation for Justice in Indonesia is quoted as saying that:

> Mass expansions of palm oil industry in Indonesia has created serious land conflict because of the land grabbing, land clearing without consent from local communities and indigenous peoples. And the likelihood of local communities and indigenous peoples' right to food is being threatened because of the massive expansions of the palm oil industry.[83]

In the same article Dr. Marcus Colchester, a senior policy advisor for the Forest Peoples Programme has commented that:

> Unfortunately, what we are seeing (in Africa) is the same kind of problems as we have already identified in Southeast Asia. That's to say the governments are handing out permits to these areas without first talking to communities, without firstly making sure their rights are recognized and secured. And so, the conflicts are brewing up in these different countries as the companies come in and take over the land and forests of these indigenous peoples, and other local communities.

2.4 Extreme Weather

As we will see in the next chapter extreme weather is a feature of climate change; so as seas warm and warming air is able to hold more moisture the amount of energy in weather systems increases, resulting in higher wind speeds, more rain and more energetic storms, hurricanes, tornadoes and cyclones. The impact of these weather events can be catastrophic, but they can be attenuated to some extent by adaptive strategies. The challenge is to be able to predict the incidence of extreme weather and plan accordingly.

In the Summary for Policymakers of its 2012 report *Managing the Risks of Extreme Events and Disasters to Advance Climate Change Adaptation* the Intergovernmental Panel on Climate Change (IPCC) states that:

[83] http://www.voanews.com/content/palm-oil-africa-indonesia-deforestation-land-grab-expansion-forests-indigenous-conflict-/1790043.html

> The impacts of climate extremes and the potential for disasters result from the climate extremes themselves and from the exposure and vulnerability of human and natural systems. Observed changes in climate extremes reflect the influence of anthropogenic climate change in addition to natural climate variability, with changes in exposure and vulnerability influenced by both climatic and nonclimatic factors.
>
> Disaster risk management and adaptation to climate change focus on reducing exposure and vulnerability and increasing resilience to the potential adverse impacts of climate extremes, even though risks cannot fully be eliminated.[84]

Hence human activities impact on both the potential causes of extreme weather events and the adaptation strategies available to manage them. The system of winds, vortices, cyclones, anticyclones and water–air interfaces that drive weather patterns across the surface of the globe is extremely complex. The variations in weather that occur at any given point on the planet's surface result from large global patterns in the atmosphere caused by the interactions of solar radiation, oceans, land mass topography and the rotation of the earth on its axis.

Wind direction and velocity are dependent on drivers such as differential solar heating creating vertical convective flows in equatorial regions and conversely, downward flows in polar regions. Infrared radiation from the sun is absorbed differently by land, water and snow surfaces. Differential warming causes variations in the temperature and pressure of overlying air masses, which rise as they warm or fall as they lose temperature. Pressure differences between masses of air generate winds, which tend to blow from high-pressure areas to areas of low pressure, deflected by the rotation of the earth by what is known as the Coriolis effect.

Fast-moving, upper atmosphere winds known as jet streams help move weather systems around the world. These predominantly flow from west to east in the boundary between the troposphere and stratosphere. The main jet streams are the polar jets, found at between 7 and 12 km above the surface of the earth, whilst the subtropical jets flow at an altitude of between 10 and 16 km (see Fig. 2.11). Jet streams tend to follow a wavy path due the Coriolis effect, with wind speeds of between 100 and 400 km/h, and measuring between 200 and 300 km wide by around 5 km deep.

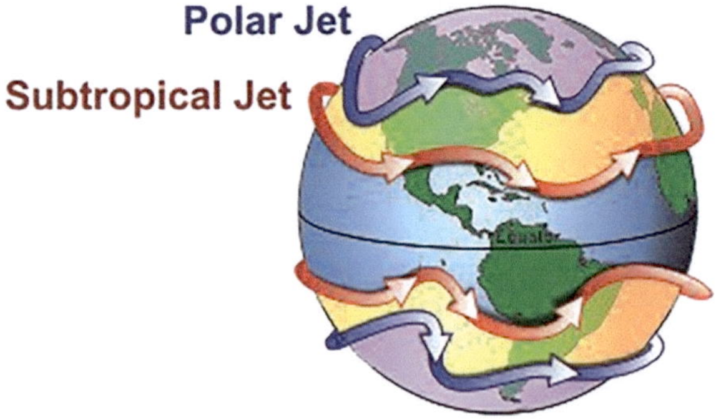

Fig. 2.11 Illustration showing typical wave pattern for jet streams (https://commons.wikimedia.org/wiki/File:Jetstreamconfig.jpg)

[84] https://www.ipcc.ch/report/managing-the-risks-of-extreme-events-and-disasters-to-advance-climate-change-adaptation/

Their path and extent of waviness depend on their strength which in turn depends on the convective forces in the atmosphere below. For example, the polar jet streams have historically been locked in by the polar ice beneath, however, in the recent decades, the area of ice has been steadily reducing, resulting in a much weaker jet stream and anomalous weather conditions in the northern hemisphere (see Extreme cold section below).

Jet streams are important in the formation of large weather systems such as cyclones and anticyclones. Cyclones rotate counter-clockwise in the Northern Hemisphere and clockwise in the Southern Hemisphere. Cyclones are also known as ‘lows’, because their centres are low-pressure areas. Clouds and precipitation are usually associated with these systems. Anticyclones, or ‘highs’, rotate in the opposite direction and are high-pressure areas, usually bringing clearer skies and more settled weather.

The boundary between two air masses having different temperatures and pressures is called a weather front. Here air velocity, temperature and humidity change abruptly, producing atmospheric instability. When there is a significant imbalance of temperatures, moisture contents and pressures in the atmosphere, storms develop, bringing precipitation and sometimes thunder and lightning.

Another major weather system that has a global impact is known as the El Niño-Southern Oscillation (ENSO) cycle which is created by fluctuations in temperature difference between ocean and land in the east central equatorial Pacific. The cold phase of this cycle is referred to as La Niña, whilst El Niño is the warm phase. The normal wind direction in equatorial Pacific is from east to west, resulting in cold water being drawn to the surface from the deeper eastern ocean and a temperature variation at the surface from 22 °C in the east to 30 °C in the west. During El Niño episodes these winds weaken resulting in warmer ocean temperatures in the east, whilst a La Niña event results in cooler ocean temperatures. El Niño episodes and La Niña events may alternate, but not always. The duration of either can be anything from 9 months up to several years, with intervals of between 2 and 7 years between.

As can be seen from Fig. 2.12, strong El Niño episodes can impact on weather across a large area, running from eastern Africa across much of Australasia to the Americas. A La Niña event typically has an almost diametrically opposite effect to an El Niño. During strong El Niño episodes, hurricane formation in the Atlantic is suppressed and the route of the seasonal cyclones is deflected to the eastern pacific. The opposite is true for La Niña, during which hurricane risk increases substantially for the USA eastern seaboard and the Caribbean. For example, the most active hurricane season ever occurred in 2005, including the catastrophic Hurricane Katrina, coinciding with the first La Niña event for five and a half years, whilst Hurricane Mitch, which occurred in 1998 at the start of a 2 year long La Niña event, reportedly led to some 11,000 confirmed deaths, mostly in Honduras and Nicaragua, with around the same number reported missing. In 2005 Hurricane Katrina led to the inundation of New Orleans, with over 1800 deaths in total and more than 80% of the city flooded. In 2012 Hurricane Sandy resulted in a storm surge that hit New York producing major flooding of streets and subways, power cuts and a cost to the US economy overall of some $65 billion. Atlantic hurricanes have also been found to have their genesis in ‘tropical waves’ which move from the western seaboard of Africa across the Atlantic, increasing in energy as they head westwards.

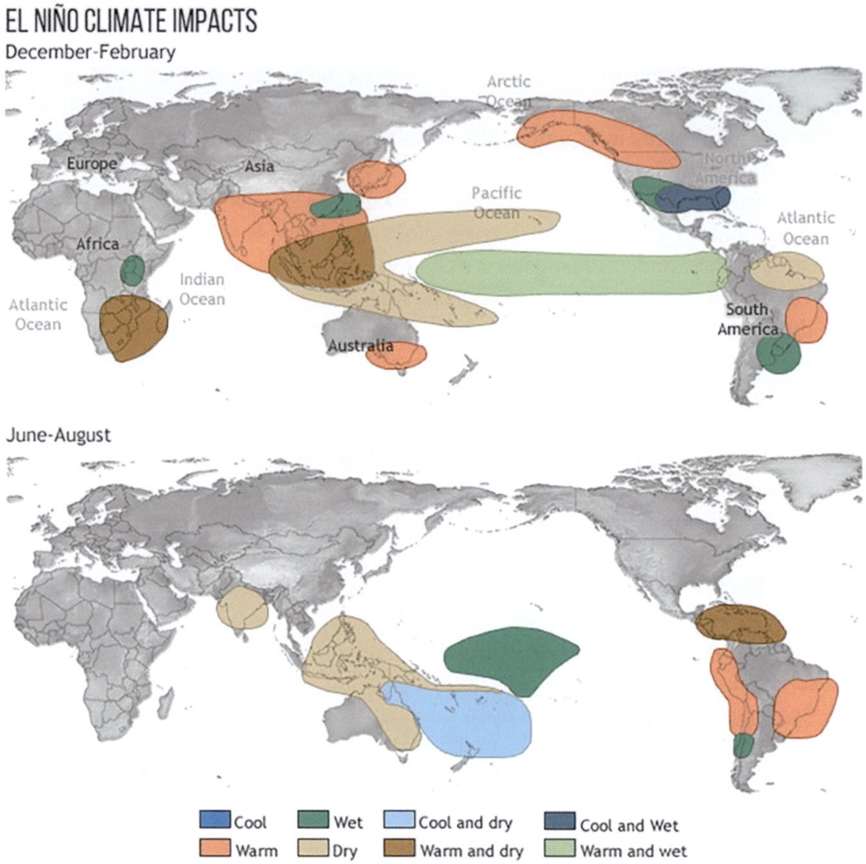

Fig. 2.12 Typical impact on weather from a strong El Niño episode (http://en.wikipedia.org/wiki/El_Ni%C3%B1o)

The location of the jet stream alters markedly depending on whether El Niño or La Niña dominates, for example, the jet stream over the USA follows a much more northerly path during a La Niña event.

Local weather is influenced by many factors, including the location's latitude, elevation and proximity to water bodies. Even the degree of urban development, which creates 'heat islands', and the amount of snow cover, which chills an overlying air mass, play important roles.

2.4.1 Storms and Flooding

Following a prolonged series of storms over Britain at the end of 2013 and beginning of 2014, the Meteorological Office reported that:

> 'The severe weather in the UK coincided with exceptionally cold weather in Canada and the USA. These extreme weather events on both sides of the Atlantic were linked to a persistent pattern of perturbations to the jet stream, over the Pacific Ocean and North America'.
>
> 'The major changes in the Pacific (sub-tropical) jet stream were driven by a persistent pattern of enhanced rainfall over Indonesia and the tropical West Pacific associated with higher than normal ocean temperatures in that region'.
>
> (Also) 'recent studies have suggested an increase in the intensity of Atlantic storms that take a more southerly track, typical of this winter's extreme weather. There is also an increasing body of evidence that shows that extreme daily rainfall rates are becoming more intense, and that the rate of increase is consistent with what is expected from the fundamental physics of a warming world'.[85]

Polar vortices are large cyclones that sit permanently over the Arctic and Antarctic at high altitude: typically within the troposphere, and measuring around 2000 km across. They tend to weaken in the summer and strengthen in winter and their boundaries are dependent upon the location and strength of the polar jet streams. For example, in the winter of 2013/14 a weakening of the jet stream resulted in it being deflected to the south over mainland North America, allowing the northern polar vortex to travel south bringing unusually cold air, snow and ice storms over much of Canada and the USA east of the Rockies. Similar conditions also occurred during the winter of 2022/23.

During the episode described above by the UK Met Office, while much of the population of North America was dealing with intense cold, some parts of the UK were inundated with flood water: either from overflowing rivers, storm surges from the sea, large waves, high water tables or some combination of these (see Fig. 2.13). There were also very high winds, in places hurricane force 12, which not only contributed to the flooding but also took down trees and power lines, damaged buildings and infrastructure, and created widespread travel chaos. There was also an increased incidence of sink-holes forming due to dissolution of underground rock from increased water flows: occurring primarily where soluble rocks such as limestone and chalk lay close to the surface. Excessive runoff also destabilised embankments, resulting in mud and rockslides.

All such events have significant political aftershocks, which can impact on the careers of politicians and even the life expectancy of governments. The focus of public scrutiny tends to evolve in a number of stages. In the immediate aftermath, the speed and effectiveness of the response of the authorities will be of particular concern. This might be followed by questioning the effectiveness of defences and, in the long term, what preventive measures should have been in place. Inevitably funding will lie at the heart of much of the debate as the questions proliferate.

As we will be looking at the causes and impacts of global warming in the next chapter, we will focus here on the causes of storm damage and flooding during extreme weather and the issues surrounding mitigation and adaptation to these. Clearly the key drivers for disruption and damage from a storm event are high winds and excessive precipitation. On-shore winds can result in the destruction of flood

[85] https://nora.nerc.ac.uk/id/eprint/505192/1/N505192CR.pdf

Fig. 2.13 Flooding of the Somerset Levels in February 2014 (https://commons.wikimedia.org/wiki/File:Confluence_of_the_Parrett_and_Tone_at_Burrowbridge_during_flooding_in_February_2014.jpg)

defences; inundation of exposed land and communities; and damage from waves and entrained detritus. The impacts can be very much worse when combined with high tides, and further still when rivers have been swollen by rain or melting snow. Rivers break their banks for a number of reasons, but primarily because more water is entering the river system than is being discharged at the coast. Water will overflow at low points along the river banks, and find its level on the adjoining land. Most rivers have their source or sources in elevated catchments and slow down, widen and meander as they reach flatter low lying land before discharging at the coast. These latter reaches are usually referred to as flood plains and may incorporate low lying sacrificial areas into which flood waters are channelled.

Because rainfall is usually highest in elevated regions, the amount of water that runs off into the river close to its source is critical. Where the land is impermeable or soils are shallow, rivers soon become not only swollen but heavily silted and polluted from soil erosion. A paper from December 2013 published in the Soil Use Management journal reported on a study of soils in the catchment of the rivers feeding the flooded Somerset Levels in southwest England:

> Field investigations between 2002 and 2011 identified soil structural degradation to be widespread in SW England with 38% of the 3243 surveyed sites having sufficiently degraded soil structure to produce observable features of enhanced surface-water runoff within the landscape. Soil under arable crops often had high or severe levels of structural degradation. Late-harvested crops such as maize had the most damaged soil where 75% of

> sites were found to have degraded structure generating enhanced surface-water runoff. Soil erosion in these crops was found at over one in five sites [6].

It is evident from this abstract that much of the upper catchment of the river Parrett, for example, that feeds the periodic floods in the Somerset Levels, is bordered by fields that provide little storm water attenuation. It is interesting to note that the Labour Government in Britain introduced in 2005 a requirement for farmers wishing to obtain a subsidy through the Single Payment Scheme to carry out a Soil Protection Review (SPR) meeting a series of Good Agricultural and Environmental Conditions (GAECs) criteria. The 2010 version of the SPR and associated guidance included requirements that specifically relate to the growing of maize and other forage crops, which traditionally has involved long periods during which the soil remained bare, leaving no roots to bind it and retain moisture.

In 2010 Defra's Rural Payments Agency (RPA) published *Cross Compliance Guidance for Soil Management* for those carrying out an SPR which includes a menu of 'Land Use Measures' for forage crops including: that ground cover crops should be sown under maize; that land be replanted within 10 days of harvesting; and that a temporary cover crop be left in the soil throughout winter.[86] Part 6 of the SPR provided an optional recommendation stating that 'in order to protect water quality it may be advisable to place six metre grassy buffer strips next to watercourses adjacent to arable land. Buffer strips can contribute to the reduction of pollution from farming activities…'. The scheme ended in December 2023 and was replaced with a sector-based set of rules.

If these measures were universally applied, then it is likely that many rivers that run through farmlands would be less prone to flooding and the accumulation of silt, as well as suffer less pollution from pesticides and fertilisers.

It is widely thought that the roots of trees help bind soil and reduce the potential for runoff and soil erosion. However, not all forests are good for erosion control: for example, where there are only thin canopy layers and minimal litter cover, along with shallow root systems and soil, attenuation of storm runoff will be minimal. One study published in 2007 concluded that on average a 10% reduction in tree cover results in a 28% increase in flood risk (141).

One of the countries, which is most closely associated with flooding is Bangladesh. This is a country of 147,570 km^2, 80% of which is alluvial flood plain, much of which lies within 5 m of sea level and, with a population in 2022 of around 170 million, one of the densest and fastest growing countries in the world. Bangladesh lies at the confluence of three major rivers: the Ganges, Brahmaputra and Meghna, which come together in the Ganges delta and empty into the Bay of Bengal, along with numerous other minor rivers and waterways. Some 20% of the land is flooded between June and September every year, fed by the monsoon rains in the mountains that provide the catchment for these river systems, along with the spring snow melt in the Himalaya. Silt deposited by these predictable annual floods

[86] https://assets.publishing.service.gov.uk/media/5a7d985140f0b65084e7625c/Cross_Compliance_Guide_to_Soil_Management_2010_edition.pdf

has over the centuries created the fertile farmland that the country depends upon for much of its subsistence crops. Despite being a regular event, these floods typically result in millions of homes being destroyed and thousands of casualties, although the figures are much higher when flooding from rivers and cyclones combine, adding high winds and storm surges to the inundation from rivers. Storm surges can also create a problem with salination, in some instances contaminating drinking water, agricultural land and fresh water habitats.

It is not unusual for commentators to blame Bangladesh floods on deforestation of the Himalaya catchment and increased volume of melt water from the glaciers. However, Thomas Hofer of the UN Food and Agricultural Organisation (FAO) in his project report *What are the impacts of deforestation in the Himalayas on flooding in the lowlands? Rethinking an old paradigm* concludes that 'Large-scale flooding has always occurred in Bangladesh. Based on these project findings there is no reason to assume that anthropogenic activities in the Himalayas have changed the frequency and dimension of flooding in Bangladesh'.[87] Indeed flooding in Bangladesh is most closely correlated to rainfall in the nearby Meghalaya hills, which experience some of the highest rainfall levels of anywhere on the planet. Although heavily forested—nearly 70% of Meghalaya State is tree covered—the shallow soil provides little attenuation to water runoff down precipitous slopes onto the Meghna floodplains of Bangladesh. Hofer suggests that, despite anomalous years, the frequency and volume of flooding in Bangladesh has not increased in some 120 years.

Where mostly impermeable surfaces border river systems, such as when waterways run through built up areas, a significant proportion of precipitation ends up in the river. Many urban areas have a rain water drainage infrastructure that empties into open waterways, sometimes combined with sewage outfalls. According to the UNESCO *Aquatic Habitats in Urban Water Management* website 'Runoff (accounts) typically for 10–20% of the average annual rainfall in rural areas. In urban areas, where surfaces are highly impervious, typical runoff volumes range between 60% and 70% of the average annual rainfall'.[88] Where river catchments are paved over this has an even more pronounced impact on downstream flooding than poor soil husbandry. Whilst building on flood plains increases the risk of flooding directly, since flood water has typically nowhere to go other than into nearby low-lying buildings, rather than soaking away into the water table. Furthermore, overburdened storm drains may back-up and cause flooding some distance from the river system.

Although the silt that is washed away from land bordering river systems and deposited downstream can contribute to the development of fertile flood plains and provide nutrients that can benefit ecosystems, there are also many negative impacts to the resultant sedimentation:

---

[87] https://www.fao.org/4/XII/0982-B2.htm

[88] http://www.aquatic.unesco.lodz.pl/index.php?p=water_cycle

- Deposition in watercourses, storm drains, reservoirs, and so on can reduce volume available to contain flow, resulting in banks being breached and necessitating remedial dredging.
- Entrained toxic substances can be absorbed by silt, poisoning fish and reducing oxygen levels.
- Excessive richness of nutrients (eutrophication) in runoff can result in choking of waterways with weeds, whilst also promoting algal growth and increasing turbidity.
- Silt can smother marine communities, suffocate corals, mangrove stands, seagrass beds, etcetera.
- Suspended material can damage fish gills and destroy their habitats.

It is important to realise that dredging is not a cure-all panacea for flooding since it 'can damage plants and animals living in the river and cause other problems such as increased erosion and higher flood risk downstream' [7]. Furthermore, disturbing the river bed during dredging can cause silt to be carried downstream, hence precautions should be taken to remove this through the placing of obstructions immediately downstream to filter or contain the silt. Hence prevention of soil erosion and attenuation of surface water runoff should always be considered as the first line of defence against river flooding.

Coastal flooding can of course occur without any form of precipitation, although the weather systems that generate the high winds that create high waves or storm surges are often associated with rain. As occurred in Southern Britain during the winter storms of 2013/2014, severe flooding generally occurs when there is a combination of high onshore winds and high tides, exacerbated by greater than usual volumes of water in river systems discharging at the coast. According to the UK Meteorological Office's website:

> The main cause of a storm surge is high winds pushing the sea water towards the coast, causing it to pile up there. There is also a smaller contribution from the low pressure at the centre of the storm "pulling" the water level up, by about 1 cm for every 1 millibar change in pressure.... For example, when Hurricane Katrina approached the US coast in 2005, it generated a storm surge of more than 8 metres in some areas. This led to widespread flooding, including almost all of the city of New Orleans where the sea defences couldn't cope with the water level.[89]

A similar storm surge caused devastation in the East of England and the Netherlands during the night of the 31st January and morning of the first February 1953 when a massive depression in the North Sea combined with high tides to elevate sea levels by 5.6 m above average. A combination of inadequate coastal defences and lack of early warning led to more than 2500 deaths including 307 in England and 1836 in the Netherlands, with more than one hundred thousand people displaced and tens of thousands of buildings destroyed. In the Netherlands around 1365 km^2 of land was inundated, including 9% of all farmland, whilst in the East of England some 647 km^2 were flooded.

[89] https://weather.metoffice.gov.uk/learn-about/weather/types-of-weather/storms/storm-surge#:~:text=The%20main%20cause%20of%20a,1%20millibar%20change%20in%20pressure

Fig. 2.14 Overview of the delta works sea defence project, Zeeland, the Netherlands (http://en.wikipedia.org/wiki/File:Deltawerken_na.png (Wikimedia commons file))

In the years following that night some five billion euros has been spent by the Dutch Government on the 'Delta Works', a hugely ambitious programme of flood defences (see Fig. 2.14).[90] These major civil engineering structures comprise a combination of dams, operable barriers and levees. Dams were built where it was acceptable for the land-side habitats to become fresh water, whilst operable barriers were built where habitats had to remain salt water or access was required for shipping.

Storm surges or tidal waves tend to be more dramatic when they occur over relatively shallow stretches of water, such as the North Sea, which has an average depth of 90 m, dropping to a minimum of 15 m over parts of the Dogger Bank, which lies some 100 km off the East coast of England. They differ from tsunamis in that the latter are caused by either a submarine earthquake, subsidence or volcanic eruption, rather than extreme weather or tidal surges.

[90] https://www.holland.com/global/tourism/getting-around/interests/land-of-water/delta-works

2.4.2 Drought and Dust Storms

The other side of the coin for a warming planet is an increase in the potential for prolonged drought episodes. Put simply, drought is a shortage of water. It may be caused by a prolonged period of abnormally low rainfall, a lowering of the water table, poor deployment of water resources, poorly planned agriculture and land management or some combination of these. Droughts can last for many years and can lead to repeated crop failures, famine, disease, economic collapse, mass migration, social unrest and war. For example, the Darfur conflict in Sudan was fuelled by decades of drought in the region, resulting in nomadic Arab Baggara people moving away from drought afflicted lands into areas occupied by non-Arab peoples.

Of course it is not just Africa that experiences drought. For example, the central plains of the USA, lying in the rain shadow of the Rockies, tend to experience prolonged periods of low rainfall. These reached epic proportions in the 1930s, between 1946 and 1956, and between 1987 and 1989; whilst California experienced drought conditions from 2011 to 2017, with further periods with little rain contributing to the catastrophic fires of 2025. The droughts of the 1930s contributed to the creation of what became known as the 'Dust Bowl', when the deep-rooted grass that held the soil together was removed using mechanised ploughs that resulted in vast areas of bare soil exposed to the powerful prevailing westerly winds. This resulted in dust storms that became known as 'black blizzards' (see Fig. 2.15), which at their height reached as far afield as New York, Washington, Atlanta and Chicago, restricting visibility in these cities at times to less than 1 m. One such storm was estimated to have dumped some 5500 tonnes of soil on the city of Chicago in 2 days. The term 'Dust Bowl' refers to an area of the Great Plains comprising, according to the US Soil Conservation Service: "the western third of Kansas, southeastern Colorado, the

Fig. 2.15 A dust storm approaching Stratford, Texas, in 1935 (http://upload.wikimedia.org/wikipedia/commons/3/34/Dust-storm-Texas-1935.png)

Oklahoma Panhandle, the northern two-thirds of the Texas Panhandle, and northeastern New Mexico".[91] An area of some 400,000 km^2, although the drought in the 1930s eventually impacted some 70% of the US land mass.

Records indicate that droughts are cyclical in this region of the USA, however, the impact of this drought was particularly bad because of the change of land use adopted by new settlers, mainly from Europe, who arrived at the beginning of the twentieth century and set up homesteads on land allocated under a series of Federal Acts. This led to a major transition from large-scale cattle farming to cultivation of cash crops such as cotton and wheat, encouraged by a prolonged period of wet weather in the 1920s and a rise in commodity prices in the years following World War I and the Russian Revolution.

All this was happening during the Great Depression, following the Wall Street crash of 1926, and many of the settlers affected lost their only source of food and income and were forced to abandon their farms. It has been estimated that some 2.5 million had left by 1940, including 200,000 who headed for California, but unemployment was high wherever they went and many ended up living in very poor conditions. There are no reliable statistics on how many died, although it is thought to be many thousands, some from malnutrition and others from 'dust pneumonia': a form of pneumoconiosis brought about by inhalation of fine dust.

As early as '1935, federal conservation programmes were created to rehabilitate the Dust Bowl, changing the basic farming methods of the region by seeding areas with grass, rotating crops, and using contour ploughing, strip ploughing, and planting "shelter belts" of trees to break the wind. Farmers were defensive when outsiders criticised their farming methods. Only when they were paid did they begin to put the new farming techniques into practice. The dollar per acre they earned often meant the difference between being able to stay a bit longer or having to abandon their land'.[92] It is these techniques that saved the same region from more severe consequences during the subsequent droughts. The drought that lasted from 1946 until 1956 impacted roughly the same area as that of the 1930s but, although crop yields in some areas fell by 50% and many left farming, there was no mass exodus of the destitute. In fact nearly one million abandoned the rural life during both of the decades 1940 to 1950 and 1950 to 1960. Farms were becoming bigger, mechanisation was displacing labour and industry was offering alternative employment.

The next major drought in the USA occurred in the late 1980s and its impact was greatest further north, resulting in the failure of economically important corn and soybean crops, major forest fires and the upper Mississippi river becoming unnavigable. It became one of the North America's most severe and most expensive environmental disasters, resulting in a cost to the US economy of around $39 billion.

[91] http://www.pbs.org/wgbh/americanexperience/features/general-article/dustbowl-drought/

[92] http://www.pbs.org/wgbh/americanexperience/features/general-article/dustbowl-drought/

2.4.3 *Inversions and Smog*

Where the dust storms that led to desolation of the region that became known as the 'Dust Bowl' were driven by strong winds blowing over exposed flat lands, the opposite can occur in areas that are ringed by elevated land. Some of the world's largest and most congested cities, such as Beijing, Athens and Mexico City, lie within basins almost entirely ringed by hills. These hills not only shelter the cities from winds but can also allow inversions to be established. A temperature inversion occurs when a layer of still cool air forms, within which the temperature is not increasing with height and hence no convective uplift is occurring. When this layer forms a lid over a basin any convection occurring beneath will not penetrate the layer, so that the air is cooled and falls back to the ground. Hence any pollutants are trapped, which can result in them reaching high concentrations and the formation of smog. Inversions tend to occur during episodes of high pressure, coincident with low wind speeds and frequently during episodes of cold weather. Of course this is also when emissions from power stations and heating systems will be at their highest.

In China, as some 15% of its territory was shrouded in smog and Beijing was reporting 'beyond index' levels of particulate matter, Chinese Premier Li Keqiang stated at the 2014 National People's Congress in Beijing, that pollution is a major problem and the government will 'declare war' on smog by removing high-emission cars from the road and closing coal-fired furnaces.

Premier Li wrote in his work report submitted at the Congress that pollution is 'nature's red-light warning against the model of inefficient and blind development,' and that 'fostering a sound ecological environment is vital for people's lives and the future of our nation'. Following the publication of lower than anticipated growth rates for China released at the time of the Congress, some commentators suggested that this was due to a drag on the economy from dealing with both the causes and effects of pollution.[93]

Temperature inversions can create conditions in which pollution levels rise and visibility is reduced due to the accumulation of airborne particulates, even where there is no basin. Cities are particularly prone to smog, not only because of the greater density of pollution sources but also because of the close proximity of buildings can impact the wind environment at street level. In 1950s London the conditions were not dissimilar to those that regularly occurred in Beijing in the 2010s. The combination of inefficient combustion engines, coal-fired power stations, coal fires and uncontrolled industrial emissions was resulting in regular episodes of smog, especially on still winter days. The winter of 1952 was very cold and so there was extensive use of coal burning across the whole of London and its suburbs. The wind had been from the East, carrying pollution across the North Sea from the continent, but on 5th December an anticyclone settled in over London creating a temperature inversion that remained for some 5 days, resulting in enormous quantities

[93] https://phys.org/news/2014-03-china-premier-declares-war-pollution.html

of pollutants being held within the still air layer close to the ground. According to the UK Met Office:

> During the period of the fog, huge amounts of impurities were released into the atmosphere. On each day during the foggy period, the following pollutants were emitted: 1,000 tonnes of smoke particles, 2,000 tonnes of carbon dioxide, 140 tonnes of hydrochloric acid and 14 tonnes of fluorine compounds. In addition, and perhaps most dangerously, 370 tonnes of sulphur dioxide were converted into 800 tonnes of sulphuric acid.[94]

Visibility for much of this period was down to a couple of metres or less, the particulates taking on a black and yellow tinge due to the combination of tarry substances and sulphur, resulting in the term 'pea-souper' being coined. Inevitably many people contracted diseases of the lung such as bronchitis and bronchopneumonia. Estimates vary, but a retrospective study from 2004 concluded that some 12,000 people died as a result of their exposure to what became known as the Great Smog of 1952 [8]. This study takes account of the 'excess deaths' from chronic obstructive pulmonary disease that occurred in the months following the smog episode.

Despite some commentators of the day attempting to attribute the Great Smog to natural causes and initial resistance from the Government because of the economic implications of mitigation, a Select Committee on Air Pollution was established, eventually resulting in the first enactment in England, Wales and Scotland of a *Clean Air Act* some 4 years later, in 1956. As well as prohibiting 'dark smoke' and emissions of grit and dust, this Act introduced the power for Local Authorities to designate 'smoke control areas' and grants for those impacted to adapt their furnaces and fireplaces to meet the requirements in the Act.

London in the 2010s still experienced high pollution concentrations, but the primary sources were from traffic and industry. For example, between the 12th and 14th March 2014 an anticyclone settled over north west Europe resulting in pollution levels in a number of cities, including London and Paris, exceeding EU Air Quality Index (AQI) limits. Very similar conditions to those that stimulated the Great Smog of 1952 resulted in a cocktail of traffic exhausts and industrial emissions being held under a temperature inversion. In Paris the concentration of particulates smaller than 2.5 microns diameter ($PM_{2.5}$) reached 185 micrograms per cubic metre, which, although not reaching Beijing levels (see Chap. 1), resulted in the authorities introducing emergency measures such as rationing car use by daily alternation of odd and even number plates, lowering speed limits and withdrawing charges from bicycle hire and public transport.

2.4.4 Extreme Cold and Snow

It is perhaps counterintuitive to link extreme cold, snow and ice storms with global warming but it is useful to examine what scientists have revealed about the potential causes of the weather that much of North America experienced over the winter of

---

[94] https://www.metoffice.gov.uk/weather/learn-about/weather/case-studies/great-smog

Fig. 2.16 Lake Michigan waterline near Chicago: winter of 2013–14 (https://commons.wikimedia.org/wiki/File:Chicago_winter_2014_lake_michigan_frozen.jpg)

2013 to 14. We saw earlier how during this time Southern Britain experienced the wettest winter since records began. At the same time much of Western Europe was experiencing temperatures around 3 °C higher than average. However the warming of the Arctic led to a weakening of the polar jet stream which allowed cool air from over the Arctic to move south towards North America. Between January and March 2014 most of North America east of the Rockies experienced a prolonged period of snow and ice storms, resulting in around 50% of the USA and most of Canada being covered with snow for several months. By the end of February, 91% of the water in the Great Lakes was frozen (see Fig. 2.16). The minimum temperature across much of the central and eastern states was between 4 °C and 5 °C lower than the 1901 to 2000 mean. In contrast a large proportion of the western coastal states were experiencing record high winter temperatures.

Referring to Fig. 2.17, It can be seen how that 'In early January, the polar vortex weakened and broke down, allowing fragments of cold air to slosh out of the bowl into mid-latitudes. The image on the left shows the weakened vortex formation on January 5, 2014. The high-pressure building up in the Arctic slowed down the jet stream, which caused it to buckle into deep folds and flow farther south than usual, introducing cold Arctic air into the central and eastern USA'.[95] The 'geopotential height' referred to in the Figure is the height that the specified atmospheric pressure (500 millibars in this case) occurs above mean sea level. The unconstrained spread of a region of low geopotential height indicates that the air beneath has cooled,

[95] http://www.climate.gov/news-features/event-tracker/wobbly-polar-vortex-triggers-extreme-cold-air-outbreak

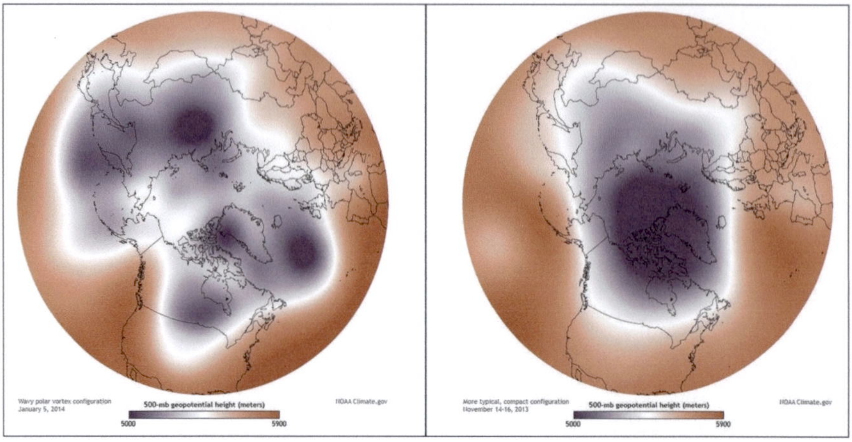

Fig. 2.17 Changes in the polar vortex and jet stream during the winter of 2013–14 (https://en.wikipedia.org/wiki/Polar_vortex)

become more dense and allowed troughs to form and hence spread the region of cold air further south than normal.

Authorities have linked the occasional weakening of the polar jet stream with a reduction in the area of sea ice in the Arctic, along with the elevation of air and sea temperatures in the region. Data from the US National Snow and Ice Data Center (NSIDC) indicate that this has been occurring for decades with an obvious and seemingly irreversible downward trend. Between 1979 and 2023 sea ice areas in the Northern Hemisphere fell on average by 24% in August and 12.5% in January.[96] The air temperature near the surface within the Arctic Circle has typically risen by around 3 °C or 4 °C since the early twentieth century.[97]

2.5 The Politics of Food Production and Agriculture

The environmental, ecological, economic and social impacts of agriculture have always been important in the survival and prosperity of humans. Indeed the very fabric of our surroundings has been sculpted by agricultural endeavour over the centuries. As Rachel Carson averred in *Silent Spring*: 'the history of life on earth has been a history of interaction between living things and their surroundings'; and in the years leading up to the publication of this landmark book the drive to increase the efficiency of food production led to a near catastrophic destruction of wildlife through the wholesale use of DDT and other pesticides. Carson's intervention is

[96] http://www.ncdc.noaa.gov/snow-and-ice/extent/sea-ice/N/1

[97] http://www.climate4you.com/Polar%20temperatures.htm

frequently credited with providing the catalyst for the modern environmental movement, as well as to DDT being outlawed across much of the planet.

Of course there remain many issues concerning the sustainability of agriculture and food production in the twenty-first century. In many ways these have become of greater concern: mostly because of the need to feed a rapidly growing population in a world in which there is widespread migration from the relative poverty of agrarian economies, whilst fertile land is being lost through urbanisation, inundation by the sea, erosion, degradation and desertification. In parallel with this the maritime environment is also being degraded and fish stocks depleted. In many parts of the world industrial scale fishing has brought the more popular types of fish with the consumer, such as cod and tuna, to levels that threaten recovery whilst indiscriminately killing unwanted sea creatures through discards and contributing to the destruction of important ecosystems such as reefs through trawling.

The IPCC Working Group II included in Table TS.AII.1 of its technical summary for its contribution to the sixth Assessment Report the following 'Climate-related representative key risks (RKRs)' relating to agriculture and ecosystem services:

- Risks to ecosystem services, people, livelihoods and key infrastructure in low-lying coastal areas and associated with a wide range of hazards, including sea level change, ocean warming and acidification, weather extremes (storms, cyclones) and sea ice loss, for example.
- Transformation of terrestrial and ocean/coastal ecosystems, including change in structure and/or functioning and/or loss of biodiversity.
- Systemic risks due to extreme events leading to the breakdown of physical infrastructure and networks providing critical goods and services.
- Food insecurity and the breakdown of food systems due to climate change effects on land or ocean resources' [9].

Rising water temperature, increased stratification, and ocean acidification all pose a risk to marine ecosystems and the services they provide for coastal livelihoods and fishing communities in particular. Whilst rising air temperatures, changes in precipitation patterns, and extreme weather events pose risks to terrestrial ecosystems and the services they provide for terrestrial livelihoods, and in particular for communities whose livelihoods depend on provisioning services.

According to a 2011 report from the EU Standing Committee on Agricultural Research:

> Many of today's food production systems compromise the capacity of Earth to produce food in the future. Globally, and in many regions including Europe, food production is exceeding environmental limits or is close to doing so. Nitrogen synthesis exceeds the planetary boundary by a factor of four and phosphorus use has reached the planetary boundary. Land use change and land degradation, and the dependence on fossil energy contribute about one- fourth of Greenhouse Gas emissions. Agriculture, including fisheries, is the single largest driver of biodiversity loss. Regionally, water extracted by irrigation exceeds the replenishment of the resource.[98]

[98] https://wayback.archive-it.org/12090/20230310195327/https://ec.europa.eu/environment/archives/eussd/food.htm

We saw earlier in this chapter that the concept of 'Planetary Boundaries' was developed by a team led by Johan Rockstrom in 2009.[99] Of the nine thresholds relating to global ecosystems postulated by this team a number relate to food production, fishing and biodiversity. For example, in a section dealing with 'interference with the global phosphorus and nitrogen cycles' the authors state that:

> Human-induced degradation of ecosystem states (e.g., overfishing, land degradation) and increase in nitrogen and phosphate flows at regional to global scales may cause undesired non-linear change in terrestrial, aquatic, and marine systems, while simultaneously functioning as a slow driver influencing anthropogenic climate change at the planetary level.

The nitrogen compounds used in food production and associated with combustion, vehicle exhausts and many more are reactive. They can be biologically, photochemically or radiatively active, resulting in potential polluting episodes. For example, one of the compounds—nitrous oxide (N_2O) has a global warming potential of 310 over 100 years: that is more than 300 times greater than carbon dioxide; whilst having a life expectancy in the atmosphere of approximately 120 years. In the USA nitrous oxide contributes around 6% of total greenhouse gas emissions, whilst some 80% of this comes from agricultural sources, including some 4% from the decomposition of manure and urine—which also contributes to methane emissions, another powerful greenhouse gas.[100]

Molecular nitrogen (N_2) on the other hand, which makes up around 78% of the earth's atmosphere, is non-reactive. In their paper examining nitrogen use in France and the development of a metric that they call the 'nitrogen food-print' Chatzinpros & Barles report that:

> Global food production is the primary cause for anthropogenic reactive nitrogen (Nr) creation and has a major influence on the global nitrogen cycle. By the end of the 20th century, about 75% of global human-driven inputs of Nr were used for agriculture but only 30% of these inputs were effectively recovered into vegetal proteins to feed humans and livestock. The non-recovered fraction is mostly lost to the environment where it contributes to the nitrogen cascade, which is defined as "the consequential transfer of Nr through environmental systems and which results in environmental change as Nr moves through or is temporally stored within each system". The nitrogen cascade is now recognized as a major crosscutting theme over all environmental problems and global change issues such as climate change, biodiversity losses, groundwater pollution, eutrophication, tropospheric ozone generation and stratospheric ozone depletion with severe effects on ecosystems and human health.[101]

Nitrogen is used in fertilisers because it forms part of the chlorophyll molecule that is responsible for producing food for the plant through photosynthesis, resulting in green, leafy growth. Phosphorus is also used in fertilisers, primarily to exploit its ability to enhance the metabolic processes through root development and flowering.

A significant proportion of the nitrogen and phosphorous in the form of nitrates and phosphates from fertilisers is carried away from fields in runoff water into the

---

[99] http://www.ecologyandsociety.org/vol14/iss2/art32/

[100] http://epa.gov/climatechange/ghgemissions/gases/n2o.html

[101] Note that the references in this extract can be found in the original paper at: http://www.biogeosciences.net/10/471/2013/bg-10-471-2013.pdf

river systems and out into the seas and oceans. Some also will percolate into groundwater and aquifers. In open waterways, both compounds can promote algal growth through photosynthesis (eutrophication) which draws in bacteria at night resulting in an oxygen deficiency (hypoxia) which can have a devastating effect on aquatic species, creating 'dead zones' that can stretch over thousands of square kilometres.

According to an Issue Brief published in 2013 by the UN Development Programme (UNDP):

> Worldwide there are now more than 500 'dead zones' covering 250,000 km^2 with the number doubling every ten years since the 1960s. The economic costs to fisheries, tourism and other coastal livelihoods are already in the many tens of billions of dollars annually and will only continue to increase in the 'business as usual' scenario.[102]

Hypoxia causes major changes in the structure and functions of ecosystems, including:

- Mass mortality of fish and benthos (marine life on the sea bed).
- Changes in species composition.
- Changes in nutritional relationships.
- Decrease in biodiversity and species richness.
- Decrease in fisheries production.
- Increase in harmful and noxious algal blooms.

One of the largest dead zones can be found at the mouth of the Mississippi river in the Gulf of Mexico. In a bad year it can stretch to around 22,000 km^2, although it has shrunk to around 5000 km^2 at times. The main factors governing size are the intensity of agricultural activity in the Mississippi catchment, the turbulence generated by storms and hurricanes locally and effectiveness of flushing mechanisms (see below). In 2012, for example, a major drought reduced production considerably in the lands bordering the Mississippi and its tributaries, resulting in a dead zone of 7400 km^2. Figure 2.18 shows an infrared image which shows the red zone suffering from hypoxia with improving oxygen levels with distance from the coast through to the blue oxygen-rich water beyond.

A 2008 Rapid Response Assessment from the UN Environment Programme (UNEP) *In Dead Water* reported that '... 10% to 15% of the oceans and seas where fish stocks have been and remain important... (are being) ... increasingly damaged by over-harvesting, unsustainable bottom trawling and other fishing practices, pollution and dead zones, and a striking pattern of invasive species infestations in the same areas. These areas ... may lose more than 80% of their tropical and cold water coral reefs due to rising sea temperatures and increasing concentrations of carbon dioxide leading to a decrease in seawater pH (acidification). These same areas are also facing rapidly growing pollution from coastal development and the slowing of flushing mechanisms due to climate change'.[103] These flushing mechanisms are a function of what is known as 'ocean thermohaline circulation' established by convection currents that occur as cool water sinks to the bottom of the ocean displacing

[102] https://www.undp.org/publications/issue-brief-ocean-hypoxia-dead-zones

[103] https://wedocs.unep.org/handle/20.500.11822/7658

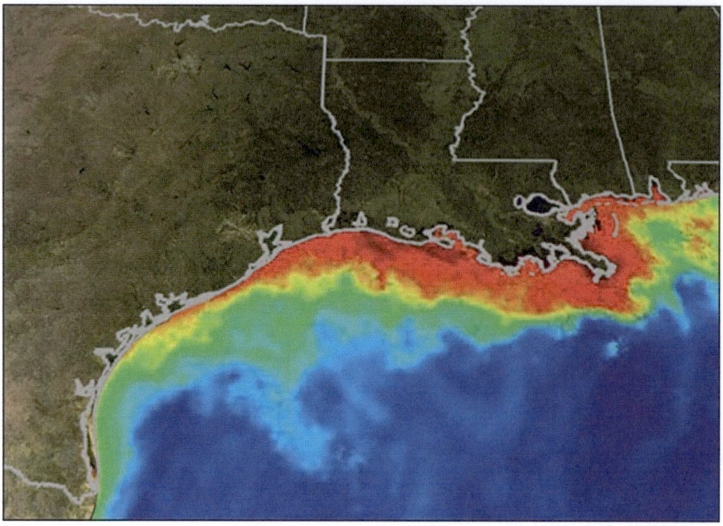

Fig. 2.18 The dead zone in the Gulf of Mexico at the mouth of the Mississippi river (http://serc.carleton.edu/microbelife/topics/deadzone/index.html)

rising warmer water. Increasing surface water temperatures are slowing down this process, which is crucial to coastal water quality and nutrient cycling: and hence the deterrence of dead zone growth.

Another example of a water body having a major problem with hypoxia and eutrophication is the Baltic Sea. In this case the sea is semi-enclosed, with only a relatively narrow channel connecting it to the North Sea, resulting in a water exchange that takes several decades. The Baltic is bordered by nine countries, including Sweden, Finland and Russia, and takes the water from rivers whose catchments include vast areas of agricultural land, as well as major industry and conurbations. Although, like the Gulf of Mexico, the area of its dead zone fluctuates, it has been estimated to have reached approximately 70,000 km^2, which represents an alarming 17% of its total surface area.[104] Because of the need for international cooperation to deal with the causes of the pollution to this important environment the Baltic Marine Environmental Protection Commission, otherwise known as the Helsinki Commission, or HELCOM, was established following the Helsinki Convention in 1992, with the contracting parties comprising all those countries bordering the Baltic Sea, along with the European Union. The resulting Baltic Sea Action Plan commits all parties to reducing the magnitude of pollution reaching the Baltic through a series of 'ecological objectives and measurable targets'.[105]

On a global scale the surface pH of water in lakes, seas and oceans is gradually reducing in direct response to the increase in carbon dioxide uptake and the

[104] http://pubs.acs.org/doi/pdf/10.1021/es802762a

[105] http://helcom.fi/baltic-sea-action-plan

formation of carbonic acid. Around 25% of carbon dioxide emissions are dissolved in the oceans. For example, the concentrations of carbon dioxide in sea water measured off the coast of Hawaii between 1989 and 2015 showed an increase in CO_2 concentration of around 15% and a reduction in pH of 7%. Lowest pH values in 2015 were approximately 7.8, whilst values below seven are considered acidic. Nevertheless this reduction in pH can impact on shell and skeleton formation and contribute to the destruction of corals, shellfish and plankton, changing the structure and dynamics of marine ecosystems and hence fish stocks.[106]

Another threat to marine ecosystems is the colonisation by species that are not native. This can occur through ships discharging their ballast tanks in regions distant from where they were filled; whilst some creatures have been shed from their temporary homes attached to ship's hulls (see section on Water pollution and ecological impacts in Chap. 8).

'The spread of invasive species causes ecosystem degradation and fragmentation. These processes are identified as the world's greatest threats to nature and key causes of species extinction'.[107] Habitats that have been impacted by pollution events, temperature rise due to climate change or trawling, for example, tend to be more vulnerable to colonisation by invasive species.

Perhaps unsurprisingly the outcome of all of the above is that fisheries are collapsing throughout the world. For example, according to the 2013 report from the International Scientific Committee for Tuna and Tuna-like Species in the North Pacific Ocean, by 2010 stocks of Bluefin tuna had fallen by 96% since before the introduction of industrial scale fishing, although a 2024 report from the same organisation found that stocks recovered significantly following the implementation of stringent management measures,[108] whilst the harvest of the Caspian Sea sturgeon fell from 27,700 tonnes in 1977 to 461 tonnes in 2000 [4, p. 274]. One landmark study from 2003 reported by Smith et al., concluded that '90% of large fish in the oceans had disappeared over the last 50 years'.

This loss has continued and means that the pressure on land-based agriculture is even greater, and with some forecasts predicting that the global human population will increase from 7.7 billion in 2019 to 9.7 billion in 2050, the challenge becomes one of great urgency (see Fig. 10.1 in Chap. 10).

Globally there is an ever-increasing reliance on meat and meat products, much of them highly processed, as primary sources of nutrition. Societies that have for centuries been almost entirely vegetarian in their diet are turning to meat consumption at the same time as they are becoming urbanised, buying cars and increasing their per capita carbon footprint. However, as the UN Food and Agriculture Organization (FAO) stated in its 2009 report on the *State of Food and Agriculture: Livestock in the Balance:*

---

106 https://oehha.ca.gov/epic/climate-change-drivers/acidification-coastal-waters

107 https://www.bio-conferences.org/articles/bioconf/full_html/2021/03/bioconf_pibidr2021_00024/bioconf_pibidr2021_00024.html/

108 https://isc.fra.go.jp/pdf/ISC24/ISC24_ANNEX13-Pacific_Bluefin_Tuna_Stock_Assessment_in_2024-FINAL.pdf

> Livestock is the world's largest user of land resources, with grazing land and cropland dedicated to the production of feed representing almost 80 percent of all agricultural land. The sector uses 3.4 billion hectares for grazing … and 0.5 billion hectares for feed crops; the latter figure corresponds to one-third of total cropland. The total land area occupied by pasture is equivalent to 26 percent of the ice-free terrestrial surface of the planet.[109]

If the 3.9 billion hectares referred to above were dedicated to growing crops that were providing vegetarian food having a suitable nutritional value for human sustenance, then according to one study reported by Kurt Schmidinger in his 2012 thesis on *Worldwide Alternatives to Animal Derived Foods,* ten times as many people could be sustained.[110]

In an effort to improve both efficiency and costs, agriculture in the developed world has become largely industrialised. This has led to farming on a large scale: some farms comprising thousands of acres of single crops (monoculture) and others creating production lines for the meat or dairy markets.

According to a 2019 report by the UN Food and Agriculture Organisation, the livestock sector has been responsible for around 10% of global water use, much of which has been used in the irrigation of feed crops,[111] although this proportion has been much higher in dry areas, such as in parts of Africa. Typically the production and processing of 1 kg of beef requires on average some 15,800 L of water, whilst pasta requires 1900, rice 2500 and potatoes only 300 L/kg.[112] These are global averages and there are significant differences between grazed and industrially farmed animals and between different countries.

Industrially farmed (intensively reared) animals require more water than grazed (extensively reared) animals whilst there are a number of other negative environmental impacts of intensive livestock production. The 2009 FAO report referred to above states that:

> The separation of industrialized livestock from its supporting land base interrupts the nutrient flows between land and livestock. This creates problems of depletion of nutrients at the source (land, vegetation and soil) and problems of pollution at the sink (animal wastes, increasingly disposed of into waterways instead of back on the land).[113]

This contributes to the eutrophication process in waterways and coastal waters and the reduction of biodiversity and fish stocks referred to earlier in this chapter.

Figure 2.19 shows an aerial photograph of a waste lagoon taking animal waste from a 'feedlot'—or concentrated animal feeding operation—in Texas. The cattle are intensively reared in pens, some of which can be seen in the upper part of the image. The lagoon measures approximately 600 m across and contains largely manure. Because it is open, like many others in the USA, for much of the time there are likely

[109] http://www.fao.org/docrep/012/i0680e/i0680e04.pdf

[110] http://www.futurefood.org/DissertationSchmidinger.pdf

[111] https://openknowledge.fao.org/server/api/core/bitstreams/22e23c47-5393-451e-b6aa-3f2c6fbc7cbe/content

[112] https://watercalculator.org/water-footprint-of-food-guide/

[113] http://www.fao.org/docrep/012/i0680e/i0680e04.pdf

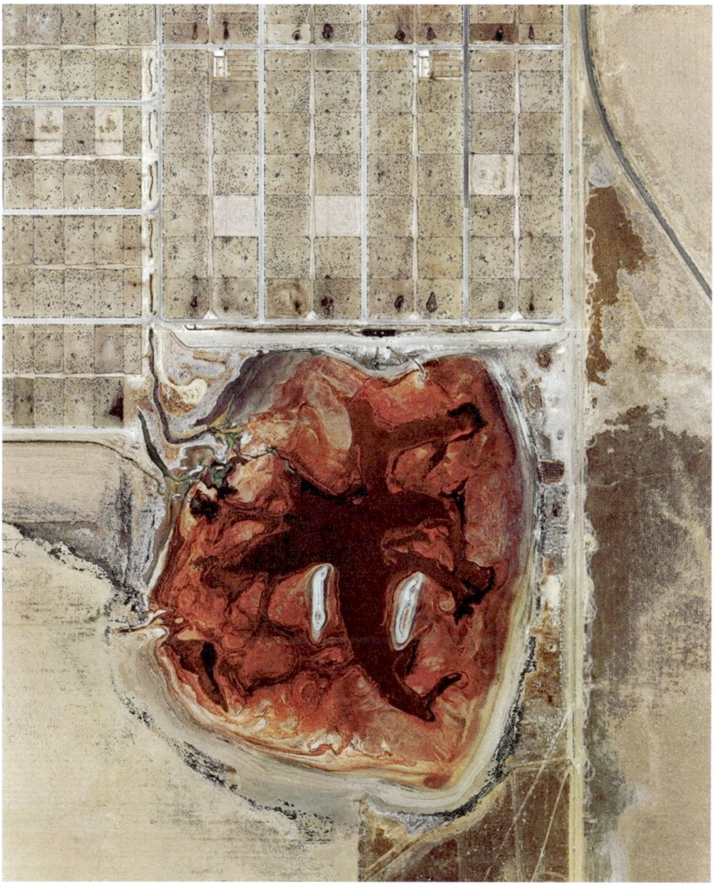

Fig. 2.19 Aerial photograph of Coronado Feeders, Dalhart, Texas (2012) (Photo: Mishka Henner)

to be malodorous emissions of chemicals such as methane and ammonia carried by the winds for many kilometres. It is important to note that in the USA the Environmental Protection Agency regulates all animal feeding operations under the 1972 Clean Water Act, which introduced a permit system for point sources of pollution, including feedlots, under a 'National Pollutant Discharge Elimination System' (NPDES).

Livestock makes a significant contribution of around 18% to global anthropogenic greenhouse gas emissions, calculated through the entire food chain. This includes 9% of global carbon dioxide emissions, 37% of global methane emissions and 65% of nitrous oxide emissions.[114] Although extensively reared livestock has a higher GHG footprint per kg of product than intensively reared animals, globally they only represent 8% of the total meat output.[115]

---

[114] http://www.fao.org/docrep/012/i0680e/i0680e04.pdf

[115] https://www.tabledebates.org/sites/default/files/2020-10/FCRN_int_vs_ext_livestock.pdf

Although there are greenhouse gas emissions associated with harvesting of arable crops and processing of the product, the cultivation of vegetative species provides a sink to carbon dioxide and other airborne pollutants. However, where forests have been cleared to make way for arable crops the carbon sequestration provided by the trees is significantly greater than that provided by the crops that have replaced them.

The intensive rearing processes associated not only with the kind of livestock farming shown in Fig. 2.19 but also with the indoor rearing of poultry and pigs, have been developed to increase efficiencies. According to the 2009 FAO report referred to above:

> Productivity gains are, however, also associated with a number of environmental concerns. The relatively low resistance to diseases of highly productive breeds, the concentration of large numbers of animals in large production units and the need to avoid disease outbreaks has led producers to use substantial amounts of drugs, often as routine preventive measures. Residues from these drugs pass into the environment, harming ecosystems and public health. In particular, the sometimes indiscriminate use of antibiotics has led to the selection of antibiotic-resistant strains of bacteria, now threatening human health in Europe and North America.[116]

Industrial scale farming has also led to a focus on breeds that are the most productive, whilst providing a product that is popular to the consumer. However, the FAO again sounds a warning in that ‘the rapid spread of intensive livestock production that utilises a narrow range of breeds has contributed to the marginalisation of traditional livestock production systems and the associated animal genetic resources’.[117]

We have already heard about how the impacts of DDT were highlighted by Rachel Carson in the mid-twentieth century. DDT is an organochlorine insecticide which, along with chlordane and other similar chemicals, persists in the environment for years, with the potential therefore to penetrate the entire food chain and accumulate in living organisms (bioaccumulation). Despite these chemicals being phased out across most of the developed world under the 2004 *Stockholm Convention on Persistent Organic Pollutants* (POPs), they are still used extensively in Southeast Asia, India and Africa, primarily in the control of insect-borne diseases such as malaria. A study by van der Berg et al. from 2012 found that 82% of the global use of DDT between 2003 and 2007 was in India, representing between 4500 and 5200 tonnes/year, with the remainder in Africa and North Korea. India was by far the largest manufacturer of DDT, with a limit of 10,000 tonnes/annum imposed by statute, whilst China, South Africa and North Korea were also reported as being producers, although China ceased production in 2008/9. Almost all of the stocks were thought to be used for disease control, apart from in North Korea where it has been reported to be used in the treatment of crops, in contravention of the Stockholm Convention.[118] India committed to phase out the use of DDT by 2024; however, it

[116] http://www.fao.org/docrep/012/i0680e/i0680e04.pdf

[117] http://www.fao.org/docrep/012/i0680e/i0680e04.pdf

[118] https://pmc.ncbi.nlm.nih.gov/articles/PMC3339467/

has continued to export product to some African countries. One 2024 study reported on extensive contamination of water resources with DDT in a number of African countries, with the highest concentrations being found in South Africa and Egypt.[119]

Resistance to pesticides has been increasing steadily since their introduction. For example, in the 1940s on average farmers lost 7% of their crop to pests, whilst since the 1980s this rate has close to doubled, despite significantly greater pesticide use.[120] In 2007 pesticide use globally was reported to run at circa 2.3 million tonnes of active ingredient per annum, with around half a million in the USA alone. It has been steadily rising since then and is expected to be more than six million tonnes globally by 2050.

Despite the phasing out of organochlorine insecticides there remain major concerns about the health and environmental impacts of some of the alternatives still in use. A number of insecticides are toxic to important pollinators, such as bees, and have been linked to 'colony collapse disorder'. Studies summarised in an assessment by the European Food Safety Authority in a report published in 2013, have linked the widely used group of insecticides known as neonicitinoids to this phenomenon.[121] One study from 2012 demonstrated that neonicitinoid exposure resulted in 'high mortality due to homing failure at levels that could put a colony at risk of collapse'.[122] These studies led to a EU-wide ban on selected noenicitinoids which commenced on 1 December 2013.[123] Although some temporary derogations were granted for sugar beet these were rescinded in 2023. The use of neonicitinoids has also been reviewed in the USA, although the Environmental Protection Agency has not introduced a ban, but changes to labelling were mandated requiring that the products not be used while bees are present.[124] The UK Government produced a *National Pollinator Strategy* in 2014 with the aim of achieving the following outcomes:

- More, bigger, better, joined-up, diverse and high-quality flower-rich habitats (including nesting places and shelter) supporting our pollinators across the country.
- Healthy bees and other pollinators which are more resilient to climate change and severe weather events.
- No further extinctions of known threatened pollinator species.
- Enhanced awareness across a wide range of businesses, other organisations and the public of the essential needs of pollinators.
- Evidence of actions taken to support pollinators.[125]

[119] https://www.sciencedirect.com/science/article/pii/S2405844024040854#:~:text=The%20highest%20DDT%20concentrations%20were,(5.62%20μg%2FL)

[120] http://www.panna.org/issues/pesticides-101-primer#2

[121] http://www.efsa.europa.eu/en/press/news/130116.htm

[122] http://www.ncbi.nlm.nih.gov/pubmed/22461498

[123] http://www.eea.europa.eu/highlights/neonicotinoid-pesticides-are-a-huge

[124] https://www.epa.gov/pollinator-protection/epa-actions-protect-pollinators

[125] https://www.gov.uk/government/uploads/system/uploads/attachment_data/file/409431/pb14221-national-pollinators-strategy.pdf

According to the Synthesis Report from the authoritative *International Assessment of Agricultural Knowledge, Science and Technology (IAAKST): Agriculture at the Crossroads* published in 2009:

> While climate change and population growth could collude to overwhelm the Earth's latent potential to grow food and bio-materials that sustain human life and well-being, both forces might be offset by smarter agriculture. Present cultivation methods are energy intensive and environmentally taxing, characteristics that in time both exacerbate demand for limited resources and damage long term productivity. Agroecosystems that both improve productivity and replenish ecosystem services behind the supply chain are desperately needed.[126]

There is little evidence that the recommendations from this report have been implemented.

It is interesting to note that genetic engineering through manipulation of the genome to synthesise the DNA of organisms has been used commercially since the early 1990s in an attempt to improve yield and productivity for a number of crops. However, the success of genetically modified organisms (GMOs) has been mixed. The IAAKST report referred to above found that 'the pool of evidence of the sustainability and productivity of GMOs in different settings is relatively anecdotal, and the findings from different contexts are variable'. One example of a detailed scientific study reported in 2008 by the University of Kansas found that genetically modified soya was some 10% less productive than the conventional equivalent. Other studies in countries such as Argentina, South Africa, Mexico, China and India reported increases in yield and financial returns for GM cotton. IAAKST report that some studies 'have also shown the potential for decreased insecticide use, while others show increasing herbicide use'.[127]

Perhaps it would be an understatement to say that the widespread use of GMO remains an extremely controversial issue. It is the impact that GMO might have on conventional crops that is the dominant concern of environmental lobby groups such as Greenpeace.[128] In this context, the IAAKST report clearly states that 'regardless of how new varieties of crop plants are created, care needs to be taken when they are released because through gene flow they can become invasive or problem weeds, or the genes behind their desired agronomic traits may introgress into wild plants threatening local biodiversity'.

Figure 2.20 shows the distribution of GM crops globally in 2018, clearly demonstrating that a de facto ban has been in place throughout Europe, Russia and most of Africa; whilst there has been significant uptake in the Americas, for example, where intensive industrial farming is endemic and Governments have been largely supportive.

We will be exploring the science behind these issues and the environmental impact of food and farming from cradle to grave in more detail in Chap. 9: Products,

---

[126] https://www.unep.org/resources/report/agriculture-crossroads-synthesis-report

[127] https://www.unep.org/resources/report/agriculture-crossroads-synthesis-report

[128] https://www.greenpeace.org/usa/sustainable-agriculture/issues/gmos/#:~:text=We%20shouldn't%20accept%20a,sovereignty%20over%20what%20to%20grow

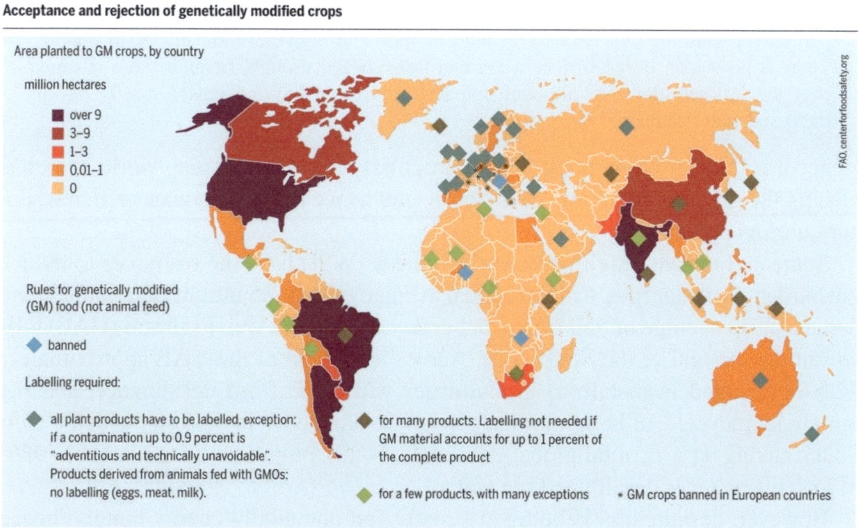

Fig. 2.20 Area of land with genetically modified crops by country in 2018 (https://commons.wikimedia.org/wiki/File%3AGenetically_Modified_Crops_Around_the_World.png)

manufacture and consumption. We will also be looking at the politics surrounding biodiversity, ecosystem resilience and invasive species later in this chapter.

2.6 The Politics of Water

We have already seen how tensions around water supply and water stress, along with disputes over transboundary river basins, can fuel conflict. We have also looked at how droughts have created great upheavals and cost the lives of thousands of people. Water is a resource that many of us take for granted, but when supplies are impacted or sources of drinking water are contaminated the repercussions can be profound.

According to the World Water Development (WWD) Report on 'managing water under uncertainty and risk' published in 2012 'agriculture accounts for 70% of all water withdrawn by the agricultural, municipal and industrial (including energy) sectors. Responsible agricultural water management will make a major contribution to future global water security'.[129]

The Technical Summary to the IPCC WGII fifth Assessment Report published in 2014 warns that:

> In many regions, changing precipitation or melting snow and ice are altering hydrological systems, affecting water resources and quality (medium confidence). Glaciers continue to

[129] https://sustainabledevelopment.un.org/content/documents/404water.pdf

> shrink in many regions due to climate change (high confidence). Climate change has caused permafrost warming and thawing in high latitude and high-elevation mountain regions. There is no evidence that surface water and groundwater drought frequency has changed over the last few decades, although impacts of drought have increased mostly due to increased water demand.[130]

Hence it is the increasing population, particularly in the developing world, which is placing the greatest strain on water supply and hence improving water efficiency in agriculture is crucial.

Some nations with high water stress indices (WSI) have the resources to reduce the burden. For example, with an annual average of only 35 mm of rain (Dubai) and demand at an average of 550 L/person per day the United Arab Emirates (UAE) falls within the top end of the WSI range. Across the whole of the UAE approximately 72% of demand is met from groundwater with 21% from desalination and the remainder recycled, although the proportion desalinated is increasing, with the Gulf States having 41% of total global desalinated water output, and a remarkable 98% of Dubai's water is desalinated sea water.

However, desalination is both very costly and potentially energy hungry, hence solar powered desalination is likely to represent the way forward where the financial resources are available, such as that built on the Arabian Gulf coast of Saudi Arabia at Al-Khafji. A key driver for this has been the 'King Abdullah Initiative for Solar Water Desalination' launched in 2010 of which the Al-Khafji plant is the first phase. The ultimate aim is for the entire water supply for Saudi Arabia to be derived from solar desalination.

Other regions are being impacted so severely by climate change that technological fixes are unlikely to provide a solution. For example, the Garnaut Climate Change Review, commissioned by the Australian Government and published in 2008, with an update in 2011, included an examination of the drought affecting large swathes of prime agricultural land in Australia. A report prepared by the Centre for Australian Weather and Climate Research for the 2011 update stated that:

> Large areas of southern Australia, in particular the southern Murray-Darling Basin, Victoria, south-west Australia and south-east Queensland experienced prolonged drought from 1997 to 2009. The dry spell in the southern Murray-Darling Basin, Victoria, and south-west Australia was unprecedented over the 110 years of reliable rainfall records. The drought resulted in declining storage levels in reservoirs, several years of severe water restrictions in cities, and years of low allocations to irrigators in the southern Murray-Darling Basin and elsewhere in Victoria.[131]

The Basin contributes some 39% towards the national income from agriculture, hence when the 2008 Garnaut Review reported predictions that economic production from the Basin could fall by some 12% by 2030, 49% by 2050 and 92% by 2100, primarily due to climate change, the alarm bells were rung. A 2.2 billion Australian dollar desalination plant in Adelaide started to produce potable water in

---

[130] https://www.ipcc.ch/report/managing-the-risks-of-extreme-events-and-disasters-to-advance-climate-change-adaptation/

[131] http://www.bom.gov.au/research/publications/cawcrreports/CTR_036.pdf

2011, subsequently meeting around 16% of the total demand across South Australia state. This compares with around 86% drawn from the river Murray at the height of the drought in 2008.

A 2022 State of the Climate report from the Australian Bureau of Meteorology confirms the long term trend in reduced rainfall in the southwest with a 15% reduction falling between April and October since 1970.[132]

2.7 The Politics of Waste

It could be said that the way a society manages its waste is a strong indicator of how developed it is and, like many other sustainability indicators, the developing world is facing many of the same problems dealing with its waste that the developed world faced as populations agglomerated and uncontrolled industrialisation took hold in the nineteenth century. Again we have to turn to China where, according to a report prepared for the World Bank East Asia Infrastructure Department in 2005, waste volumes were projected to increase by a factor of 2.5 by 2030 compared with 2004, when it overtook the USA as the world's largest waste generator.[133] Globally some 2.24 billion tonnes of solid waste were produced in 2020, which according to the World Bank, will rise to 3.88 billion tonnes by 2050.[134] As can be seen from Fig. 2.21, it is the developed world that produces the most waste per capita.

We saw in Chap. 1 some of the impacts of uncontrolled waste in the form of plastics entering the waterways, seas and oceans of the world; whilst earlier in this chapter, we examined the challenges involved in the alternatives to landfill through incineration and energy from waste facilities. It will be seen in later chapters that almost every aspect of anthropogenic activity involves the generation of some form of waste. In this section however we will be looking at the politics around the management of waste globally, whilst in Chap. 9, we will be exploring the impacts of waste and the blight caused by plastic waste in particular.

Waste takes many forms and, although some waste products, such as animal ordure, are natural and biodegradable, those that are most difficult to deal with tend to be anthropogenic. Although there are many ways to classify waste it is usual to consider firstly whether it is hazardous, since this will instigate a requirement to separate it from other wastes and delineate the processes required to ensure safe handling and disposal. Globally there are different systems in place for classifying

---

[132] http://www.bom.gov.au/state-of-the-climate/2022/

[133] https://documents.worldbank.org/en/publication/documents-reports/documentdetail/237151468025135801/waste-management-in-china-issues-and-recommendations

[134] https://www.worldbank.org/en/topic/urbandevelopment/brief/solid-waste-management#:~:text=In%202020%2C%20the%20world%20was,3.88%20billion%20tonnes%20in%202050

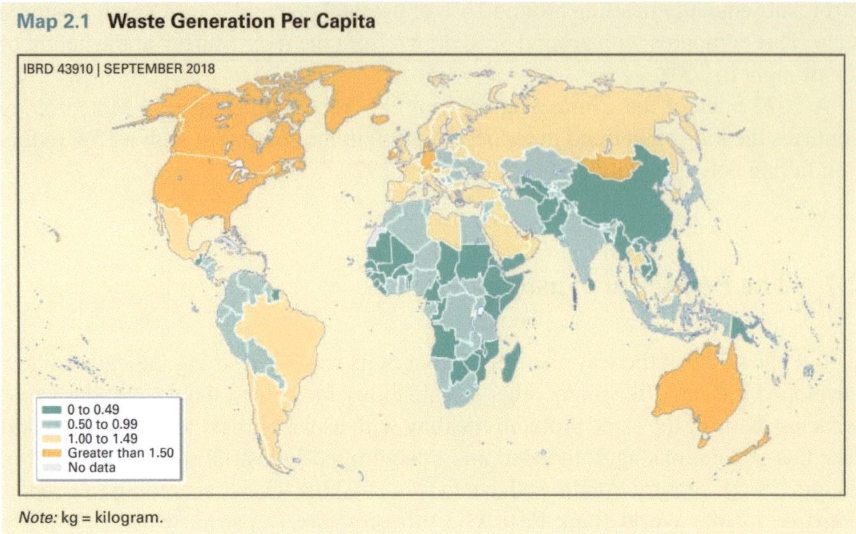

Fig. 2.21 Waste generation for 2018, measured in kilograms per person per day (https://commons.wikimedia.org/wiki/File:Waste_generation_per_day_per_capita,_September_2018.jpg)

waste, such as the *European List of Wastes*[135] and the US *Resource Conservation and Recovery Act* definitions.[136]

Of course it is primarily growth that is to blame for increased waste generation. Take for example the Maldives: an archipelago in the Indian Ocean that has become a magnet for tourists, on which its economy is strongly dependent. Tourists generate some 3.5 kg of solid waste per day which is more than five times that of the average native Maldivian. In an attempt to cater for this waste a landfill site was created on the island of Thilafushi, itself effectively an artificial island (see Fig. 2.22). With some 400 tonnes of rubbish being transported by boat to the island every day there has been very little management of its disposal, with hazardous and household waste frequently mixed, significant quantities lost to the ocean and uncontrolled burning resulting in potential harm to those exposed to the resultant smoke.

The volume of solid waste has grown worldwide as the consumer society has taken hold and cheap disposable products have been developed with built in obsolescence, requiring greater packaging and using materials that are difficult to recycle; whilst per capita wealth has increased, with a consequent increase in the demand for consumables.

Up until quite recently most major cities around the planet deposited much of their solid waste on nearby land that was deemed least desirable for other uses. Although most of these dumps are grouped under the general heading of 'landfill

135 http://ec.europa.eu/environment/waste/framework/list.htm

136 https://www.epa.gov/laws-regulations/summary-resource-conservation-and-recovery-act

Fig. 2.22 The waste island of Thilafushi in the Maldives (Image: http://www.atlasobscura.com/places/thilasfushi-the-rubbish-island)

Fig. 2.23 Waste pickers at Olusosun landfill site, Lagos Nigeria (Image: Mumtahina Rahman; From: https://commons.wikimedia.org/wiki/File:Waste_disposal-landfill.jpg)

sites' they vary widely in how they are managed and their environmental impact. Figure 2.23 for example shows the site at Olusosun in Lagos, Nigeria.

This is one of the largest landfill sites in the world, with some 12,000 tonnes of waste deposited every day. Extraordinarily there are around a thousand homes located on the site, many of the occupants being employed as waste pickers, collecting the most valuable items for trade. However, with very little control of the wastes deposited, the dump is a major source of pollution from which, according to a survey carried out by a team from the Afe Babalola University in Nigeria and reported in 2023: 'A major source of air and groundwater pollution, the Olusosun landfill has a detrimental impact on the health of those who live there. Solid waste, groundwater interactions, and contaminated migration into the nearby neighbourhood were studied. It was observed that the degradation of waste products in dump sites releases harmful leachate into the groundwater'.[137]

Similar sized sites were closed in 2012 at Gramacho in Rio de Janeiro in Brazil and in 2011 at Bordo Poniente in Mexico City. The Gramucho site was replaced with a new recycling plant, although the waste pickers who earned a living from recycling waste from the site, as an alternative to a life of crime in many cases, have largely not been catered for. Closure of the Bordo Poniente plant led initially to rubbish accumulating in the streets of Mexico City since alternative sites were inadequate or barred because of political disputes. As with Gramucho some 1500 waste pickers had their livelihood removed overnight.

This was highlighted by the United Nations Environment Programme's Global Partnership on Waste Management (GPWM) Framework which summarises the impact of poor waste management as follows:

> Besides having serious economic and environmental and health implications, unsound waste management also has a social dimension. Like most environmental hazards, deficiencies in waste management also disproportionately affect poorer communities more as wastes are often dumped in land adjacent to slums. As waste is often considered as a source of income for waste pickers, exposure to hazardous substances in improperly separated wastes such as lead, mercury and infectious agents from healthcare facilities is common. Crude practices to recover valuable materials, e.g., from e-waste, setting free dioxins, other persistent organic pollutants (POPs), and other harmful emissions, have heavy impacts on the health of workers in the informal waste sector and further contribute to air, land and water contamination.[138]

The GPMW Framework document quotes an extraordinary statistic from a 2008 briefing note produced by the Public Private Initiative Advisory Facility (PPIAF) that 'In developing countries about 1% of the urban population—at least 15 million people—survive by salvaging recyclables from waste'; which included around six million people in China alone.[139] Estimates for the global total have been steadily increasing: by 2022 the number quoted was 20 million, although the figure for China varies greatly. This is largely an unorganised industry, bypassing whatever employment laws exist, and with a significant proportion of the workers being children.

Like Olusosun many landfill sites have polluted the neighbouring ground and water resources. According to a 2014 report from the US-based National Centre for

[137] https://neptjournal.com/upload-images/(3)D-1443.pdf

[138] http://www.unep.org/gpwm/Portals/24123/images/Essential%20Documents/GPWMFrameworkFINAL%2022022012%20wo%20track%20changes.pdf

[139] http://www.ppiaf.org/sites/ppiaf.org/files/publication/Gridlines-44-Informal%20Recycling%20-%20MMedina.pdf

Atmospheric Research (NCAR), approximately 41% of the world's waste was being burned in unregulated open fires, equivalent to some 1.1 billion tonnes per annum.[140] 'The study concluded that the fires produce emissions equivalent to as much as 29% of officially reported human-related global emissions of small particulates (less than 2.5 microns in diameter), as well as 10% of mercury and 64% of a group of gases known as polycyclic aromatic hydrocarbons (PAHs)'. Like many of these sites both Thilafushi and Olusosun are subject to sporadic fires, which frequently smoulder underground for months or years and break out at the surface from time to time.

The authors also estimated that the burning of waste increases carbon dioxide emissions globally by 5% compared with currently reported anthropogenic emissions. Indeed 'In certain developing countries—such as Lesotho, Burundi, Mali, Somalia, and Sri Lanka—the trash burning produces more carbon dioxide than is tallied in official inventories (for all sources). This discrepancy can be important in international negotiations over reducing greenhouse gas emissions'.

A significant amount of rubbish burning and seepage into groundwater occurs at illegal sites, where rubbish has been fly tipped. In the Campania region of southern Italy between the towns of Acerra, Nola and Marigliano the branch of the mafia known as the Camorra has been responsible for industrial scale dumping of waste, which was first exposed in the 1970s. A mixture of wastes has been dumped in thousands of illegal sites, including a wide range of hazardous substances, such as asbestos, radioactive waste and cabling. The latter has been burnt to extract the copper, thus releasing dioxins into the atmosphere. Many of the dumps are burning continuously, with occasional breakouts of underground fires without warning, resulting in the area becoming known as 'the land of fire'. Illegal dumping is endemic throughout Italy: as far back as 2004 it was estimated that there were some 5000 sites. A map produced in December 2013 shows areas where illegal sites have been identified, including the Tollo area of the Abruzzo region, the Salento area of the Puglia region and Caltanisseta in Sicily.[141]

Studies published in the Lancet Oncology in 2004[142] and by the Sbarro Institute for Cancer Research at Temple University, Philadelphia in 2014[143] indicate that rates for a number of cancers, as well as for birth defects, amongst the residents of the triangular area formed between Acerra, Nola and Marigliano are significantly higher than the national average. This led to the area becoming known also as the 'triangle of death'. This is a fertile area, with a long history of agriculture, and much of the illegal dumping has been on or adjacent to agricultural land, resulting in grave concerns about contamination of crops.

According to Antonio Giordano, director of the Sbarro Institute, his '…father's revelations (in the 1970s) created political turmoil, because they revealed for the

[140] https://opensky.ucar.edu/islandora/object/archives%3A6471

[141] http://italychronicles.com/map-italy-show-illegal-toxic-waste-dumping/

[142] https://www.uonna.it/lancet-journal-acerra.htm

[143] http://news.temple.edu/news/2014-06-03/temple-researcher%E2%80%99s-work-spotlights-cancer-threat-italy

first time the link between the industries, the Camorra mafia illegally dumping their toxic waste and the politicians who were silently protecting them both'.[144]

This kind of activity is neither restricted to the mafia nor to Italy, for example, Europol—the European Union's law enforcement agency—has highlighted waste trafficking as a low-risk high profit margin activity for criminal gangs (low risk to the gangs that is) that is 'one of the fastest growing new areas of criminal and organised crime activity, in which some groups are estimated to earn 4 billion euros per annum'.

> Europol reported in its 2011 Organised Crime Threat Assessment (OCTA) how illicit waste trafficking is often facilitated through cooperation with legitimate businesses, including those in the financial services, import/export and metal recycling sectors, and with specialists engaged in document forgery to acquire permits. Toxic waste is trafficked from Southern to South East Europe and the Western Balkans (Romania, Hungary and Albania), as well as other Member States. Italy has also become a transit point for e-waste (second-hand electrical and electronic equipment) en route to Africa and Asia. There is evidence of corruption in the public and private sectors, in relation to the issuing of certificates by laboratory technicians. Intermediate storage sites are often used to disguise the ultimate destination of waste, which makes it difficult to identify the source companies.[145]

For England the Environment Agency (EA) produces an annual waste crime report—*Cracking down on waste crime*.[146] The report covering 2012 to 2013 claimed that some 1300 illegal waste sites had been closed down or brought into regulation between end of April 2012 and March 2013. The statistics reported therein indicate that, at 30%, the greatest proportion of illegal waste stopped was from construction and demolition, whereas hazardous waste only represented some 2%. The equivalent report in 2022 states that some 18% of waste was being managed illegally, which represents some 30 million tonnes of waste.[147]

However, what the EA statistics do not reveal are the legal sites that gradually become a nuisance to local residents, such as the dump at Cornwall Drive in Bromley which has gained widespread notoriety. This site appeared at the end of a local man's garden when he leased it to waste management company Waste4Fuel which managed remarkably to obtain a license for a waste transfer station despite the site's green field location in close proximity to a residential area.

Waste4Fuel obtained a license to handle general non-hazardous waste, storing no more than 5500 tonnes on the site at any one time. However it has been estimated that the heap grew to some 25,000 tonnes by July 2014, presenting an ever growing problem with rats, flies, dust, smoke and fires, which regularly required the attention

---

[144] http://news.temple.edu/news/2014-06-03/temple-researcher%E2%80%99s-work-spotlights-cancer-threat-italy

[145] https://www.europol.europa.eu/media-press/newsroom/news/europol-warns-of-increase-in-illegal-waste-dumping#:~:text=Europol%20reported%20in%20its%202011,in%20document%20forgery%20to%20acquire

[146] https://www.gov.uk/government/uploads/system/uploads/attachment_data/file/288605/LIT_8777_8cc7d6.pdf

[147] https://www.gov.uk/government/speeches/crackdown-on-waste-crime-time-to-stop-trashing-our-future

of the fire brigade. All of these factors resulted in local residents having to keep windows closed and being unable to enjoy their gardens. However after years of legal disputes the site was eventually cleared in 2017.

It is clear from the above that 'land-filling … has the potential to devalue property, contaminate land and pollute surface and groundwater, threatening human health, ecosystem health, biodiversity and potable water supplies' [4] (Chap. 10). Good waste management must therefore aim at reducing the amount of waste that is either uncontrollably burnt or deposited in landfill. Across EU countries, for example, landfill sites are supposed to be closely controlled and monitored under the provisions of the 1999 *Landfill Directive*. This sets out criteria for the characteristics of waste that are acceptable for landfill. For example, municipal, inert and non-hazardous wastes would mostly be acceptable, whilst only hazardous wastes that have been rendered stable and non-reactive by such processes as vitrification or solidification can be considered for landfill, provided their long term leachability meets certain criteria. Landfill sites are classified as handling either inert, non-hazardous or hazardous waste. All waste must be pre-treated to either: reduce volume, reduce hazardous nature, facilitate handling or enhance recovery.

The Landfill Directive pays particular attention to biodegradable waste and the requirements for methane control and monitoring. All EU states were required to reduce the biodegradable content of waste going to landfill by 35% of the 1995 benchmark by 2020.

> The following wastes may not be accepted in a landfill:
>
> - liquid waste;
> - flammable waste;
> - explosive or oxidising waste;
> - hospital and other clinical waste which is infectious;
> - used tyres, with certain exceptions;
> - any other type of waste which does not meet the acceptance criteria laid down in Annex II (of the Directive).[148]

In England, for example, the Environment Agency has published guidance on *Waste acceptance at landfills*.[149] All landfill sites operate under a permit system that requires operators to follow designated waste acceptance procedures (WAP). These are based on the *List of Wastes (England) Regulations* 2005 which in turn implements the *European Waste Catalogue* referred to above. The requirements have remained largely unchanged since Britain withdrew from the European Union.

In order to encourage alternatives to landfill the UK Government introduced a landfill tax in 1996. This introduced a levy to be paid by the landfill operator at a rate per tonne that depends on the type of waste, with a significant premium being charged for what are broadly defined as 'active' materials: being materials that are potentially more polluting, whilst also offering significant opportunity for recovery

---

[148] http://europa.eu/legislation_summaries/environment/waste_management/l21208_en.htm

[149] https://wte.org

or re-use, such as wood, metals and plastics. Those qualifying for the lower tax are listed in the *Landfill Tax (Qualifying Materials) Order* 2011—and subsequent amendments.

The European Commission has instigated proceedings against a number of EU States for infringements of the Landfill Directive. Serial offenders include Greece, Italy and Spain. We have already heard about Italy's problem with illegal waste dumping in areas such as the 'triangle of death' in Campania, but Greece also has a problem with illegal dumping and was taken to the EU Court of Justice in 2005 over hundreds of illegal landfill sites, and then again in 2010 and 2013, resulting in millions of euros in fines accumulating.

For centuries humans have burnt their waste as a more hygienic and frequently easier alternative to burying it. However, in early civilisations where populations formed into clusters, such as in ancient Athens, and in the absence of open areas for fires, waste was largely dumped in the streets, such that eventually street levels rose on a bed of accumulated rubbish. It was not until 320 BC that a law was introduced in Athens forbidding this practice. In ancient Rome laws requiring property owners to clear rubbish from in front of their buildings were introduced and open pits were provided outside of the city walls for its disposal.

Incineration of waste within an enclosed furnace dates from 1874 when a 'destructor' was commissioned in Nottingham, England. The first waste incinerator to be commissioned in the USA was built on Governor's Island New York City in 1885 and, by the beginning of the twentieth century, some 15% of major American cities were incinerating their solid waste. As early as 1920 an incinerator in Atlanta was producing steam for the generation of electricity and in the 1930s, there were some 700 incinerators across the USA, although this fell to 265 by 1966 due to problems associated with emissions. However, energy from waste (or waste to energy) plants were not widely adopted in the USA until the 1978 *Public Utility Regulatory Policies Act* (PURPA) produced a guaranteed market for the electricity produced, reinforced by the Energy Security Act of 1980 which resulted in loan guarantees and tax incentives for the construction of waste to energy plants. This was all undone by the 1986 Tax Reform Act which removed their tax free status, meaning that few waste to energy plants were built in the US until 2004 when electricity from these plants qualified for Renewable Electricity Production Tax Credits.

According to the US-based Energy Recovery Council (subsequently the Waste to Energy Association): 'In 2010, 86 (waste to energy) plants operate in 24 states and have capacity to process more than 97,000 tons of municipal solid waste per day'.[150] By 2025 this had shrunk to 75 plants, compared with 522 in Europe. As with the Kings Lynn scheme described earlier in this chapter, there remains considerable controversy over the construction and operation of incineration plants, particularly amongst those within the local community.

[150] https://wte.org

Incineration experienced boom years across most of the developed world during the first half of the twentieth century. In the UK as early as 1912, there were 338 incinerators, 80 of which were generating electricity. However, according to Eurostat statistics, by 2012 only 16% of municipal waste was being incinerated,[151] with only 24 energy from waste plants in England burning some four million tonnes of waste per annum. Defra figures quoted in a 2014 UK Government report state that 'A 2010 survey found only 2% of commercial and industrial waste was incinerated with energy recovery in England'.[152]

In Denmark, where winters are cold, the first incinerator, built in 1903, was designed to generate steam for district heating, since when every city in Denmark has been provided with its own energy from waste plant, resulting in some 50% of municipal waste providing fuel for combined heat and power generation by 2012. This has resulted in Denmark having to import waste from other countries and its recycling rates being very low.[153]

Japan on the other hand disposes of more than 70% of its waste by incineration, representing some 1700 incinerators, 24.5% of which recover energy from the combusted waste. Japan has focussed on incineration of waste because of its population density and lack of suitable sites for landfill. However by 1997 dioxin emissions reached some 8 kg-TEQ per annum, 90% of which were from incineration (where TEQ refers to toxic equivalence for a mixture of dioxins weighted to take account of their relative toxicity). At that time concentrations of airborne dioxin were reported as being some three times higher than in equivalent locations in the USA and Europe and subsequently measures were introduced to reduce emissions, which consequently fell rapidly: to around 0.3 kg-TEQ by 2007.

Refuse collection can also be a political hot potato. In most countries local government is responsible for collecting household waste and keeping public areas free of rubbish. There are clearly public health impacts around the accumulation of uncollected refuse—food waste in particular—whilst excessive provision of waste bins can be unsightly where space is limited.

The rash of different coloured wheelie bins that decorates front gardens and alleys across the British Isles has evolved as a result of a policy to separate recyclable from general waste and increase collection intervals, whilst making it easier to handle the resultant heavier bins. Getting the balance right between bin size and collection interval is very important since if the interval is too great then the customer is more likely to dispose of surplus waste to municipal tips or through fly-tipping, whilst food waste may act as a nursery for maggots or a magnet for rats and other pests.

---

[151] https://www.politico.eu/article/denmark-devilish-waste-trash-energy-incineration-recycling-dilemma/

[152] https://www.gov.uk/government/uploads/system/uploads/attachment_data/file/284612/pb14130-energy-waste-201402.pdf

[153] https://www.politico.eu/article/denmark-devilish-waste-trash-energy-incineration-recycling-dilemma/

In early 2014 the Department for Communities and Local Government (CLG) for England and Wales was pressing local authorities to introduce weekly refuse collections, although most councils at the time were operating an alternate weekly collection service, with weekly collections for food waste where provided. Most councils were resisting the Government's strictures and hence Government considered introducing an amendment to the 1990 Environmental Protection Act to require weekly general waste collections. By 2023 this had not happened.

The proliferation of wheelie bins tends to be at its worst for terraced houses with narrow frontages and no rear access for refuse vehicles or refuse collection personel. Each residence may have four or five receptacles for waste: in the UK, these comprise typically black and blue wheelie bins for general and recyclable waste respectively; occasionally smaller bins for bottles and food waste with separate collections for garden waste—although this latter service may be subject to a charge.

2.8 The Politics of Biodiversity and Ecosystem Resilience

Many of the human activities referred to so far in this chapter have an impact on the natural world, with which we cohabit in the biosphere in a state of continuous tension. As populations grow and become more urbanised this tension can only increase. According to the Summary of the 2012 UN Environment Programme 5th Global Environmental Outlook (GEO5):

> Up to two thirds of species in some taxa are threatened with extinction; species populations are declining, since 1970, vertebrate populations have fallen by 30 per cent; and since 1970 conversion and degradation has resulted in declines of 20 per cent of some natural habitats. Climate change will have profound impacts on biodiversity, particularly in combination with other threats.
>
> Habitat loss and degradation, including from unsustainable agriculture and infrastructure development; unsustainable exploitation; pollution and invasive alien species remain predominant threats to terrestrial and aquatic biodiversity. All contribute to a decline in ecosystem services, which may lead to increasing food insecurity and endanger poverty reduction and the improvement in human health and well-being.[154]

GEO6, published in 2019, confirmed all of these impacts, whilst confirming that the decline had continued.

All of these impacts come at a financial cost and the Economics of Ecosystems and Biodiversity (TEEB) study published its eponymous interim report in 2008 under the auspices of the EC to evaluate and highlight these costs on the basis that:

> Recent pressure on commodity and food prices shows the consequences of this loss (of biodiversity) to society. Urgent remedial action is essential because species loss and ecosystem degradation are inextricably linked to human well-being. Economic growth and the conversion of natural ecosystems to agricultural production will, of course, continue. We cannot—and should not—put a brake on the legitimate aspirations of countries and indi-

[154] https://www.unep.org/geo/geo-resources/geo-5

> viduals for economic development. However, it is essential to ensure that such development takes proper account of the real value of natural ecosystems. This is central to both economic and environmental management.[155]

Despite its age this report remains valid. It provides the results of a study that draws from work carried out on the 'Cost of Policy Inaction' and the GLOBIO model,[156] concluding that 'cumulative welfare losses could be equivalent to 7% of annual consumption by 2050'. However, it was recognised that this figure was conservative since it excluded 'numerous known loss categories, for example, all marine biodiversity, deserts, the Arctic and Antarctic; some ecosystem services are excluded as well (disease regulation, pollination, ornamental services, and many more), while others are barely represented (e.g., erosion control), or under-represented (e.g., tourism); losses from invasive alien species are also excluded'.

The Millennium Ecosystem Assessment (MA) referred to earlier in this chapter concluded that approximately 60% of ecosystem services have been degraded through loss of biodiversity, pollution, climate change and bad land use. TEEB, which was rebranded as the Natural Capital Coalition (NCC) in 2014, has 'calculated the economic impact of the degradation of natural systems at approximately 2 to 5 trillion US dollars per year'.[157] Smith et al. [4, pp. 262–263] compare economic impact of inaction with the cost of preventative action, providing an estimate for 'an annual budget to restore global habitat/ecosystem services (of) around an additional US$110 to 130 billion per annum'. See also Section dealing with *Social and economic impacts* in Chap. 3.

Madagascar is among the most biologically diverse places on the planet, with extraordinarily high numbers of endemic flora and fauna: '98 per cent of Madagascar's land mammals, 92 per cent of its reptiles, 68 per cent of its plants and 41 per cent of its breeding bird species do not exist anywhere else'.[158]

Madagascar might be said to represent a microcosm of global ecosystems, perhaps exemplifying many of the negative impacts that humans are having on the natural environment. Population growth has been near exponential: with total numbers rising from 4.2 million in 1940 to 8.75 million in 1980 to 15.26 million in 2000 and 29.96 million in January 2023.

It has been estimated that since the arrival of humans around 2000 years ago Madagascar has lost around 90% of its forest cover, some 70% of which was lost between 1895 and 1925 under French colonial rule. The French introduced coffee as a cash crop, accelerating the widespread slash and burn of forests, or *tavy,* for growing not only coffee but also subsistence rice and for grazing cattle; whilst

---

155 http://www.teebweb.org/publication/the-economics-of-ecosystems-and-biodiversity-an-interim-report/

156 https://www.cbd.int/financial/doc/copi-2008.pdf

157 https://www.theoep.org.uk/natural-capital-committee#:~:text=The%20NCC%20defined%20natural%20capital,land%2C%20minerals%20and%20oceans"

158 https://www.earthrangers.com/EN/CA/omg-facts/diversity-of-madagascar/#:~:text=It%20is%20found%20off%20the,exist%20nowhere%20else%20on%20Earth!

precious hardwoods such as rosewood and ebony were exploited for trade, and other trees used in the manufacture of charcoal. As recently as June 2014 reportedly some 3000 tonnes of rosewood logs were seized by authorities in Singapore, indicating a burgeoning illegal trade in rainforest timber logged in Madagascar, much of which ends up in China.

> The timber appears to have originated from Madagascar's fast-dwindling stockpile of rosewood logs, which accumulated after a surge in illegal logging that followed the 2009 coup displacing former president Marc Ravalomanana from power. National parks and forest reserves were ransacked during the ensuing political crisis, but global outcry led the interim government to establish an export ban on rainforest logs.[159]

As with many indigenous populations that live in forested areas, clearing trees to make way for subsistence crops has been an integral part of the Malagasy way of life for centuries:

> *Tavy* is the lifeblood of Malagasy culture and the rural economy. Typically an acre or two of forest is cut, burned, and then planted with rice. After a year or two of production the field is left fallow for 4–6 years before the process is repeated. After 2–3 such cycles the soil is exhausted of nutrients and the land is likely colonized by scrub vegetation or alien grasses. On slopes, the new vegetation is often insufficient to anchor soils, making erosion and landslides a problem.[160]

This widespread soil erosion has left a legacy of poor and infertile soils which can no longer sustain the growing indigenous population of what has become one of the poorest nations on the planet. As well as the decimation of forests, more than 300 animal species, many unique to Madagascar, are threatened with extinction. According to a paper published in Science on *Averting Lemur Extinctions amid Madagascar's Political Crisis*, of 101 lemur species on the island, 90 are threatened with extinction, including 22 that are critically endangered: 'Madagascar's political turmoil (since 2009) has left existing environmental laws that would have protected forests and wildlife unenforced, and the international community has withdrawn most of its funding for conservation programmes'.[161]

However, as we saw earlier in this chapter, the Madagascar government has signed up to the World Bank's Wealth Accounting and Valuation of Ecosystem Services (WAVES) programme which it is hoped 'will help Madagascar use environmental accounting to generate the economic information that it needs to improve decision-making. In particular, it will develop macroeconomic indicators that take account of the use of natural capital and will develop accounting systems in the mining, water resources and forestry sectors'.[162]

---

[159] https://news.mongabay.com/2014/06/singapore-intercepts-massive-illegal-shipment-of-madagascar-rosewood/#:~:text=Coup%20leaders%20sell%20out%20Madagascar's%20forests%2C%20people&text=In%20the%20absence%20of%20governance,critical%20economic%20incentive%20for%20conservation

[160] http://rainforests.mongabay.com/20madagascar.htm

[161] https://www.science.org/doi/10.1126/science.1245783

[162] https://www.wavespartnership.org/en/about-us

Once deforestation has stripped the topsoil of its binding roots, and soils have been degraded by years of mismanaged cultivation, then whole tracts of land can become barren wildernesses. When combined with long-term drought and exposure to prevailing winds then a phenomenon such as that which created the 'Dust Bowl' that blighted the Great Plains of North America during the 1930s can develop. Some of the more mountainous regions of the world have highly erodible soils on steeply sloping land that are being washed away, in the case of Ethiopia, for example, at a rate of some two billion tonnes a year.

In parts of Africa and Asia desertification is a major threat to hundreds of millions of people, many of whom live in extreme poverty. Something like one third of the entire land area of China is desert, representing an area of 3 million km^2. A significant portion of this area has become desert in recent years: for example, at 1.3 million km^2 by far the largest desert area in China, the Gobi grew by some 65,000 km^2 between 1994 and 2008; and is steadily closing in on Beijing. Dust storms periodically add to Beijing's pollution problems and sometimes stretch as far as the Korean peninsula.

The authorities in China have recognised the problem of desertification, and its impact on agriculture in particular, for many decades, launching an afforestation programme in 1978, scheduled to continue until 2050. This mammoth project has the official title of the 'Three-North Shelterbelt Programme', although it is more popularly known as the 'Green Great Wall of China'. The ambition has been to plant some 350,000 km^2 of forest along the southern edge of the great northern desert areas of China. This would increase the tree cover from 5% in 1978 to 15% by the project completion. An overview of progress by the Chinese State Forestry Administration reported that tree cover had increased to 12.4% of the land area of China by the end of 2012.[163] However, there have reportedly been many problems with the effectiveness of the programme, such as the planting of trees in arid and semi-arid lands. For example, some 53,000 hectares of trees planted in the Minqin area of Ganzu province have either died or become stunted. This area has an annual rainfall of only 110 mm and since the tree planting programme commenced groundwater level has fallen by between 10 and 40 m.[164]

Another problem has occurred through the blanket planting of single tree species:

> In Ningxia, for example, 70 percent of the trees planted were poplar and willow. In 2000, one billion poplar trees were lost to a disease (*Anoplophora*), wiping out 20 years of planting efforts. The vulnerability to disease, compounded with its rapid exhaustion of soil and water resources, makes single-species forests "green deserts," according to agro-botanist Zhang Yan. Single-species forestry helps increase the official tally of trees, but it fails to improve the environment.[165]

The Three-North Shelterbelt is one in a line of hugely ambitious afforestation programmes that have achieved varying degrees of success in recent history. For

[163] http://www.china.org.cn/environment/2013-08/12/content_29689240.htm

[164] https://www.theepochtimes.com/article/desertification-in-china-1521959

[165] https://www.theepochtimes.com/article/desertification-in-china-1521959

example, President Roosevelt launched the Great Plains Shelterbelt project in 1934 in response to the disastrous events that followed the formation of the Dust Bowl. By 1942 some 220 million trees had been planted over a 160 km wide swathe stretching from the Canadian border to southern Texas. Perhaps slightly more ambitiously, following the Second World War, Stalin launched his 'Plan for the Transformation of Nature'. Launched in 1948 it included a plan to plant around 600,000 km^2 of forest, but collapsed in 1953 following his death. It is interesting to note that both Nazi Germany and fascist Italy launched major afforestation plans, in part to fuel the industrialisation requirements in the build-up to war.

References

1. Carson, R. (1962). *Silent spring*. Houghton Mifflin Harcourt.
2. Meadows, D.H., Meadows, D.L., Randers, J. and Behrens III, W. (1972) The Limits to Growth. Potomac Associates/Universal Books.
3. Tainter, J. A. (1988). *The collapse of complex societies*. Cambridge University Press.
4. Smith, M. H., Hargroves, K., & Desha, C. (2010). *Cents and sustainability*. Earthscan.
5. Gilbert, G. (2011). *Global impacts of chernobyl radioactive fallout*. Dorrance Publishers.
6. Palmer, R., & Smith, R. (2013). Soil structural degradation in SW England and its impact on surface-water runoff generation. *Soil Use and Management, 29*, 567–575. https://doi.org/10.1111/sum.12068
7. Scottish Environmental Protection Agency. (n.d.). *Floods, dredging and river changes*. SEPA.
8. Bell, M. L., Davis, D. L., & Fletcher, T. (2004). A retrospective assessment of mortality from the London Smog Episode of 1952: The role of influenza and pollution. *Environmental Health Perspectives, 112*(1), 6–8.
9. Intergovernmental Panel on Climate Change. (2022). *Impacts, adaptation and vulnerability* (Working Group II Contribution to the Sixth Assessment Report of the Intergovernmental Panel on Climate Change). IPCC.

Chapter 3
Climate Change and the Interconnectedness of All Things

The two greatest problems of our times—overcoming poverty in the developing world and combating climate change—are inextricably linked. Failure to tackle one will undermine efforts to deal with the other: ignoring climate change would result in an increasingly hostile environment for development and poverty reduction, but to try to deal with climate change by shackling growth and development would damage, probably fatally, the cooperation between developed and developing countries that is vital to success.

A Blueprint for a Safer Planet, Nicholas Stern, 2009

3.1 Introduction

Virtually everything in the universe is either the result of historical energy exchange or part of an ongoing thermodynamic process. The First Law of Thermodynamics defines the conservation of energy which is equally applicable to the entire universe or the smallest living organism. The total amount of energy in the universe is constant, although energy can be transferred from one part of the universe to another. For its own part, the earth benefits from energy transmitted as broad spectrum radiation from the sun and an atmosphere that creates conditions at the planet surface which support an extraordinary variety of flora and fauna.

The earth has been continuously evolving ever since the so-called big bang and has experienced numerous changes in climate, geology, topography and life forms in the process. However, ever since humans have been exploiting natural resources and have become increasingly industrialised and urbanised, there have been impacts that run the risk of catastrophic and irreversible damage to a civilisation that has experienced close to exponential growth in modern times.

In harnessing the potential energy in various fuels to either produce useful energy directly or kinetic energy in the generation of electricity and the movement of goods and people, humans have unwittingly triggered the emission of vast quantities of

P. Appleby, *The Overburdened Planet*,
https://doi.org/10.1007/978-3-032-01510-5_3

waste products that are polluting the planet and its atmosphere. This has been augmented by emissions such as those associated with increasingly intensified agriculture, the regular burning of forests and crops, livestock rearing, the burning and degradation of waste and the release of previously trapped methane and other greenhouse gases.

3.2 Greenhouse Effect and Global Warming

The earth's atmosphere comprises a mixture of gases which vary in composition and concentration across the surface of the planet and with distance from the planet's surface; as well as with time, both in the short and the long terms. The main constituents in the air we breathe close to the planet's surface are oxygen (21%), nitrogen (78%) and argon (0.9%). In addition, there are trace quantities of carbon dioxide (0.04%) and other gases. Proportions of water vapour, pollutant gases and particulates vary with climate and the availability of a source.

Historically scientists have divided the atmosphere into a number of layers, distinguished by specific physical characteristics, such as temperature gradient or gaseous composition (see Fig. 3.1).

The layer closest to the surface of the earth is known as the troposphere and extends to an average of 12 km above sea level or approximately 9 km at the poles and 17 km at the equator. As this layer bears the weight of the rest of the atmosphere, its average density is the highest and it contains some 80% by mass of the gases in the atmosphere. It also contains most of the water vapour, carbon dioxide and other so-called greenhouse gases, along with the major systems that impact on weather. In this layer the temperature steadily reduces with increased height to around minus 55 °C at the tropopause. Above this the stratosphere stretches to between 50 and 55 km from the planet's surface and contains the ozone layer, which lies in a belt which stretches 15 to 35 km from the surface of the earth. At between 2 and 8 parts per million, the concentration of ozone in this region is 7–30 times greater than the average in the atmosphere.

Ozone is formed by the action of ultraviolet radiation from the sun on the oxygen atoms in this region, resulting in a process of splitting and recombining that divides O_2 into separate O molecules that are then combined with other O_2 molecules to form O_3 (ozone) resulting in a temperature rise back up to just below freezing at the stratopause. However, in the mesosphere, the temperature falls again to around minus 85 °C until the ultra-low density region known as the thermosphere which stretches from the mesopause at 80 km to the exobase which lies at between 350 and 800 km above the earth's surface, where temperatures can reach as high as 1500 °C, marking the start of the transition region to outer space known as the exosphere. In this region lie extremely low densities of helium, nitrogen, oxygen and carbon dioxide which disappear entirely at around 10,000 km from the earth.

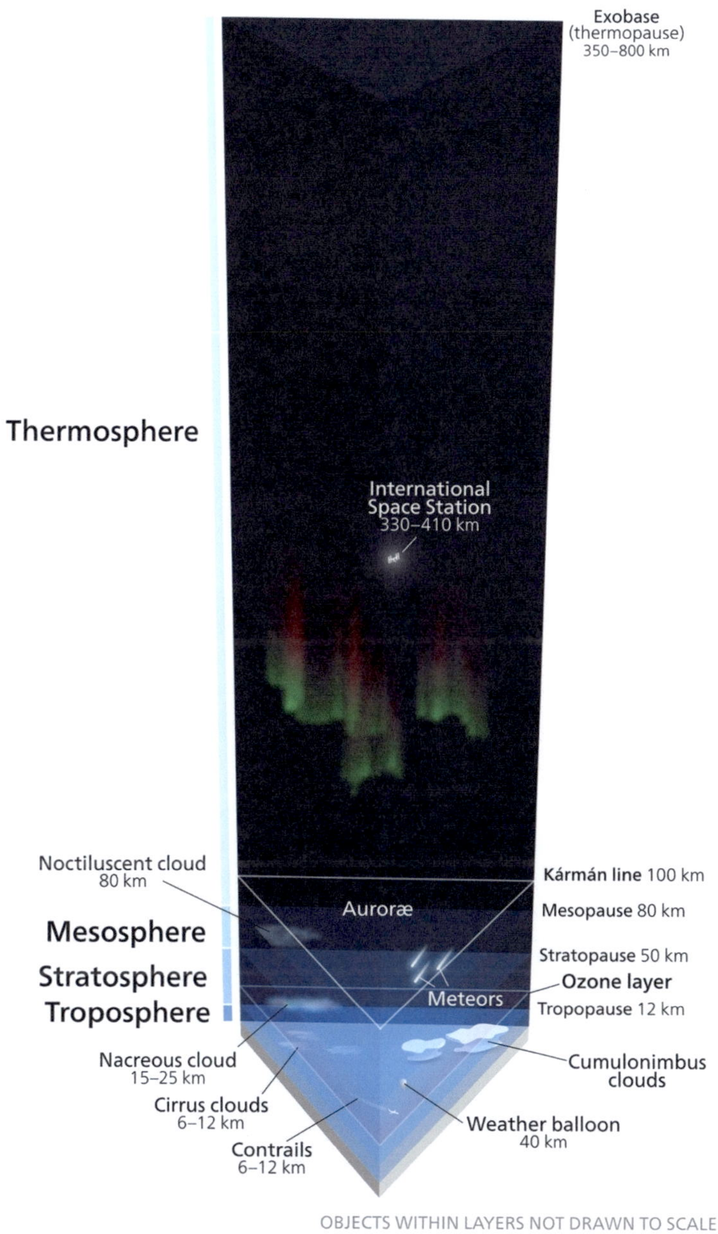

Fig. 3.1 Earth's atmosphere (Image: Kelvinsong; http://en.wikipedia.org/wiki/Atmosphere_of_Earth#mediaviewer/File:Earth%27s_atmosphere.svg)

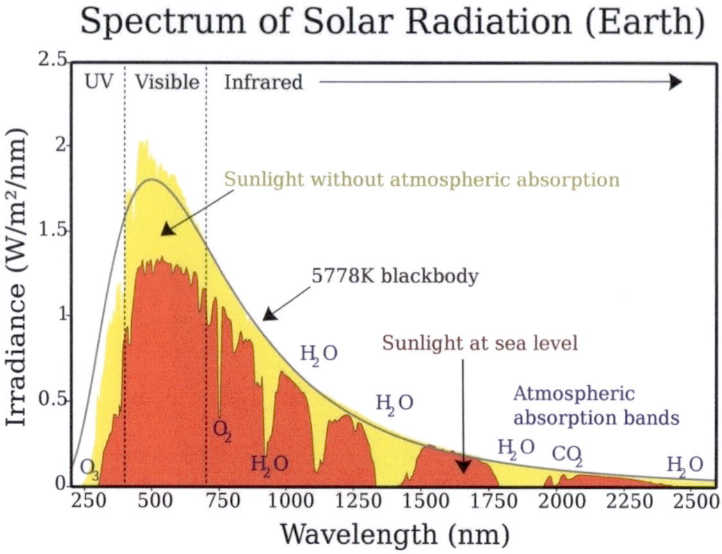

Fig. 3.2 Spectrum of solar radiation with and without atmospheric absorption (Image: Robert A Rohde. From: http://en.wikipedia.org/wiki/Sunlight#mediaviewer/File:Solar_spectrum_en.svg)

Radiation from the sun lies within a broad spectrum from ultraviolet radiation at the lower wavelengths through the visible spectrum to infrared radiation (also known as near-infrared) (see Fig. 3.2).

The figure shows the impact that the atmosphere has on incident radiation from the sun, indicating how much water vapour (H_2O), carbon dioxide (CO_2) and ozone (O_3) reduce solar intensities at sea level within specific wavelength bands. What this spectrum does not show is the infrared radiation transmitted from the earth back into space and filtered by water vapour, CO_2 and other gases, creating what has become known as the 'greenhouse effect'. The greater the filtering by 'greenhouse gases' in the atmosphere the higher average air temperatures at the earth's surface become, resulting in global warming. Water vapour is the dominant greenhouse gas, although because it is variable in concentration, as well as being weather dependant, its role in global warming fluctuates. Of the other gases that find their way into the atmosphere from terrestrial sources carbon dioxide is by far the most important. However other gases, such as methane, nitrous oxide and fluorinated gases, are much more powerful absorbers of infra-red radiation, and although present in far smaller quantities, when their concentrations in the atmosphere are converted to 'carbon dioxide equivalent' their contribution to global warming in 2015 was around 35% (see Fig. 3.3).

The 'F-gases' in Fig. 3.3 included the ozone depleting 'Montreal gases' which in this context are chlorofluorocarbons (CFCs), hydrochlorofluorocarbons (HCFCs) and methyl chloroform (CH_3CCl_3): all controlled under legislation driven by the Montreal protocol as ozone depleting gases; whilst the acronyms 'PFC, HFC, SF6' are included within the list of Kyoto greenhouse gases: perfluorocarbons,

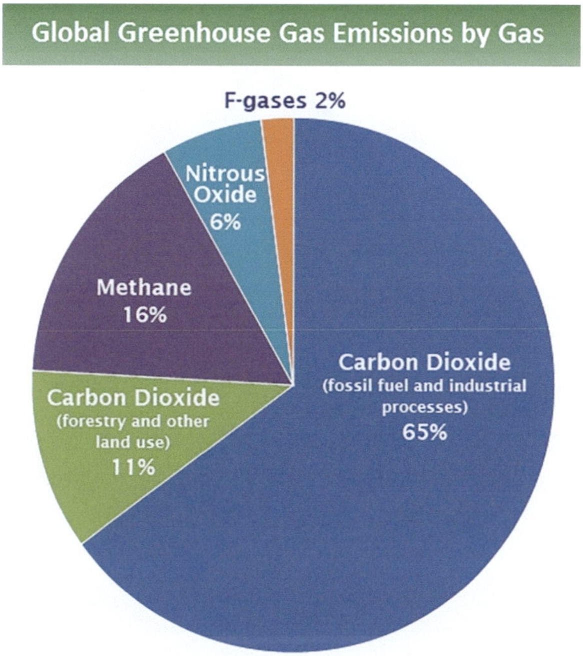

Fig. 3.3 Global greenhouse gas emissions by gas (2015) (https://commons.wikimedia.org/wiki/File:Global_emissions_gas_2015.png)

hydrofluorocarbons and sulphur hexafluoride, respectively. These are all anthropogenic gases primarily associated with industrial materials such as refrigerants, blown insulants and degreasing agents. Figure 3.4 shows the proportion of GHGs for each source as reported in 2016.

The concentration of these gases expressed as the impact they would have on warming the planet if they were carbon dioxide (CO_{2eq}) is greater than their emissions as a percentage of the total. This is because they absorb far more infrared radiation than carbon dioxide does. The relationship between the absorption properties of a given greenhouse gas to those for CO_2 is popularly known as 'global warming potential' or GWP, values for which have been estimated for these gases over time horizons of 20 and 100 years and published in Appendix 8A, Chapter 8 of the IPCC Working Group 1 (WG1) Assessment Report (AR5) on the *'Physical Science Basis' of Climate Change 2013*.[1]

The strict definition of GWP as adopted by the IPCC is 'An index, based on radiative properties of greenhouse gases, measuring the radiative forcing following a pulse emission of a unit mass of a given greenhouse gas in the present-day atmosphere integrated over a chosen time horizon, relative to that of carbon dioxide'; where radiative forcing (RF) is the 'the change in the net, down-ward minus upward,

[1] http://www.ipcc.ch/pdf/assessment-report/ar5/wg1/WG1AR5_Chapter08_FINAL.pdf

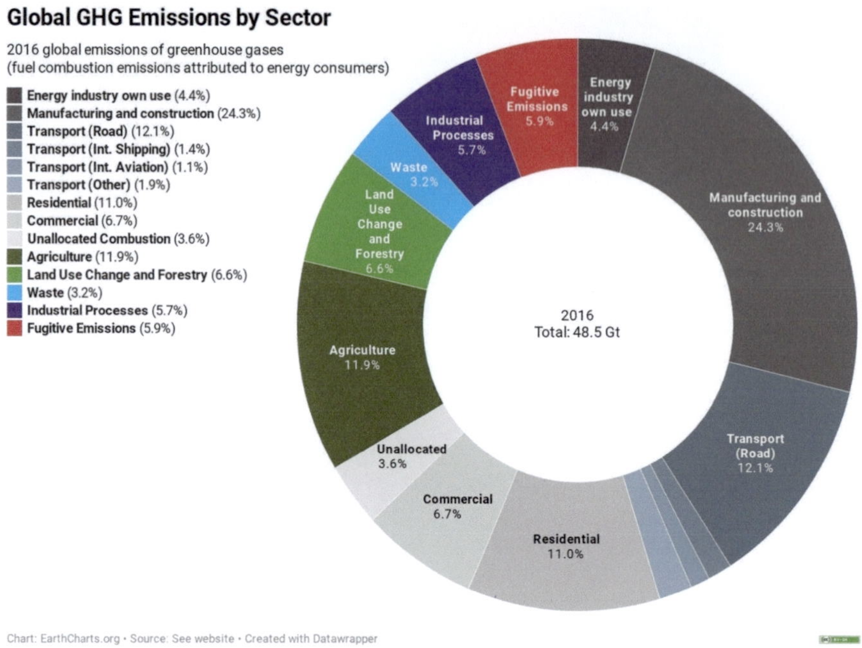

Fig. 3.4 Global greenhouse gas emissions by source (2016) (https://commons.wikimedia.org/wiki/File:Global_GHG_Emissions_by_Sector_2016.png)

radiative flux (expressed in W/m^2) at the tropopause or top of atmosphere due to a change in an external driver of climate change, such as, for example, a change in the concentration of carbon dioxide or the output of the Sun,' calculated using a pre-industrial base year of 1750, at which time the only greenhouse gases present in the atmosphere were water vapour, carbon dioxide, methane and nitrous oxide, apart from small quantities of halocarbons associated with marine phytoplankton and algae.[2] GWP is dependent to varying degrees on the time horizon for which it has been determined. For example, the GWP for methane over a 20 year horizon is 84 and over 100 years is 28. On the other hand, nitrous oxide is listed has having a GWP of 264 over 20 years and 265 over 100 years. Some of the industrial chemicals have GWP values which are orders of magnitude higher than these: for example, refrigerant gas HFC-23 has a GWP over 100 years of 12,400, whilst the value for sulphur hexafluoride is 23,500 for the same time horizon. Fortunately the emission rates of these gases is very small compared with carbon dioxide, particularly since their life expectancy (adjustment time) in the atmosphere is 222 and 3200 years, respectively.

In contrast, water vapour has by far the highest concentration of any greenhouse gas in the atmosphere and, having a GWP of 2 or 3 times that for CO_2, it is an

[2] https://www.ipcc.ch/site/assets/uploads/2018/02/WG1AR5_AnnexIII_FINAL.pdf

important 'climate forcer' in the short term, however, its average lifetime is only 10 days and it is unevenly distributed due to the continual processes of evaporation, condensation and precipitation with changes in temperature and weather patterns.

Another important constituent of the atmosphere is black carbon or soot. This is the product of incomplete combustion of fossil fuels taking the form of very fine particles, having mass median diameters of less than 2.5 microns (PM2.5). As an effective absorber of infra-red radiation black carbon has a GWP of between 450 and 1500 but a life expectancy of only 4–12 days. Black carbon is found throughout the atmosphere and can impact air quality and respiratory health when in the lower atmosphere whilst also depositing on surfaces such as snow and ice where it increases the rate of absorption of solar radiation and hence melting rates (see Sect. 3.3) and on foliage, where it can similarly increase radiation absorption and damage the health of plants.

According to Renee Cho of the Earth Institute at Columbia University 'Black carbon, like all particles in the atmosphere, also affects the reflectivity, stability and duration of clouds and alters precipitation. Depending on how much soot is in the air and where black carbon sits in the atmosphere, it has different effects. If it absorbs heat at the level where clouds are forming, they will evaporate. When it lies above lower stratocumulus clouds that block the sun, it stabilises them and thus has a cooling effect. Because black carbon interacts with other components of particulate matter, such as sulphates and nitrates that reflect sunlight and cool the atmosphere, scientists do not know exactly how much black carbon itself directly contributes to global warming'.[3]

The 2013 IPCC WG1 AR5 referred to above offers effective radiative forcing (ERF) as a more accurate method of estimating global warming since, unlike RF, it takes account of the short term impacts of atmospheric temperatures and water vapour, including rapidly changing cloud formations. Because the atmosphere over equatorial regions is warmer than over polar regions, for example, it can contain between 20 and 30 times more water and hence the impact of water vapour on warming in the equatorial regions is proportionately greater.

An important factor in the determination of GWP and anthropogenic climate change is the combined effect of the differing times that greenhouse gases remain in the atmosphere and their relative effectiveness in causing radiative forcing. Carbon dioxide, for example, has a 'turnover' of only 4 years '…because of the rapid exchange between the atmosphere and the ocean and terrestrial biota (flora and fauna). However, a large part of that CO_2 is returned to the atmosphere within a few years. Thus, the adjustment time (life expectancy) of CO_2 in the atmosphere is actually determined by the rate of removal of carbon from the surface layer of the oceans into its deeper layers. Although an approximate value of 100 years may be given for the adjustment time of CO_2 in the atmosphere, the actual adjustment is faster initially and slower later on'.[4]

[3] https://blogs.ei.columbia.edu/2016/03/22/the-damaging-effects-of-black-carbon/

[4] https://www.ipcc.ch/site/assets/uploads/2018/02/WG1AR5_AnnexIII_FINAL.pdf

The concentration of greenhouse gases in the atmosphere is dependent upon the difference between the emission rate from sources and the sink rate either into reservoirs or through chemical reaction with other gases. The earth itself is an important sink, whether via flora and fauna (biota), into soils and peat, or through absorption by water bodies.

Thematic Focus Element 7 in the Technical Summary to the IPCC WG1 Assessment Report 5 describes the action of these sinks as follows:

> The human caused excess of CO_2 in the atmosphere is partly removed from the atmosphere by carbon sinks in land ecosystems and in the ocean, currently leaving less than half of the CO_2 emissions in the atmosphere. Natural carbon sinks are due to physical, biological and chemical processes acting on different time scales. An excess of atmospheric CO_2 supports photosynthetic CO_2 fixation by plants that is stored as plant biomass or in the soil.
>
> The uptake of anthropogenic CO_2 by the ocean is primarily a response to increasing CO_2 in the atmosphere. Excess atmospheric CO_2 absorbed by the surface ocean or transported to the ocean through aquatic systems (e.g., rivers, groundwaters) gets buried in coastal sediments or transported to deep waters where it is stored for decades to centuries. The deep ocean carbon can dissolve ocean carbonate sediments to store excess CO_2 on time scales of centuries to millennia.[5]

Soils are also a terrestrial sink for methane due to the presence of methanotrophic bacteria (that metabolise methane), although the greatest loss (>90%) occurs in the atmosphere, where methane combines with hydroxyl radicals (OH) to form a methyl radical (CH_3) and water vapour. Further reactions result in the formation of ozone, carbon dioxide and more water vapour, all of which continues to contribute to radiative forcing and hence global warming, although with a lower GWP than the original methane.

Nitrous oxide has no significant terrestrial sinks but is subject to photolysis in the atmosphere, where it enters the stratospheric ozone depletion cycle as NO (nitric oxide). Similarly halogenated halocarbons such as CFCs are removed by photolysis in the mid to upper stratosphere, whilst hydrogen bearing halocarbons such as HCFCs react with OH in a similar manner to methane.

As well as comprising a mixture of gases and vapours the atmosphere also contains a variety of aerosols: some formed naturally, some anthropogenic. Natural aerosols include clouds and mist in which water vapour condenses around nuclei. Fine dust and smoke in suspension from desert storms, fires and volcanic eruptions can create particulate aerosols, whilst sulphur dioxide from volcanic eruptions, coal and oil burning can condense to form sulphuric acid or sulphate aerosols. The pollution from very large volcanic eruptions has historically penetrated the stratosphere, resulting in an accumulation of sulphate aerosols in a region that has become known as the stratospheric aerosol layer or 'Junge layer', which lies at an altitude of around 30 km. The SO_2 from coal and oil fired power stations tends to congregate in the troposphere and is less widely distributed.

[5] https://www.ipcc.ch/site/assets/uploads/2018/02/WG1AR5_TS_FINAL.pdf

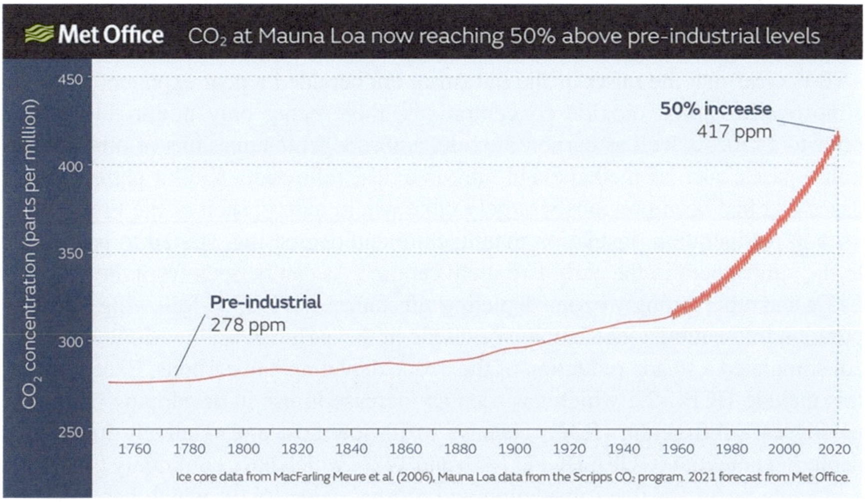

Fig. 3.5 Graph showing CO_2 fluctuations between 1750 and 2021, derived from Antarctic ice core samples and Mauna Loa (Hawaii—British Crown Copyright Met office; From: https://www.carbonbrief.org/met-office-atmospheric-co2-now-hitting-50-higher-than-pre-industrial-levels/)

The resultant aerosols from all of these sources filter sunlight, resulting in a cooling or negative radiative forcing effect that has a significant impact on global warming, although unevenly distributed across the planet's surface.

3.2.1 Global Warming Projections

In its Synthesis Report published on 1 November 2014, drawing together the evidence for climate change gathered by Working Group 1 for its 5th Assessment Report, the IPCC stated that

> Anthropogenic greenhouse gas emissions have increased since the pre-industrial era driven largely by economic and population growth. From 2000 to 2010 emissions were the highest in history. Historical emissions have driven atmospheric concentrations of carbon dioxide, methane and nitrous oxide, to levels that are unprecedented in at least the last 800,000 years, leading to an uptake of energy by the climate system.[6]

Figure 3.5 shows a composite graph showing the carbon dioxide concentrations determined from ice core samples taken from a number of sites in Antarctica; along with readings that have been taken at the Mauna Loa Observatory located at 3400 m above sea level in Hawaii. Much of this work has been coordinated by the Global Monitoring Division of the US National Oceanic and Atmospheric Administration

[6] https://archive.ipcc.ch/pdf/assessment-report/ar5/syr/SYR_AR5_FINAL_full_wcover.pdf

(NOAA) and the Commonwealth Scientific & Industrial Research Organization (CSIRO) in Victoria, Australia.

It is clear that the onset of the industrial era heralded a near exponential rise in atmospheric carbon dioxide concentrations, there being only minor fluctuations prior to 1750. As well as carbon dioxide, anthropogenic emissions of other greenhouse gases, such as methane and nitrous oxide, followed a similar pattern whilst gases that had been previously largely unknown in nature, such as the Freon gases used in refrigeration, insulation manufacture and degreasing, started to be emitted to the atmosphere in the early twentieth century. As can be seen from the plots for CFCs and other strongly ozone depleting substances in Fig. 3.6, following its incorporation into national legislation worldwide in the early 1990s, the Montreal protocol stimulated a steady reduction of the ozone depleting halocarbons. Exceptions to this include HCFC-22, which has seen an increase in use in developing countries, mostly as a refrigerant (R22), because of its low cost and relatively low ozone depleting potential (ODP); HCFC 141b and 142b, which have commonly be used as a blowing agent for foam insulation and nitrous oxide (N_2O) which has followed close to a straight line increase during the 35 years shown in the graph due to a combination of fossil fuel combustion and increase in fertiliser use in agriculture.

- CH3Cl: Methyl Chloride—low ODP, solvent
- CFC-12: Chlorofluorocarbon R12—ozone depleting, refrigerant gas
- CFC-11: Chlorofluorocarbon R11—ozone depleting, refrigerant gas
- HCFC-22: Hydrochlorofluorocarbon R22—low ODP, refrigerant gas
- CCl_4: Carbon tetrachloride—ozone depleting, solvent and fire extinguishing

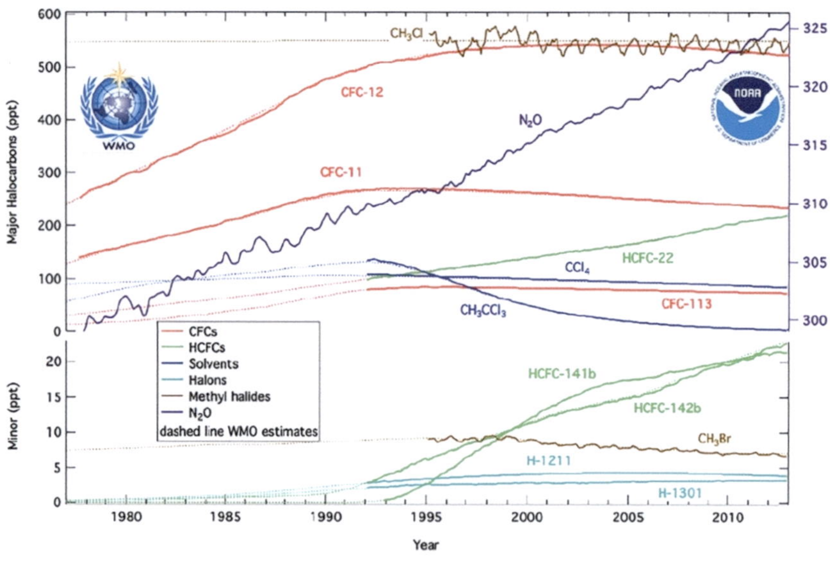

Fig. 3.6 Graph showing fluctuations in common halocarbons in the atmosphere between 1978 and 2013 (From: https://gml.noaa.gov/hats/graphs/graphs.html)

- CH_3CCl_3: Methyl chloroform—low ODP, solvent, degreasing agent
- CFC-113: Chlorofluorocarbon R12—ozone depleting, refrigerant gas
- HFC-134a:—Hydrofluorocarbon R134a—refrigerant gas
- Halon-1211: Bromochlorodifluoromethane—ozone depleting—fire extinguishing
- Halon-1301: Bromotrifluoromethane—ozone depleting—fire extinguishing
- HCFC 141b: 1, 1-dichloro-1-fluoroethane—ozone depleting—foam insulation
- HCFC 142b: 1-Chloro-1, 1 difluoroethane—ozone depleting—refrigerant, feedstock for PVDF and foam insulation

3.2.2 Temperature Change

In its Synthesis Report for AR5 referred to above the IPCC gave the following stark conclusion from the work carried out by the hundreds of scientists from many countries who had contributed to its research:

> Warming of the climate system is unequivocal, and since the 1950s, many of the observed changes are unprecedented over decades to millennia. The atmosphere and ocean have warmed, the amounts of snow and ice have diminished, and sea level has risen.[7]

If possible the warnings only grew louder with the publication of Assessment Report 6. For example, the Summary for Policymakers states that:

> Global surface temperature will continue to increase until at least the mid-century under all emissions scenarios considered. Global warming of 1.5C and 2C will be exceeded during the 21st century unless deep reductions in CO_2 and other greenhouse gas emissions occur in the coming decades.

The graphs that are reproduced in this section are a sample of the extensive graphics used in the report produced by Working Group 1 for the 2021 6th Assessment Report of the Intergovernmental Panel on Climate Change—The Physical Science Basis.[8] There is no attempt here to summarise the entire coverage of this report, which comprises nearly 2400 pages but to give a flavour of its salient points and some of the key findings from its exhaustive analysis and projections derived from the numerous climate models used.

In order to produce the consensus that led to the IPCC 6th Assessment Report, numerous computer simulations were carried out to predict the impacts of global warming on surface temperatures and sea level rise. The Coupled Model Intercomparison Project (CMIP) has been the main vehicle for this, Phase 6 of which produced the results summarised in Fig. 3.9. These used five 'Shared Socioeconomic Pathways' (SSP1-1.9, SSP1-2.6, SSP2-4.5, SSP3-7.0 and SSP5-8.5) which represent a number of scenarios along with associated radiative forcing

---

[7] https://archive.ipcc.ch/pdf/assessment-report/ar5/syr/SYR_AR5_FINAL_full_wcover.pdf

[8] https://www.ipcc.ch/report/ar6/wg1/downloads/report/IPCC_AR6_WGI_FullReport_small.pdf

predictions for the year 2100, relative to pre-industrial values, ranging from 1.9 to 8.5 W/m^2, respectively. Radiative forcing (RF) is the change of net radiative flux (incoming minus outgoing radiation) at the top of the atmosphere compared with pre-industrial levels and is used for evaluating and comparing the strength of various mechanisms affecting the earth's radiation balance and hence causing climate change.

The SSPs are a more sophisticated version of the Representative Concentration Pathways (RCP) used in previous phases of the CMIP. These were developed by a number of separate groups from a common database for greenhouse gas emissions from all major sources. Each developed a scenario based on assumptions through to 2100 and beyond based on assumed economic activity, energy sources, population growth and hence anthropogenic greenhouse gas emissions and the resultant changes in radiative forcing. Where the models for SSPs for 1.9, 2.6, 4.5 and 7.0 W/m^2 assume degrees of mitigation, SSP5-8.5 is commonly referred to as the fossil fuel-based scenario, whilst SSP1-1.9 depends on sustainable development and low consumption (see Box 3.1).

A number of different combinations of SSP and RF have been modelled, although the SSPs shown in figures below have been adopted by the IPCC authors as the most likely scenarios. Box 3.1 summarises the assumptions used in the developing the models for SSP and RF.

Box 3.1 Shared Socioeconomic Pathways (SSPs), Risk Categories for Temperature Rise and Illustrative Mitigation Pathways (IMPs)

Shared Socioeconomic Pathways:

SSP1—Sustainable development pathway—low consumption, low population growth.

SSP2—Continue with current development patterns.

SSP3—Policy focused primarily on security, barriers to international trade, low population growth in rich countries, high in poor, with steady growth globally through twenty-first century.

SSP4—Policy focus on elite, high consumption, low to medium economic growth, low population growth in rich countries, high in poor, but with total plateauing after 2050.

SSP5—Fossil fuel-based development, free markets in control, high consumption and economic growth, but with population growth tailing off after 2050.

For a detailed narrative for each SSP go to https://www.carbonbrief.org/explainer-how-shared-socioeconomic-pathways-explore-future-climate-change/

Risk Categories:

C1—Average temperature rise limited to 1.5 K by 2100 with >50% likelihood with no or limited overshoot.

C2—Average temperature rise limited to 1.5 K by 2100 with >50% likelihood with high and prolonged overshoot (0.1–0.3 K over several decades).
C3—Average temperature rise limited to 2 K by 2100 with >67% likelihood with no overshoot.
C4—Average temperature rise limited to 2 K by 2100 with potential continuing rise thereafter.
C5—Average temperature rise limited to 2.5 K by 2100.
C6—Average temperature rise limited to 3 K by 2100.
C7—Average temperature rise limited to 4 K by 2100.
C8—Average temperature rise Exceeds 4 K by 2100 and continues rising thereafter.

Illustrative Mitigation Pathways (IMPs):

IMP-Ren—Emphasis on renewables.
IMP-Neg—CO_2 removal.
IMP-LD—Low demand for resources.
IMP-GS—A gradual strengthening of mitigation measures over time.
IMP-SP—Sustainable development strategies, including reducing inequalities.

It is interesting to note that the SSP3 scenario assumes a steady global population growth to 12.62 billion people by 2100. This compares with the UN 2022 'median-variant' projection of 10.4 billion, which assumes a decline in fertility for countries where large families have historically been prevalent; and a 'high-variant' projection of 14.5 billion if the birth rate is an average of half a child higher per family.[9] At the other end of the scale, the other scenarios assume initially a rise in population, peaking at around 9.6 billion by mid-century with SSP1 and SSP5, dropping back to 2012 levels by the end of the century. This compares with the UN's 'low-variant' projection of a population rising to 9 billion by 2050 and falling to 6.5 billion by 2100, based on an average birth rate of half a child less per family than the median.

Figure 3.7 shows the energy mix and consumption on which the various SSP models have been run. It can be seen that for the worst case scenario (SSP5) nearly 50% of the global energy consumption would be derived from coal in 2100, a large proportion of which might be used to fuel the rapidly growing economies of China and India in this scenario.

The resultant atmospheric carbon dioxide, methane and nitrous oxide concentrations have been predicted by modelling emissions from power generation and other greenhouse gas sources, including those associated with changes in land use and agriculture. As can be seen from Fig. 3.9, SSP5 is projected to produce the greatest increase in atmospheric carbon dioxide emissions, but along with methane and

[9] https://www.ipcc.ch/report/ar6/wg1/downloads/report/IPCC_AR6_WGI_FullReport_small.pdf

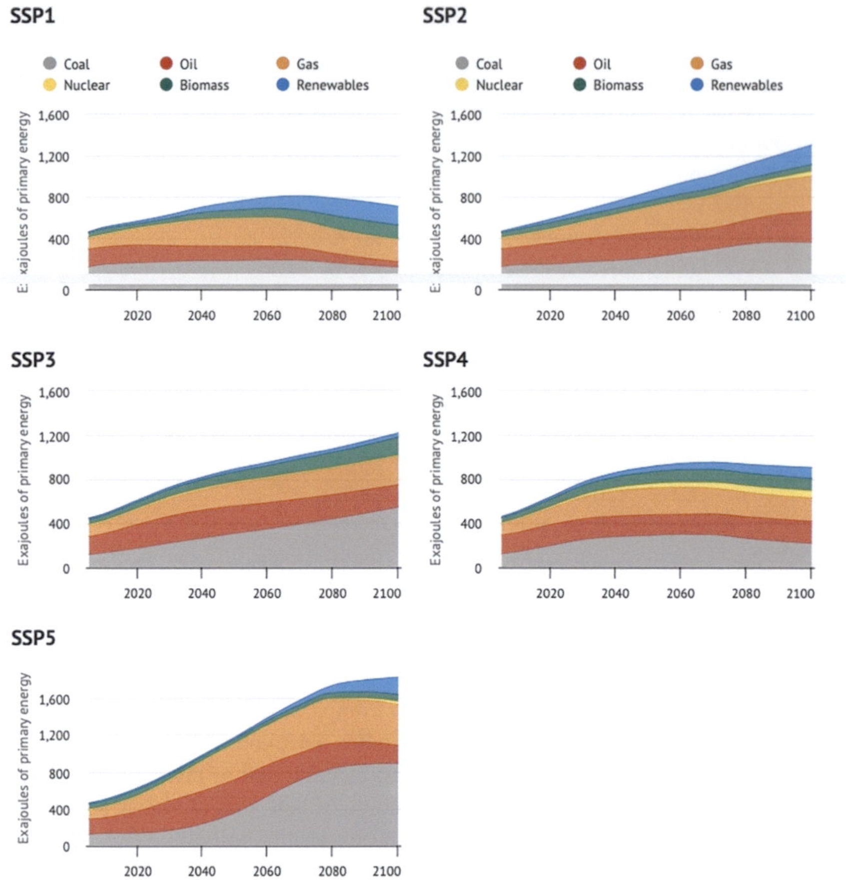

Fig. 3.7 Energy mix for SSP scenarios up to 2100. 1 Exajoule = 10^{18} Joules or 277,777,778 MWh (Image by Carbon Brief; From: https://www.carbonbrief.org/explainer-how-shared-socioeconomic-pathways-explore-future-climate-change/)

nitrous oxide showing some indication of abatement towards the end of the twenty-first century, whilst the SSP3 projections for all three continue to rise. The SSP1 and 2 projections, on the other hand, show the levels of all three dropping below 2015 rates.

The IPCC has introduced another layer of categorisation into AR6 which divides outcomes of climate change according to the risk of temperature rise during the twenty-first century and beyond. For example, C1 is based on the temperature rise above pre-industrial levels being 1.5 K in 2100 with a likelihood of greater than 50% and with no or limited overshoot prior to this.[10] The highest category is C8 which corresponds to a temperature rise of more than 4 K in 2100 and continuing to

[10] http://www.carbonbrief.org/what-happens-if-we-overshoot-the-two-degree-target-for-limiting-global-warming

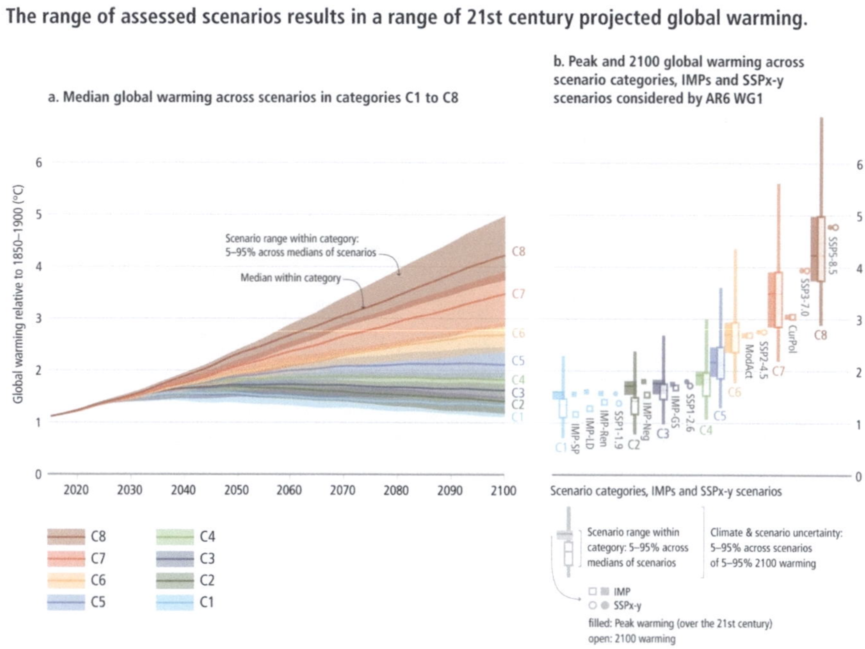

Fig. 3.8 Relationship between categories, SSPs and IMPs (Illustrative Mitigation Pathways) and projected temperature rise during twenty-first century (From: [2])

rise thereafter (See Box 3.2). The relationship between these categories and SSPs is shown in Fig. 3.8b which is reproduced from AR6.

The Illustrative Mitigation Pathways (IMPs) indicated in the graph at 3.8b are based on strategies that 'illustrate pathways that achieve deep and rapid emissions reductions through different combinations of mitigation strategies'.

Box 3.2

Figure SPM Box Fig. 1: Figure Box SPM.1 Projected global mean warming of the ensemble of modelled scenarios included in the climate categories C1–C8 and IMPs (based on emulators calibrated to the WGI assessment), as well as five illustrative scenarios (SSPx-y) as considered by AR6 WG I. The left panel shows the p5–p95 range of projected median warming across global modelled pathways within a category, with the category medians (line). The right panel shows the peak and 2100 emulated temperature outcomes for the categories C1 to C8 and for IMPs, and the five illustrative scenarios (SSPx-y) as considered by AR6 WG I. The boxes show the p5–p95 range within each scenario category as in panel-a. The combined p5–p95 range across scenarios and the climate uncertainty for each category C1–C8 is also shown for 2100 warming (thin vertical lines). {Table SPM.1, Fig. 3.11, WGI Figure SPM.8}

Note—to see the references in this box refer to [2].

Box 3.3 Caption for Figure SPM.4 from the IPCC Report [1]

Figure SPM.4 | Future anthropogenic emissions of key drivers of climate change and warming contributions by groups of drivers for the five illustrative scenarios used in this report. The five scenarios are SSP1-1.9, SSP1-2.6, SSP2-4.5, SSP3-7.0 and SSP5-8.5.

Panel (a) Annual anthropogenic (human-caused) emissions over the 2015–2100 period. Shown are emissions trajectories for carbon dioxide (CO_2) from all sectors ($GtCO_2$/year) (left graph) and for a subset of three key non-CO_2 drivers considered in the scenarios: methane (CH_4, $MtCH_4$/year, top-right graph); nitrous oxide (N_2O, MtN_2O/year, middle-right graph); and sulphur dioxide (SO_2, $MtSO_2$/year, bottom-right graph, contributing to anthropogenic aerosols in panel (b).

Panel (b) Warming contributions by groups of anthropogenic drivers and by scenario are shown as the change in global surface temperature (°C) in 2081–2100 relative to 1850–1900, with indication of the observed warming to date. Bars and whiskers represent median values and the very likely range, respectively. Within each scenario bar plot, the bars represent: total global warming (°C; 'total' bar) (see Table SPM.1); warming contributions (°C) from changes in CO_2 ('CO_2' bar) and from non-CO_2 greenhouse gases (GHGs; 'non-CO_2 GHGs' bar: comprising well-mixed greenhouse gases and ozone); and net cooling from other anthropogenic drivers ('aerosols and land use' bar: anthropogenic aerosols, changes in reflectance due to land-use and irrigation changes, and contrails from aviation) (see Figure SPM.2, panel c, for the warming contributions to date for individual drivers). The best estimate for observed warming in 2010–2019 relative to 1850–1900 (see Figure SPM.2, panel a) is indicated in the darker column in the 'total' bar. Warming contributions in panel (b) are calculated as explained in Table SPM.1 for the total bar. For the other bars, the contribution by groups of drivers is calculated with a physical climate emulator of global surface temperature that relies on climate sensitivity and radiative forcing assessments. {Cross-Chapter Box 1.4, 4.6, Fig. 4.35, 6.7, Figs. 6.18, 6.22 and 6.24, 7.3, Cross-Chapter Box 7.1, Fig. 7.7, Box TS.7, Figs. TS.4 and TS.15}.

Note—to see the references in this box refer to [1].

The 'global surface temperature' referred to in Fig. 3.9 represents the key criterion on which UN global warming mitigation strategies are based. At the time of writing, a consensus has gathered momentum that if this value exceeds 1.5 °C above pre-industrial levels, then the impacts will be disastrous. Hence a lot of attention is being paid to the C1 and IMP scenarios that set out a route map through to 2100 which should achieve the temperature profile shown in Fig. 3.10, with surface temperatures being held below a 1.5° rise.

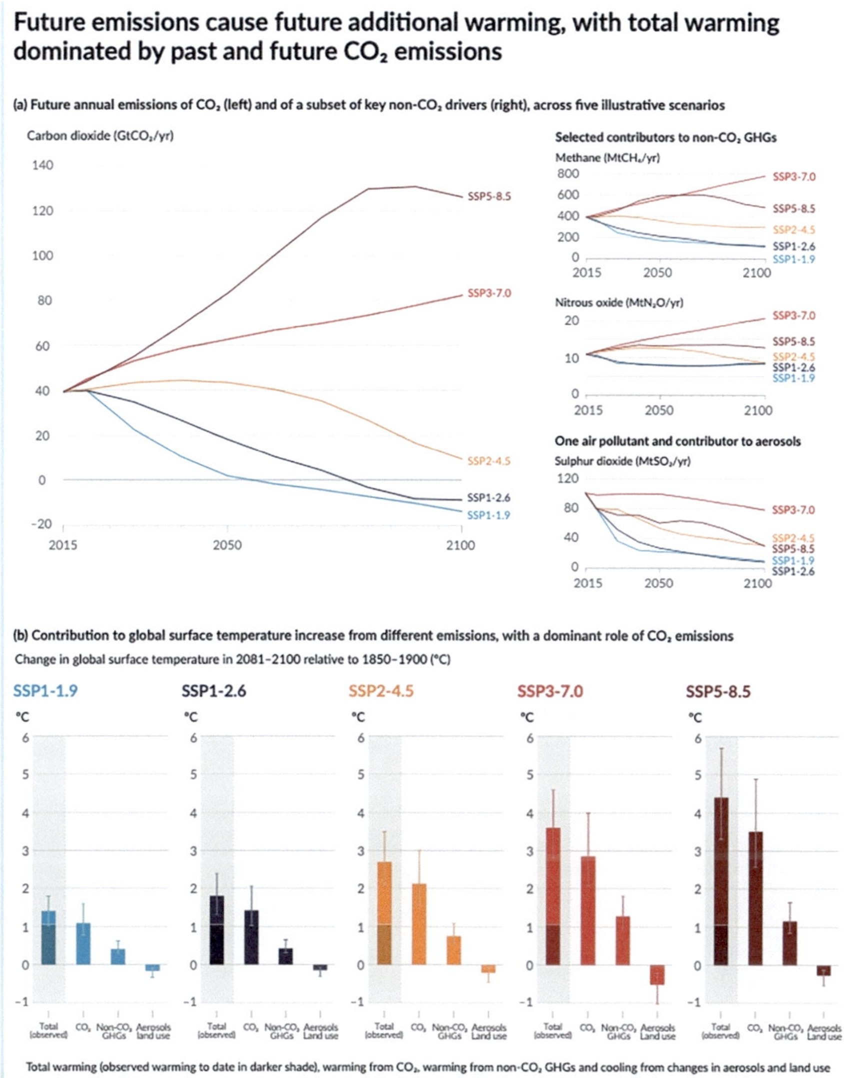

Fig. 3.9 Emissions of CO_2 and other greenhouse gases and aerosols under Shared Socioeconomic Pathways and their contribution to changes in global surface temperatures (From [1])

It is the impacts of carbon dioxide and other greenhouse gases, along with corresponding temperature rise, that are concern of course. These include impacts on: global mean sea level (GMSL); atmospheric circulation, such as jet streams and weather patterns; the water cycle (humidity and precipitation), cryosphere (ice, snow and permafrost), ocean temperature; carbon and other biogeochemical cycles,

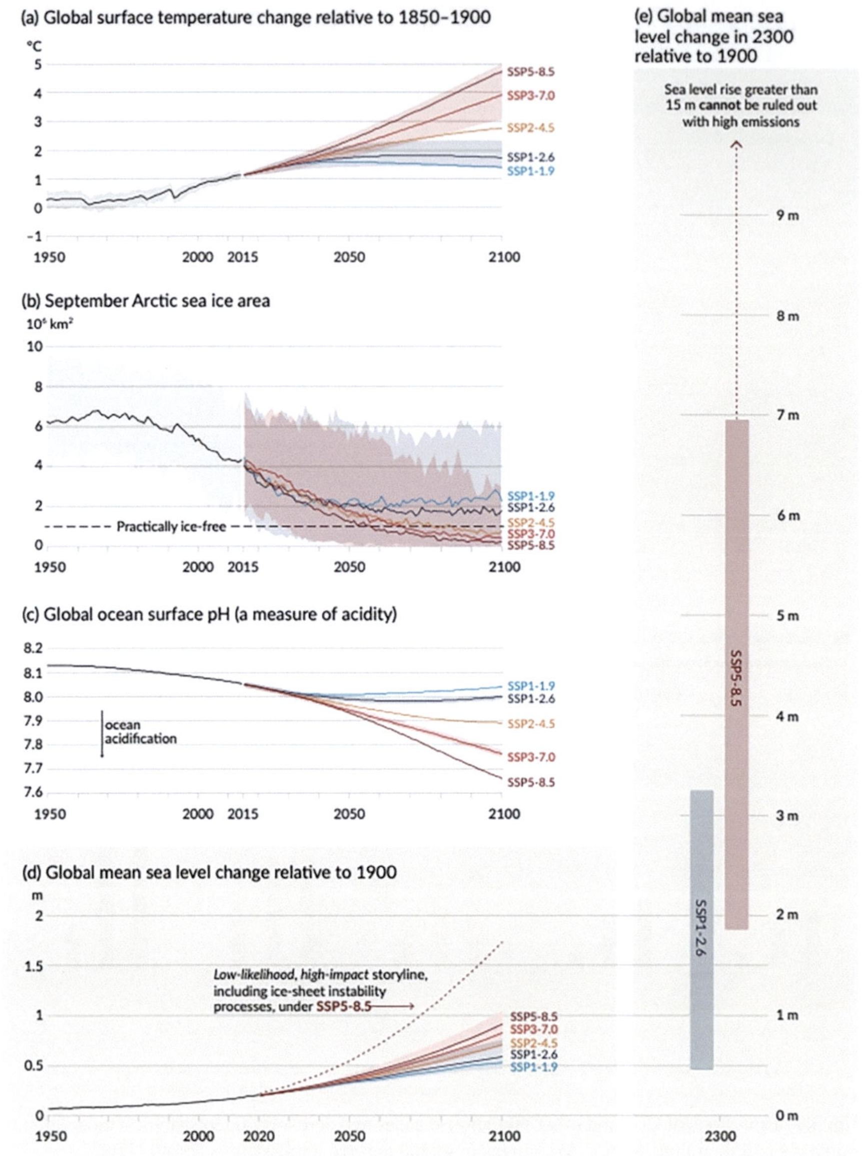

Fig. 3.10 Observed and projected changes in global surface temperature, sea ice area, ocean surface pH, sea level change up to 2100 and 2300 (From: [1])

such as the uptake of carbon dioxide and other greenhouse gases by ecosystems and surface water, including ocean acidification.

The IPCC Synthesis report for AR5 concludes with a high degree of confidence that, associated with global climate change, there is a 'risk of severe ill-health and disrupted livelihoods resulting from storm surges, sea level rise, and coastal flooding; inland flooding in some urban regions and periods of extreme heat'. The same report refers to 'systemic risks due to extreme weather events leading to breakdown of infrastructure networks and critical services; risk of food and water insecurity and loss of rural livelihoods and income, particularly for poorer populations'; as well as a 'risk of loss of ecosystems, biodiversity, and ecosystem goods, functions and services'.[11]

This theme is developed in the AR6 WG1 SPM which states that 'A warmer climate will intensify very wet and very dry weather and climate events and seasons, with implications for flooding or drought (high confidence), but the location and frequency of these events depend on projected changes in regional atmospheric circulation, including monsoons and mid-latitude storm tracks. It is very likely that rainfall variability related to the El Niño–Southern Oscillation is projected to be amplified by the second half of the twenty-first century in the SSP2-4.5, SSP3-7.0 and SSP5-8.5 scenarios'.[12]

The remainder of this chapter will examine these issues in more detail, referring extensively to the IPCC Assessment Reports.

3.3 Sea Level Rise, Snow and Ice Retreat

According to the US Geological Survey water covers around 71% of the earth's surface, but only approximately 0.13% of its volume. Even so this represents some 1386 million km^2 of water, 96.5% of which is held in the oceans. Perhaps surprisingly only 1.74% of the earth's water is reported to be held in the ice caps, glaciers and permanent snow, although this figure is based on estimates published in 1993.[13]

Chapter 13 of the IPCC Fifth Assessment Report from Working Group I sets out the mechanisms by which climate change results in sea level rise:

> Changes in ocean currents, ocean density and sea level are all tightly coupled such that changes at one location impact local sea level and sea level far from the location of the initial change, including changes in sea level at the coast in response to changes in open-ocean temperature. Although both temperature and salinity changes can contribute significantly to regional sea level change only temperature change produces a significant contribution to global average ocean volume change due to thermal expansion or contraction. Regional atmospheric pressure anomalies also cause sea level to vary through atmospheric loading. All of these climate-sensitive processes cause sea level to vary on a broad range of space and time scales from relatively short-lived events, such as waves and storm surges, to sustained changes over several decades or centuries that are associated with atmospheric and ocean modes of climate variability.[14]

[11] https://archive.ipcc.ch/pdf/assessment-report/ar5/syr/SYR_AR5_FINAL_full_wcover.pdf

[12] https://www.ipcc.ch/report/ar6/wg1/downloads/

[13] http://water.usgs.gov/edu/earthhowmuch.html

[14] https://www.ipcc.ch/report/ar6/wg1/downloads/report/IPCC_AR6_WGI_FullReport_small.pdf

Box 3.4 Caption for Figure SPM.8 from the IPCC Report [1]
Figure SPM.8 | Selected indicators of global climate change under the five illustrative scenarios used in this Report.

The projections for each of the five scenarios are shown in colour. Shades represent uncertainty ranges—more detail is provided for each panel below. The black curves represent the historical simulations (panels a, b, c) or the observations (panel d). Historical values are included in all graphs to provide context for the projected future changes.

Panel (a) Global surface temperature changes in °C relative to 1850–1900. These changes were obtained by combining Coupled Model Intercomparison Project Phase 6 (CMIP6) model simulations with observational constraints based on past simulated warming, as well as an updated assessment of equilibrium climate sensitivity (see Box SPM.1). Changes relative to 1850–1900 based on 20-year averaging periods are calculated by adding 0.85 °C (the observed global surface temperature increase from 1850–1900 to 1995–2014) to simulated changes relative to 1995–2014. Very likely ranges are shown for SSP1-2.6 and SSP3-7.0.

Panel (b) September Arctic sea ice area in 106 km^2 based on CMIP6 model simulations. Very likely ranges are shown for SSP1-2.6 and SSP3-7.0. The Arctic is projected to be practically ice-free near mid-century under intermediate and high GHG emissions scenarios.

Panel (c) Global ocean surface pH (a measure of acidity) based on CMIP6 model simulations. Very likely ranges are shown for SSP1-2.6 and SSP3-7.0.

Panel (d) Global mean sea level change in metres, relative to 1900. The historical changes are observed (from tide gauges before 1992 and altimeters afterwards), and the future changes are assessed consistently with observational constraints based on emulation of CMIP, ice sheet, and glacier models. Likely ranges are shown for SSP1-2.6 and SSP3-7.0. Only likely ranges are assessed for sea level changes due to difficulties in estimating the distribution of deeply uncertain processes. The dashed curve indicates the potential impact of these deeply uncertain processes. It shows the 83rd percentile of SSP5-8.5 projections that include low-likelihood, high-impact ice sheet processes that cannot be ruled out; because of low confidence in projections of these processes, this curve does not constitute part of a likely range. Changes relative to 1900 are calculated by adding 0.158 m (observed global mean sea level rise from 1900 to 1995–2014) to simulated and observed changes relative to 1995–2014.

Panel (e) Global mean sea level change at 2300 in m relative to 1900. Only SSP1-2.6 and SSP5-8.5 are projected at 2300, as simulations that extend beyond 2100 for the other scenarios are too few for robust results. The 17th–83rd percentile ranges are shaded. The dashed arrow illustrates the 83rd percentile of SSP5-8.5 projections that include low-likelihood, high-impact ice sheet processes that cannot be ruled out.

Panels (b) and (c) are based on single simulations from each model and so include a component of internal variability. Panels (a), (d) and (e) are based on long-term averages, and hence the contributions from internal variability are small. {4.3; Figs. 4.2, 4.8, 4.11; 9.6; Fig. 9.27; Figs. TS.8 and TS.11; Box TS.4 Fig. 1}.

Refer to [1] for Figures referenced above.

According to Chapter 9 of the IPCC AR6 WG1 full report and as shown in Fig. 3.10d 'It is virtually certain that GMSL (global mean sea level) will continue to rise until at least 2100, because all assessed contributors to GMSL are … virtually certain to continue contributing throughout this century. Even under the best case scenario of SSP1-1.9 sea level rise is projected to be around 0.5 m compared with 1900 level and 0.38 m relative to the period 1995–2014'. Whereas for SSP5-8.5 GMSL could rise by up to a metre compared with 1900, with a 'high-impact' projection of up to 1.75 m when combined with ice sheet instability (low-likelihood). This latter projection factors in a considerable contribution from melting of the Greenland icecap. As can be seen from Fig. 3.10d, this modelling has been extended to project potential increase in GMSL up to 2300 with a potential for reaching an apocalyptic 6.8 m.

'Ocean thermal expansion (38%) and mass loss from glaciers (41%) dominate the total (rise in sea level) from 1901 to 2018'. The contribution of melt water from Antarctica and Greenland was four times higher between 2010 and 2019 than 1992 to 1999.[15]

However, in its 2014 Assessment Report, the IPCC states that 'based on current understanding, only the collapse of marine-based sectors of the Antarctic ice sheet, if initiated, could cause GMSL (global mean sea level) to rise substantially above the likely range during the 21st century'. The mass of ice in Greenland and Antarctica is being eroded as the accumulation of surface snow falls below the outflow of melt water to the ocean. The West Antarctica ice sheet is particularly sensitive to subsurface ocean warming, with models indicating significant ice loss if atmospheric carbon dioxide concentrations remain above 350 ppm for millennia. Even if the contribution from Antarctic meltwater to sea level rise turns out to be insignificant in the long term:

> The available evidence indicates that global warming beyond a threshold would lead to the near-complete loss of the Greenland ice sheet over a millennium or longer, causing a global mean sea level rise of approximately 7m. Studies with fixed present-day ice sheet topography indicate that the threshold is greater than 2°C but less than 4°C (medium confidence) of global mean surface temperature rise above pre-industrial.[16]

[15] https://www.ipcc.ch/report/ar6/wg1/downloads/report/IPCC_AR6_WGI_FullReport_small.pdf

[16] https://www.ipcc.ch/report/ar6/wg1/downloads/report/IPCC_AR6_WGI_FullReport_small.pdf

Fig. 3.11 Greenland ice sheet in August 2014 (https://earthobservatory.nasa.gov/images/84607/dark-snow-project; Photo: NASA Earth Observatory)

Clearly the impact of sea level rise will vary enormously depending on the accuracy of these predictions and the effectiveness of mitigation strategies. The studies referred to by the IPCC mention enormous ranges and there appears to be very little agreement between the modelling approaches used. Although there seems to be broad agreement that catastrophic melting of either the Greenland or Antarctic ice sheets before the end of the twenty-first century is unlikely, there is growing evidence that the irreversible mechanisms that will set in train the eventual complete deglaciation of Greenland could already be occurring.

Part of this process is driven by changes in the albedo, or reflectivity, of the Greenland ice sheet. The 'Dark Snow' project, that has been examining changes in albedo across the Greenland ice cap since 2000, has found significant reductions in reflectivity in recent years. During the summer of 2014 climate scientists Jason Box and Johnny Ryan mapped the extent of dark ice and snow and studied its composition:

> Around their camp in southwestern Greenland, Ryan and Box observed that snow and ice were darkened by a combination of dust, algae, and soot from wildfires. Most of the soot and dust was likely deposited thousands of years ago, Ryan noted, "so what happened in the past is having a direct effect on how the ice sheet behaves today".[17]

Figure 3.11 shows one of the images obtained during their field study, showing the condition of ice during the summer of 2014. The low albedo results in greater absorption of sunlight and hence accelerates melting, resulting in a higher density of particulates in the ice, reducing albedo still further until either all of the ice has melted or it snows.

[17] https://earthobservatory.nasa.gov/images/84607/dark-snow-project

It is interesting to note that some of the black carbon responsible for this darkening of the ice has been deposited on the Greenland ice sheet by wild fires in the Northwest Territories of Canada and in Siberia. The increasing incidence of these has in turn been attributed largely to hotter and drier summers in the region, primarily due to climate change.

This has been recognised by the IPCC in its AR6 WG1 report which refers to '… the effects of BC (black carbon or soot) deposition on snow and glaciers, concluding that there is high confidence that darkening of snow through the deposition of BC and other light-absorbing particles enhances snowmelt in the Arctic'.

As we saw earlier the melting and consequent shrinking of glaciers and ice sheets is responsible for something like 40% of sea level rise. According to the Randolph Glacier Inventory (RGI) based on satellite imagery obtained over a period from 1999 to 2010 globally, there are approximately 198,000 glaciers having a surface area greater than 0.1 km^2, having a total area of 726,000 km^2 or around 0.5% of the total land surface. This compares with some 14 million km^2 of ice sheet, representing around 9.5% of the total land surface area. Hence, the long-term contribution of glacier melt to sea level rise will be relatively small compared with ice sheets.

We have seen that a worst case scenario for global sea level rise could arguably result in an increase of between 1.2 and 1.5 m by the end of the century. Indeed Anders Levermann, one of the contributors to the IPCC section on sea level change, in an on-line forum that took place between the publications of AR5 and AR6, concludes that a 1.5 m rise represents '…the upper limit of global mean sea level that coastal protection might need for the coming century'.[18]

It is important to understand what the implications of a 1.5-m sea level rise are for coastal communities around the world. Firstly, this rise will not be uniform: the world maps published in the IPCC reports indicate significant variations due to differences in temperature and salinity (thermosteric and halosteric conditions respectively) and changes in wind forcing. The potential impact of meltwater from Greenland and the Antarctic ice sheets and glaciers is extremely complex. For example, meltwater forcing in the sub-polar North Atlantic causes changes to the Atlantic Meridional Overturning Circulation (AMOC), which in turn causes dynamic changes of sea level in the North Atlantic, particularly in its north-western region. 'The combination of this dynamic sea level rise and the global mean sea level rise makes the north eastern North American coast vulnerable to some of the fastest and largest sea level rises during the (21st) century'.[19] It is clear that there are huge variations in projections of both global mean sea level (GMSL) rise and regional levels. These projections are not only dependent upon the SSP scenarios assumed in the models used but also on the contributions of such factors as marine ice cliff instabilities and marine ice sheet instabilities about which there are significant disagreements in the research community. For example, one panel of experts reviewing AR5 commented that:

[18] https://e360.yale.edu/features/yale_e360_forum_on_ipcc_report_2013

[19] https://www.ipcc.ch/report/ar6/wg1/downloads/report/IPCC_AR6_WGI_FullReport_small.pdf

> By 2100, GMSL is projected to rise by 0.28–0.55 m (likely range) under SSP1-1.9 and 0.63–1.01 m (likely range) under SSP5-8.5 relative to the 1995–2014 average (medium confidence). Under the higher CO_2 emissions scenarios, there is deep uncertainty in sea level projections for 2100 and beyond associated with the ice-sheet responses to warming. In a low-likelihood, high-impact storyline and a high CO_2 emissions scenario, ice-sheet processes characterized by deep uncertainty could drive GMSL rise up to about 5 m by 2150.[20]

The process by which ice sheets are eroded by warming seas is particularly evident for the West Antarctic Ice Shelf which is both thinner than the East Shelf and sits on ground which in places falls well below sea level. Hence ice sheets are being warmed from below by the Circumpolar Deep Water creating what has become known as the "weak underbelly" of the West Antarctic Ice Sheet. Marine ice cliff instabilities, on the other hand, may potentially result in a catastrophic collapse. Near vertical ice cliffs are formed where ice sheets reach the ocean, with a portion of their height below sea level. 'When glaciers retreat into an overdeepening basin—or ice shelf collapse exposes a tall ice cliff—cliffs become structurally unstable at a threshold cliff height, leading to runaway cliff failure and ice sheet disintegration' [3].

The AMOC is one of an interconnected group of thermohaline circulation systems that transport energy through the earth's oceans as shown in schematic form in Fig. 3.12. Thermohaline circulation is driven by differences in temperature and salinity and involves the circulation of huge volumes of water over millennial time scales: for example, the complete cycle within the AMOC takes approximately a thousand years. It is also responsible, in combination with wind forcing, for the generation of surface currents, such as the Gulf Stream in the North Atlantic. The surface flow of warm water in the AMOC is represented by the red line that flows from the South Atlantic up the east coast of America cooling in the Nordic and Labrador Seas and thence returning in the deep ocean as shown by the blue line that ends up in the Weddell Sea off the coast of Antarctica.

As sea levels rise, the risk of inundation, flooding, land loss and population displacement increases. As we saw in the previous chapter, these are all aggravated by the increase in extreme weather events such as hurricanes, cyclones and storm surges. Maps are available on-line based on a simple topographical modelling of contours and sea level rise [4]. These do not, however, reflect tidal variations, coastal erosion or sea defences: existing or proposed. It is clear, however, that if the sea level rises by 1.5 m, in the absence of suitably designed coastal defences, all land that is currently less than 1.5-m above sea level is at risk..[21]

> Most countries in South, Southeast, and East Asia appear to be highly threatened because of the widespread occurrence of densely populated deltas, often associated with large growing cities. Africa also appears highly threatened owing to the low levels of development combined with expectations of rapid population growth in coastal areas: Egypt and

[20] https://www.ipcc.ch/report/ar6/wg1/downloads/report/IPCC_AR6_WGI_FullReport_small.pdf

[21] https://www.researchgate.net/publication/44683423_Sea-Level_Rise_and_Its_Impact_on_Coastal_Zones

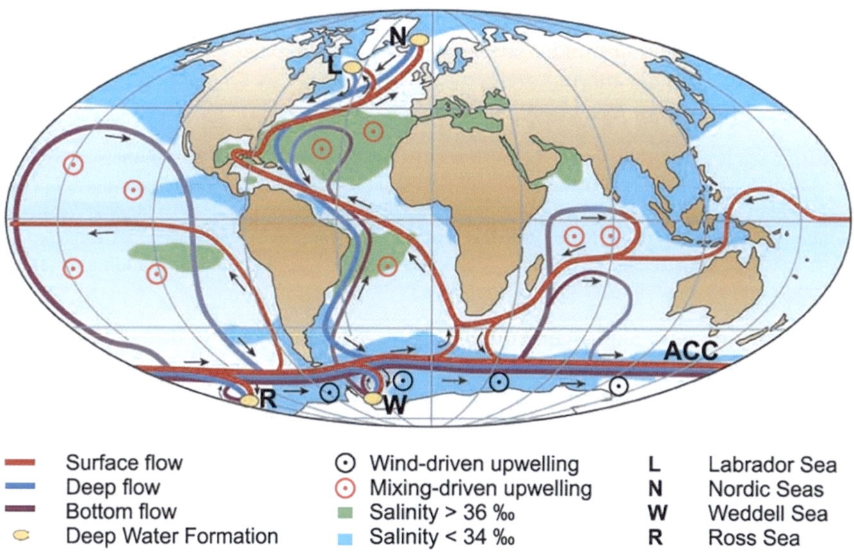

Fig. 3.12 Schematic of ocean circulation. *ACC* Antarctic Circumpolar Current. (From: http://www.eoearth.org/view/article/150290/)

> Mozambique are two "hotspots" for potential impacts. However, the small island states experience the largest relative increase in impacts, including regions of high islands like the Caribbean. Low islands such as the Maldives or Tuvalu face the real prospect of submergence and complete abandonment during the 21st century [4].

The 1600 islands in the Indian Ocean that form the Maldives archipelago would be inundated completely by a sea level rise of just over 1 m. The Maldives Government are so alarmed by this prospect that they are examining the potential evacuation of all the 345,000 islanders. Clearly the greatest impacts will be felt by all coastal communities: for example, a report published by the Stockholm Environment Institute in 2013 under its ClimateCost programme estimated the economic impacts of flooding in Europe in the twenty-first century in the absence of further adaption [5].[22] For a conservative 37-cm sea level rise by the 2080s, it was estimated that some 250,000 people would experience flooding, with an associated cost of 25 billion euros per annum at 2013 prices, the displacement of a total of 438,000 people and loss of 35% of fresh water wetlands by the 2080s. With the world on a trajectory towards a 1 m average sea level rise by 2100 it was suggested that the annual cost to European countries could increase to 156 billion euros by the 2080s.

[22] https://books.google.co.uk/books?id=bnPdBgAAQBAJ&pg=PA26&lpg=PA26&dq=Stockholm+Environment+Institute+in+2013+under+its+ClimateCost+programme&source=bl&ots=gefQHSpybX&sig=ACfU3U1ZVDHZLQMqd_DIcugxf-N3aLTnNA&hl=en&sa=X&ved=2ahUKEwijo4DsjsmDAxVTVEEAHR7OBwUQ6AF6BAgsEAM#v=onepage&q=Stockholm%20Environment%20Institute%20in%202013%20under%20its%20ClimateCost%20programme&f=false

Rising sea levels not only increase the incidence of inundation, erosion, submergence and associated dry and wetland loss and displacement of coastal communities, but also damage marine ecosystems. According to Chapter 5 of the IPCC Working Group II 5th Assessment Report:

> Rising sea level impacts marine ecosystems by drowning some plants and animals as well as by inducing changes of parameters such as available light, salinity, and temperature. The impact of sea level is related mostly to the capacity of animals (e.g., corals) and plants (e.g., mangroves) to keep up with the vertical rise of the sea. Mangroves and coastal wetlands can be sensitive to these shifts and could leak some of their stored compounds, adding to the atmospheric supply of these greenhouse gases.[23]

Coral reefs are particularly vulnerable to climate change, mainly through rising sea water temperatures and acidification. They are constructed from calcium carbonate which is mostly secreted by coral polyps and encrusting macro-algae. There is a wide variety of coral polyps which live in symbiosis with zooxanthellae, which provide corals with their colouration. Zooxanthellae are single cell microalgae that survive through photosynthesis, absorbing carbon dioxide and secreting oxygen, hence requiring shallow, clean and clear water to thrive.

Coral reefs 'occupy less than 0.1% of the ocean floor yet play multiple important roles throughout the tropics, housing high levels of biological diversity as well as providing key ecosystem goods and services such as habitat for fisheries, coastal protection, and appealing environments for tourism'.[24]

Coral reefs are subject to numerous stresses, many of which, such as temperature rise and acidification (see below), are linked to climate change, but also pollution, fishing, storm damage, sedimentation, nutrient overloading, disease and predation. Historically temperature rise of the tropical eastern Pacific Ocean has occurred because of El Niño Southern Oscillation events. Bleaching of coral reefs is the result of a stress response by the zooxanthellae to any of these which, when they die or lose their pigment, results in the coral appearing to be bleached. If the stress is prolonged, then the coral also can die and the reef may be lost.

According to the IPCC WG2 5th Assessment Report 'mass coral bleaching coincided with positive temperature anomalies over the past 30 years, sometimes followed by mass mortality', whilst also highlighting that the 'persistence of coral reefs depends on the balance between the production and erosion of calcium carbonate and on coral settlement, both of which are affected by ocean acidification'. However, 'globally, the primary climate-related driver appears to be ocean warming rather than ocean acidification, cyclonic activity, and changes in freshwater input'.[25]

Although based on observations and projections from the early 2000's IPCC's 5th Assessment Report draws a dramatic picture of the likely bleaching events globally through to the end of the twenty-first century. Projected figures are based on the climate modelling scenario A1B used for the IPCC 4th Assessment, which

---

[23] https://archive.ipcc.ch/pdf/assessment-report/ar5/wg2/WGIIAR5-Chap5_FINAL.pdf

[24] https://archive.ipcc.ch/pdf/assessment-report/ar5/wg2/WGIIAR5-CCboxes_FINAL.pdf

[25] https://archive.ipcc.ch/pdf/assessment-report/ar5/wg2/WGIIAR5-Chap5_FINAL.pdf

results in similar temperature change projections to the RCP 6.0 scenario from the 5th Assessment and between SSP2-4.5 and SSP3-7.0 under the AR6 scenarios. These projections, based on the output from four ensemble runs, indicate that '99% of the reef locations will experience at least one severe bleaching event between 2090 and 2099', with a steep increase from the 2020 onwards.[26]

Sea level rise and storm surges also impact on vegetated coastal habitats and wetlands, including mangrove forests, seagrass meadows, kelp forests and saltmarshes. Wetlands comprise land that is inundated for at least part of the year and represent some 4–6% of total land area globally, which can be found on coastal margins, and edging rivers, estuaries, deltas, and inland water bodies. Remarkably wetlands store between 20% and 25% of the world's soil carbon.

In 1990 there were around 240,000 km^2 of mangrove forest in the world, although it has been estimated that the area has been dropping by 1% annually due primarily to the major growth of shrimp farms in the intertidal zone, particularly in Southeast Asia. The Forest Resource Assessment for 2020 estimated the total mangrove forest area at 147,860 km^2.[27] Estimates vary for the extent of seagrass globally, ranging from 160,000 to 600,000 km^2, although it is thought that the total area fell at the rate of 10% per decade between 1970 and 2000, although the rate accelerated to such an extent that by 2022 it was 7% per annum.[28] The area of coastal saltmarshes has been estimated variously at between 22,000 and 400,000 km^2, although probably the most accurate estimate of 55,000 km^2 was produced by McOwen et al in 2017 from a large dataset published in the Biodiversity Data Journal.[29] Campbell et al. estimated the rate of loss of saltmarsh between 2000 and 2019 to be 0.28% per year or about 719 km^2.[30] Kelp forests are ubiquitous to rocky shorelines in cold and temperate regions, occupying some 25% of the total coast globally: or around 87,000 km.[31] Figure 3.13 shows estimated distribution of mangrove forests, seagrass and saltmarshes globally assembled from UN Environmental Programme sources in 2012.[32]

Degradation of these important ecosystems results in reduced carbon absorption due to photosynthesis and increased greenhouse gas emissions as methane and carbon dioxide is released. A 2012 analysis by Pendleton et al. of 'global "Blue Carbon" emissions from conversion and degradation of vegetated coastal ecosystems' provides a range of estimated carbon released by land-use change in these ecosystems globally, equivalent to a maximum of 0.28 Petagrams of carbon (PgC) (280 million tonnes) annually.[33]

---

[26] https://archive.ipcc.ch/pdf/assessment-report/ar5/wg2/WGIIAR5-Chap5_FINAL.pdf

[27] https://www.mangrovealliance.org/news/30-years-of-global-forest-data/

[28] https://www.wri.org/insights/understanding-seagrass#:~:text=Seagrasses%20are%20found%20in%20159,almost%20the%20size%20of%20France

[29] https://www.ncbi.nlm.nih.gov/pmc/articles/PMC5515097/

[30] https://www.nature.com/articles/s41586-022-05355-z

[31] http://www.kelpecosystems.org/about/

[32] https://www.mangrovealliance.org/news/30-years-of-global-forest-data/

[33] https://www.mangrovealliance.org/news/30-years-of-global-forest-data/

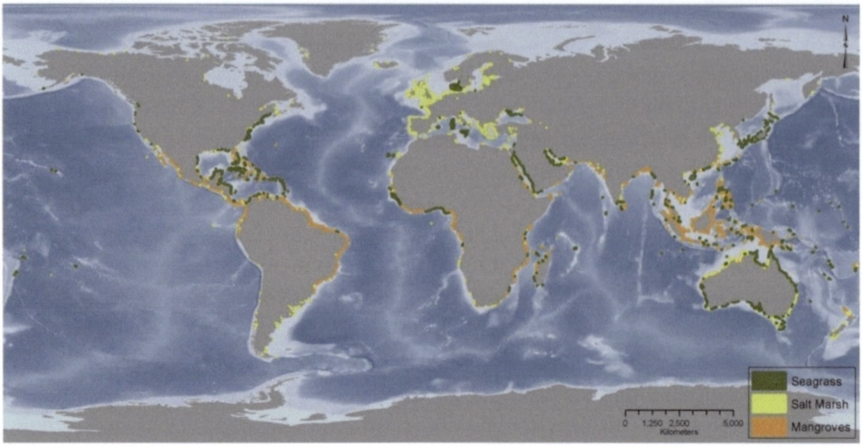

Fig. 3.13 Global distribution of seagrasses, tidal marshes and mangroves (From: http://journals.plos.org/plosone/article?id=10.1371/journal.pone.0043542#pone-0043542-t001)

Mangrove forests grow on sedimentary shorelines in the tropics, typically on intertidal deltas and estuaries. Their aerial roots are particularly suited to salt water, although they are tolerant to a wide range of salinities, and they provide a nursery for sea creatures that end up populating coral reefs and the oceanic pelagic zone. Studies published in 2013 by the UK Cambridge Coastal Research Unit have concluded that ancient mangrove forests have responded well to rising sea level and in fact their roots trap sediment at a rate that is likely to keep up with projected sea level rise.[34] Mangrove forests are also extremely resilient to storm damage from wave penetration. A 2012 report from the same Cambridge-based Unit analysed field studies from around the world and observed that attenuation of wave height per 500 m of mangrove forest depth ranged from 50% to 99%. In the Indian Ocean tsunami on Boxing Day 2004, for example, many lives were saved and homes protected by the mangrove forests that lay in the path of the enormous waves that crashed into shorelines around South Asia, although destroying most of the forest in the process.

Seagrass meadows are far more vulnerable to sea and air temperature changes, turbidity, fishing and storm damage than mangrove forests.

> Seagrass meadows provide ecosystem services that rank among the highest of all ecosystems on earth. The direct monetary outputs are substantial since highly valued commercial catches such as prawns and fish are dependent on these systems. Seagrasses provide protective shelter for many animals, including fish, and can also be a direct food source for manatees and dugongs, turtles, some herbivorous fish and sea urchins. The roots and rhizomes of seagrasses also stabilise sediments and prevent erosion while the leaves filter suspended

[34] https://www.cam.ac.uk/research/news/mangroves-could-survive-sea-level-rise-if-protected

> sediments and nutrients from the water column. Seagrass meadows are thus linked to other important marine habitats such as coral reefs, mangroves, salt marshes and oyster reefs.[35]

There are 72 known species of seagrass, of which 13 are at risk of extinction or endangered. Temperature sensitivity varies with species, so as sea temperatures rise cold water species are being driven polewards. However, species that are dying through exposure to higher water temperatures are not being replaced by species that are more tolerant to higher temperatures. Increasing storm intensities will accelerate die back through uprooting plants, increasing turbidity and smothering plants in sediment.

Kelp is a seaweed that grows in dense forests in nutrient rich coastal waters down to a depth of 40 m. It is relatively short-lived, typically up to 3 years, but the largest species can grow at 30 cm/day to a maximum length of 60 m. As with seagrass, ocean temperature rise is causing the forests to head polewards, but its overall impact is to reduce the size of forests globally.[36] In some areas kelp forests have been decimated through grazing by sea urchins, in particular where a decline in sea otter population has removed the urchin's main predator, resulting in 'sea urchin barrens'.

Salt marsh response to climate change is similar to that of mangrove forests and it is thought that its degradation and loss can be attributed mostly to land reclamation, coastal development and dredging, rather than sea level and temperature rise or storm damage.[37]

3.4 Ocean Acidification and Oxygen Content

The oceans of the world are not only warming, expanding and rising but are also absorbing carbon dioxide and becoming more acidic. The oceans absorb approximately 30% of the CO_2 from the atmosphere. According to Chapter 3 of Assessment Report 5 by the IPCC Working Group 1 headed 'Oceans':

> When CO_2 reacts with seawater it forms carbonic acid (H_2CO_3), which is highly reactive and reduces the concentration of carbonate ion and can affect shell formation for marine animals such as corals, plankton, and shellfish. This process could affect fundamental biological and chemical processes of the sea in coming decades.[38]

Records indicate that '…the pH of surface waters has decreased as a result of ocean uptake of anthropogenic CO_2 from the atmosphere. There is high confidence that the pH decreased by 0.1 since the pre-industrial era'.[39] The pH scale runs from 0 to 14, with the lower numbers indicating an acidic liquid and higher representing a basic

[35] https://iucn.org/sites/default/files/import/downloads/report_seagrass_and_climate_change.pdf

[36] https://archive.ipcc.ch/pdf/assessment-report/ar5/wg1/WG1AR5_Chapter03_FINAL.pdf

[37] https://archive.ipcc.ch/pdf/assessment-report/ar5/wg2/WGIIAR5-Chap5_FINAL.pdf

[38] https://archive.ipcc.ch/pdf/assessment-report/ar5/wg1/WG1AR5_Chapter03_FINAL.pdf

[39] https://archive.ipcc.ch/pdf/assessment-report/ar5/wg1/WG1AR5_Chapter03_FINAL.pdf

or alkaline liquid. For example, battery acid typically has a pH of zero and liquid drain cleaner a pH of 14. Pure water has a pH of approximately 7, which means that it is neither alkaline/basic nor acidic, whereas the salt water found in our seas and oceans is typically slightly alkaline, with a global average that has fallen from 8.2 to 8.1 since the start of the Industrial Revolution.

The pH of ocean water is not only dependent upon the uptake of atmospheric carbon dioxide, however, but also the production and respiration of dissolved and particulate organic matter as well as the formation and dissolution of calcium carbonate minerals. It is these interactions that produce the impacts on corals, plankton and shellfish referred to above.

Ocean acidity is also impacted to some extent by natural processes such as volcanic activity and methane hydrate releases from benthic organic matter on the ocean floor. Atmospheric sulphur dioxide and nitrogen oxides associated with power generation, industrial and agricultural activities have been found to impact on the pH of coastal waters, primarily through the so-called acid rain. Sulphuric acid and nitric acids contribute the reduction in pH of coastal waters in proximity to these sources, with the greatest contribution from the absorption of atmospheric carbon dioxide.

Under the SSP scenarios used for climate change projections in IPCC AR6, it is predicted that after a period of relative stability in the 1950s average global ocean surface pH levels will fall under all scenarios, with a maximum drop under the SSP5-8.5 scenario of 0.45 by 2100 compared with 1950. As might be expected the reductions in pH for SSP1-2.6 and -1.9 are considerably lower, with signs of recovery toward the end of the twenty-first century. According to the IPCC 5th Assessment Report (WGII Chapter 5):

> Estimates of future atmospheric and oceanic carbon dioxide concentrations indicate that, by the end of this century, the average surface ocean pH could be lower than it has been for more than 50 million years.[40]

Also important to the health of water bodies is the oxygen content, the depletion of which can result in 'dead zones' where there is no oxygen whatsoever (anoxia) or very low oxygen levels (hypoxia). We saw in Chap. 2 how rivers and coastal waters can become hypoxic through eutrophication when plant nutrients fed with fertiliser in the form of nitrates and phosphates in runoff water from fields produce phytoplankton and algal blooms that consume oxygen in the water through photosynthesis and decomposition. The increase in the size and incidence of dead zones is partly due to the intensification of farming, that in turn is due to increased population, which also increases the discharge from sewage treatment plants. It is also partly due to atmospheric nitrogen deposition linked to the burning of fossil fuels and vehicle exhausts.

Whereas the dead zones created by eutrophication and hypoxia are localised and created primarily by waterborne and atmospheric pollution, oceanic oxygen minimum zones (OMZ) or 'shadow zones' occur globally and exist at a depth between

[40] https://archive.ipcc.ch/pdf/assessment-report/ar5/wg1/WG1AR5_Chapter03_FINAL.pdf

200 and 1000 m below the surface. According to the IPCC 2014 Synthesis Report 'oxygen minimum zones are progressively expanding in the tropical Pacific, Atlantic, and Indian Oceans, due to reduced ventilation and O_2 solubility in warmer, more stratified oceans, and are constraining fish habitat'.[41] Chapter 3 of the detailed report concludes that 'the long-term deoxygenation of the open ocean thermocline is consistent with the expectation that warmer waters can hold less dissolved oxygen (solubility effect), and that warming-induced stratification leads to a decrease in the transport of dissolved oxygen from surface to subsurface waters'.[42] The ocean thermocline is a layer in which temperature changes more rapidly with depth than above or below, separating the more turbulent upper layers from the calm water below. In warm oceans, the depth of the thermocline coincides closely with that of the oxygen minimum zone, whilst in polar regions, the thermocline barely exists. The trend during the twenty-first century is for 'species richness and fisheries catch potential to increase, on average, at mid- and high latitudes (high confidence) and to decrease at tropical latitudes and in semi-enclosed seas (medium confidence)'.[43]

3.5 Ocean Ecosystem Impacts

It is clear from the above that climate change impacts ocean ecosystems and will continue to do so for the foreseeable future. The IPCC concludes that: 'overall, climate change will lead to large-scale shifts in the patterns of marine productivity, biodiversity, community composition, and ecosystem structure. Regional extinction of species that are sensitive to climate change will lead to a decrease in species richness'.[44]

Increased carbon dioxide concentration, warming of surface layers, and the input of fresh water from snow melt and increased precipitation combine to increase stratification, resulting in reduced mixing and lower oxygen concentrations in the pelagic zone (open sea). This drives out the larger fish, which head to cooler and more oxygenated waters, allowing the smaller sea creatures on which they would normally predate to thrive.

This is referred to as the 'trophic cascade effect' whereby interference in any one link in the predatory chain can impact all of the others. An important cause of this interference relates to the timing of individual life cycles (phenology) which in some instances is being impacted by climate change. This can result in a disjuncture between the feeding needs of the predator and the availability of its prey, for example, thus impacting on the entire ecosystem in which these creatures sit (phenological shift).

---

[41] https://archive.ipcc.ch/pdf/assessment-report/ar5/syr/SYR_AR5_FINAL_full_wcover.pdf

[42] https://archive.ipcc.ch/pdf/assessment-report/ar5/wg1/WG1AR5_Chapter03_FINAL.pdf

[43] https://archive.ipcc.ch/pdf/assessment-report/ar5/syr/SYR_AR5_FINAL_full_wcover.pdf

[44] https://archive.ipcc.ch/pdf/assessment-report/ar5/wg2/WGIIAR5-Chap6_FINAL.pdf

Researchers have discovered that some parts of the food chain, such as phytoplankton, use 'temperature unrelated cues' to time their annual cycle, whilst their predators experience a phenological shift due to temperature rise. For example, 'the timing of spawning in *Macoma balthica* (the Baltic clam), a marine intertidal bivalve, is under selection by temporal variation in both its food resources, phytoplankton and predation risk by juvenile shrimps (*Crangon crangon*). Over the past two-decades, *Macoma* spawning phenology has advanced by 1 day/year in the Dutch Waddensea. However, there has been no advancement in 'the timing of the phytoplankton bloom, leading to a mismatch between *Macoma* development and resources'.[45] This has led to a significant reduction in Baltic clam populations in the Waddensea, which are also subject to predation by red knot and other intertidal feeding birds as well as bottom feeding (demersal) fish and crabs.

Primary production of phytoplankton through the conversion of carbon dioxide to organic carbon by photosynthesis is crucial to a healthy pelagic ecosystem worldwide. It forms the base for the marine food chain and is used as a key parameter in earth system models (ESMs) that couple atmosphere, cryosphere and hydrosphere with climate and carbon cycles to predict biogeochemistry changes in oceans. Multi-model predictions for net primary production (NPP) of phytoplankton reproduced in WG2, Chapter 6 of the IPCC 5th Assessment Report indicate an increase in NPP in higher latitudes but decreases in oceans that it is predicted will become increasingly stratified such as the tropical Pacific, Atlantic and Indian Oceans as well as the North Atlantic.[46] The studies reported in AR5 suggest that these impacts will be significantly greater by the end of the twenty-first century under the RCP8.5 'business as usual' scenario (SSP5-8.5 under the AR6 scenarios).

Changes in net primary production (ΔPP) are typically reported in terms of grams of carbon fixed per square metre of water surface area per year (gCm^{-2} $year^{-1}$). The AR5 projections are based on a number of different modelling techniques using data from the RCP scenarios, but the overall picture for all scenarios is for a mean reduction in NPP over the course of the twenty-first century.

> Most climate change scenarios foresee a shift or expansion of the ranges of many species of plankton, fish, and invertebrates toward higher latitudes, by tens of kilometers per decade, contributing to changes in species richness and altered community composition. Organisms less likely to shift to higher latitudes because they are more tolerant of the direct effects of climate change or less mobile may also be affected because climate change will alter the existing food webs on which they depend.[47]

All of this will have a significant impact on commercial fishing, in terms of both catch potential and the average body size of fish. Both of these will be impacted not only by the NPP of phytoplankton but also by water temperature, acidification, sea ice retreat, pollution, hypoxia and overfishing. Despite all of these factors it is still projected that catch sizes will increase in some biomes, with shifting distributions

[45] https://archive.ipcc.ch/pdf/assessment-report/ar5/wg2/WGIIAR5-Chap6_FINAL.pdf

[46] https://archive.ipcc.ch/pdf/assessment-report/ar5/wg2/WGIIAR5-Chap6_FINAL.pdf

[47] https://archive.ipcc.ch/pdf/assessment-report/ar5/wg2/WGIIAR5-Chap6_FINAL.pdf

in search of cooler water polewards and at greater depth, resulting in potential invasions or extinctions.

3.6 Permafrost Thaw

Ground that remains frozen for 2 or more years consecutively is defined as permafrost.[48] This ground may be located at high latitudes, at high altitudes or beneath glaciers, lakes or seas. Most of the terrestrial permafrost formed globally can be found in the northern hemisphere across Russia, Scandinavia, Greenland, Canada and Alaska. Figure 3.14 shows the estimated distribution in the circumpolar northern region of various types of permafrost from data published in 1998.

As temperatures rise in the higher latitudes, then ground that has been frozen year-round for perhaps centuries is thawing. Much of this ground comprises peat which, as Hodgkins et al., report in their 2014 paper *'Changes in peat chemistry associated with permafrost thaw increase greenhouse gas production'*, becomes a significant source of methane and carbon dioxide as it thaws:

> As the Arctic warms, peat is expected to release more CH_4 (methane) and CO_2 as permafrost thaws and exposes more peat, and as peat experiences faster decomposition due to increasing temperatures. The production of CH_4 relative to CO_2 is also likely to increase due to increasing anaerobic conditions caused by peat waterlogging associated with thaw and subsidence. Our results suggest that estimates of CH_4 climate feedbacks based on these effects alone are likely to be conservative: as permafrost thaws, overall carbon release, as well as the proportion released as CH_4, is likely to increase on a per-volume basis as thaw-induced subsidence and shifts in plant community increase organic matter lability (instability) and favor increasingly methanogenic conditions.[49]

Measurements taken at between 10 and 20 m below the surface in a number of boreholes in the circumpolar northern permafrost region indicate that between 1977 and 2010 mean annual temperatures increased by between 1 °C and 2 °C.[50] In parts of Russia, permafrost that was originally 10–15 m thick disappeared between 1975 and 2005, whilst the southern boundary of the permafrost region moved north by some 80 km.

It has been estimated that between 1400 and 1600 Pg of carbon is stored in high latitude permafrost and soils.[51] This compares with the cumulative total of carbon found in the earth's atmosphere in 2011 of between 445 and 585 Pg and the approximately 1000 Pg of carbon calculated to be the trigger for a 2 °C rise in global mean surface temperature[52].

---

[48] https://archive.ipcc.ch/pdf/assessment-report/ar5/wg1/WG1AR5_Chapter04_FINAL.pdf

[49] http://www.pnas.org/content/111/16/5819.full

[50] https://archive.ipcc.ch/pdf/assessment-report/ar5/wg1/WG1AR5_Chapter04_FINAL.pdf

[51] http://www.permafrostcarbon.org/

[52] https://archive.ipcc.ch/pdf/assessment-report/ar5/wg1/WG1AR5_Chapter12_FINAL.pdf

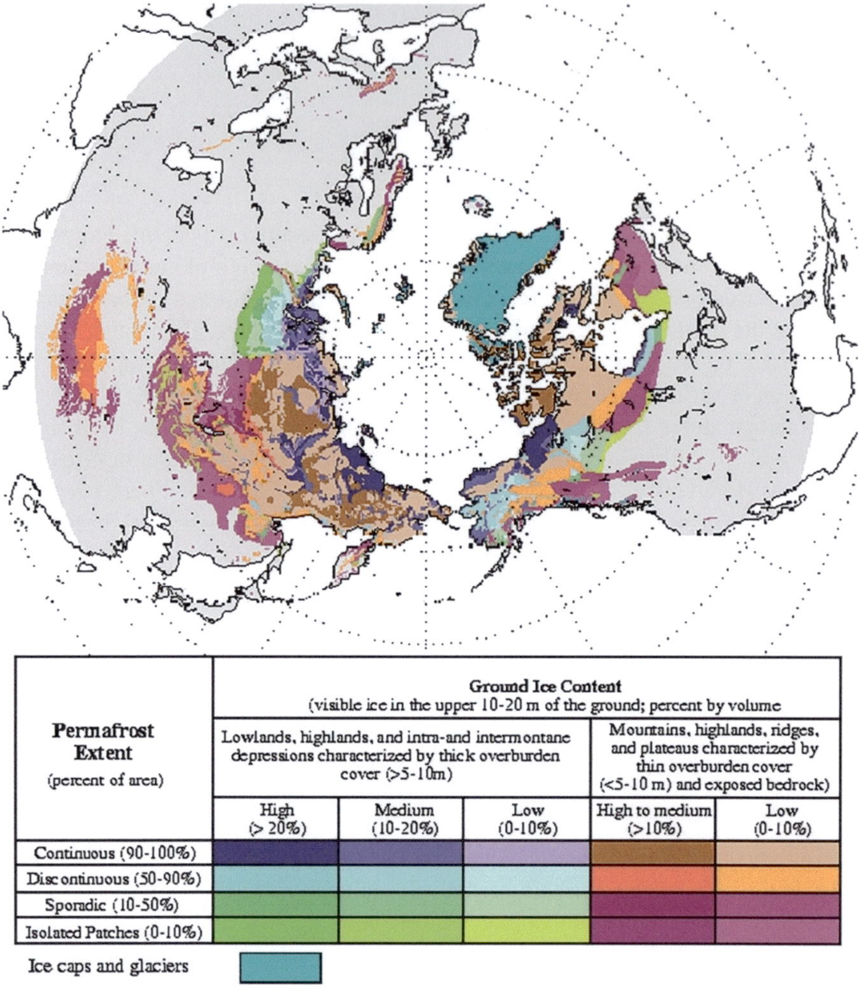

Fig. 3.14 Circumpolar northern zonation of various permafrost types (Image: Brown et al. From: https://commons.wikimedia.org/wiki/File:Circum-Arctic_Map_of_Permafrost_and_Ground_Ice_Conditions.png)

Projections for permafrost degradation reported in Chapter 12 of the IPCC WG1 AR5 Report suggest that between 26% and 93% of 'near-surface' permafrost could be lost by the end of the twenty-first century compared with the average area estimated for 1986 to 2005. The bottom end of the range being that predicted for RCP2.6 and the top for RCP8.5. 'Near-surface' refers typically to the top 3.5 m of permanently frozen ground which, once thawed, would be responsible for the majority of methane and carbon dioxide emissions. Bearing in mind that in northern Siberia permafrost has been measured at a maximum thickness of around 1500 m, and 740 m in northern Alaska, the risk of a total thaw is negligible. Chapter 12 also refers to estimated contributions by the methane emissions associated with permafrost degradation to temperature rise by the end of the twenty-first century ranging from very little up to 0.69 °C.[53]

This analysis has been further supported in the modelling reported in Chapter 9 of WG1 AR6 which includes a figure that illustrates the vast range of projected volumes of permafrost loss to a depth of 3 m, but with a worst case scenario close to a total loss by the end of the twenty-first century.[54]

3.7 Freshwater Resources

In Chap. 2, we examined the politics of water and examples of water stress and drought from around the world. In years to come pressure on freshwater supplies will come from both population growth and climate change, with the most vulnerable populations in the developing world likely to be most affected.

Projections reported in the IPCC Fifth Assessment Report indicate that as climate change takes hold, although there will be regional variations, the incidence of 'new or aggravated water scarcity' is likely to increase, with an overall decrease in 'renewable groundwater resources' by the end of the twenty-first century:

> Broadly, water resources are projected to decrease in many mid-latitude and dry subtropical regions, and to increase at high latitudes and in many humid mid-latitude regions (high agreement, robust evidence). Even where increases are projected, there can be short-term shortages due to more variable streamflow (because of greater variability of precipitation) and seasonal reductions of water supply due to reduced snow and ice storage. Availability of clean water can also be reduced by negative impacts of climate change on water quality; for instance, the quality of lakes used for water supply could be impaired by the presence of algae producing toxins.[55]

An analysis by Schewe et al., referred to extensively in the IPCC chapter on Freshwater Resources mentioned above, using a large ensemble of global hydrological models (GHMs) and the five RCP scenarios, demonstrates that: '… climate change is likely to exacerbate regional and global water scarcity considerably. In particular, the ensemble average projects that a global warming of 2 °C above present (approximately 2.7 °C above pre-industrial) will … increase the number of people living under absolute water scarcity (defined as less than 500 m^3 per capita per year) by another 40% (according to some models, more than 100%) compared with the effect of population growth alone'.[56]

According to the UN Water for Life Programme 'hydrologists typically assess scarcity by looking at the population-water equation. An area is experiencing water

---

[53] https://archive.ipcc.ch/pdf/assessment-report/ar5/wg1/WG1AR5_Chapter12_FINAL.pdf

[54] https://www.apicultural.co.uk/exponential-increase-in-global-honey-bee-populations-as-a-driver-of-wild-bee-declines

[55] https://archive.ipcc.ch/pdf/assessment-report/ar5/wg2/WGIIAR5-Chap3_FINAL.pdf

[56] http://www.pnas.org/content/111/9/3245.full.pdf+html

stress when annual water supplies drop below 1,700 m^3 per person. When annual water supplies drop below 1,000 m^3 per person, the population faces water scarcity, and below 500 cubic metres "absolute scarcity"'.[57]

We will be discussing the extremes of water scarcity and drought later in this chapter, but in summary, the key impacts of climate change on freshwater supply in the long term are likely to include: greater evaporation as temperatures rise; changes in groundwater recharge from snowmelt and precipitation; increased saltwater contamination of groundwater due to sea level rise; increased soil erosion and consequent sediment load from more intense rainfall and damage to freshwater ecosystems from evaporation, pollution and saltwater intrusion.

3.8 Terrestrial Ecosystems and the Biosphere

The earth's biosphere contains all living plants and animals found in the atmosphere, on and below land and in the oceans, as well as the detritus produced therefrom. It is divided into biomes that are characterised by the typical communities of plants and animals associated with individual regions. The biosphere, along with the ecosystems of which it is formed, is particularly sensitive to climate change. Analysis of historical ecosystems reconstructed from fossils (the palaeoecological record) has shown that past episodes of climate change have resulted in 'biome shifts', wherein entire communities of plants and the animals that depend on them change location in order to adapt and survive.

One example of a biome shift was described earlier under the heading of Permafrost thaw. This is sometimes referred to as the 'tundra–boreal biome shift' or the spread of northern latitude (boreal) forest into the tundra permafrost. Earth system models (ESM), referred to in the IPCC 5th Assessment Report, predict a northward shift of Arctic vegetation with climate warming, as the boreal biome migrates into what is currently tundra. These predictions are supported by observations of 'shrub expansion in tundra, increased tree growth at the tundra–forest transition, and tree mortality at the southern extent of the boreal forest' in recent decades. Vegetation changes, associated with a biome shift and augmented by degradation through wild fires, will modify the ecosystem carbon balance, through permafrost thawing and associated carbon dioxide and methane emissions, feeding back into additional global warming.[58]

According to the Executive Summary to Chapter 4 of the IPCC Fifth Assessment Report from Working Group 2:

> Many plant and animal species have moved their ranges, altered their abundance, and shifted their seasonal activities in response to observed climate change over recent decades

[57] Link no longer available

[58] https://archive.ipcc.ch/pdf/assessment-report/ar5/wg2/WGIIAR5-Chap4_FINAL.pdf

(high confidence). They are doing so now in many regions and will continue to do so in response to projected future climate change (high confidence).[59]

Climate change will be a powerful stressor on both terrestrial and freshwater ecosystems. Forecasts reported by IPCC indicate with high confidence that this will result in an increase in species extinctions in the second half of the twenty-first century and beyond, and especially under the RCP6.0 and 8.5 scenarios, broadly confirming the assertion made in its 4th Assessment Report 'that approximately 20–30% of the plant and animal species assessed to date are at increasing risk of extinction as global mean temperatures exceed a warming of 2–3°C above preindustrial levels.'

The cumulative impacts of anthropogenic habitat modification, overexploitation, pollution and invasive species will significantly increase the stress from climate change. 'Under the fastest rates and largest amounts of projected climate change (RCP8.5), many species will be unable to move fast enough to track suitable environments, which will greatly reduce their chances of survival'.[60] Under the scenarios that assume mitigation will slow down and reduce climate change impacts (RCP2.6) species loss will be lower due to the increased time that should be available to adapt or move to a more suitable location.

Apart from temperature rise, animals and plants will be impacted by rising surface carbon dioxide concentrations and nitrogen deposition. The response of plants to elevated CO_2 levels varies with species and concentration, for example some trees and woody plants, including some crops, are more productive or faster growing under elevated CO_2 levels; although certain plants, such as clover, generate higher levels of cyanide and become toxic to herbivores when exposed to greater airborne concentrations of carbon dioxide.

'Nitrogen (N) deposition describes the input of reactive nitrogen from the atmosphere to the biosphere both as gases, dry deposition and in precipitation as wet deposition'.[61] Nitrogen enters the atmosphere as NO_x in the products of combustion from power stations, industry and wildfires, the water courses in the run-off from fields and soil via fertilisers. It is a nutrient that promotes plant growth via transpiration from the soil and absorption from the air. It is a major constituent of plant tissue and facilitates conversion of carbon dioxide into carbohydrates and amino acids—the key components of plant proteins. Too much nitrogen results in toxicity which can cause plant dehydration or grazing animals to become ill from reduced oxygen transfer, the symptoms of which range from reproductive difficulties to death.

Change in land use, including deforestation and the wholesale exploitation of forests, peatlands and savannahs for agriculture, grazing and logging, has contributed approximately one third of the carbon dioxide emissions into the atmosphere since 1750. The percentage of land area covered by trees fell from around 85% in 1750 to 31% in 2020, whilst agricultural land areas have risen from a few percent to

[59] https://archive.ipcc.ch/pdf/assessment-report/ar5/wg2/WGIIAR5-Chap4_FINAL.pdf

[60] https://archive.ipcc.ch/pdf/assessment-report/ar5/wg2/WGIIAR5-Chap4_FINAL.pdf

[61] http://www.apis.ac.uk/overview/pollutants/overview_N_deposition.htm

over 38% in the same period. According to Chapter 4 of IPCC WG2 AR5 'the natural carbon sink provided by terrestrial ecosystems is partially offset at the decadal time scale by carbon released through the conversion of natural ecosystems (principally forests) to farm and grazing land and through ecosystem degradation (high confidence)'.[62]

As well as contributing to climate change there is a well-established association between changes in temperature and precipitation and the health and survival of trees: 'extensive tree mortality and widespread forest dieback (high mortality rates at a regional scale) linked to drought and temperature stress have been documented recently on all vegetated continents'.[63]

This trend is likely to continue since 'most models suggest that rising temperatures, drought, and fires will lead to forests becoming a weaker sink or a net carbon source before the end of the (21st) century'.[64]

Peat not only provides the substrate for tundra but also for the peat swamp areas of Southeast Asia, with deposits scattered across much of the planet, although they cover only some 3% of the total land surface area. However, they hold the equivalent of around 50% of atmospheric carbon (as CO_2 equivalent) and one-third of the world's total soil carbon stock.

Peat has been used as a fuel for centuries, but more recently deforestation and agriculture have increasingly disturbed peatlands, releasing the methane and carbon dioxide trapped therein. Historically peatlands have provided a net sink for atmospheric carbon, whilst in recent times they have become a net source. For example, according to the IPCC: 'deforestation, drainage, and burning in Indonesian peat swamp forests can … (contribute) significantly to global GHG (greenhouse gas) emissions, especially during periods of intense drought associated with ENSO (El-Niño Southern Oscillation) when burning is more common'.[65]

With changing patterns of precipitation and overexploitation or degradation of land and natural resources there has been a steady expansion of dry and water stressed regions of the planet. There are many definitions of desert, but, generally speaking they are regions which suffer extreme water stress and which are sparsely populated with animals and plants. They include extremely hot regions, such as the Sahara, and extremely cold, such as Antarctica. The percentage of land area that can be defined as desert is approximately 33%.

The UN Convention to Combat Desertification (UNCCD) prefers the term 'drylands' which it defines as having a maximum ratio of precipitation to evapotranspiration of 0.65: that is, significantly lower rainfall than water loss to the atmosphere and demand from plants.[66] Drylands not only include hyper-arid, arid and semi-arid regions (deserts, semi-deserts and grasslands, respectively), but also dry sub-humid

[62] https://archive.ipcc.ch/pdf/assessment-report/ar5/wg2/WGIIAR5-Chap4_FINAL.pdf

[63] https://archive.ipcc.ch/pdf/assessment-report/ar5/wg2/WGIIAR5-Chap4_FINAL.pdf

[64] https://archive.ipcc.ch/pdf/assessment-report/ar5/wg2/WGIIAR5-Chap4_FINAL.pdf

[65] https://archive.ipcc.ch/pdf/assessment-report/ar5/wg2/WGIIAR5-Chap4_FINAL.pdf

[66] http://www.un.org/en/events/desertification_decade/whynow.shtml

areas, such as rangelands—lands that are primarily used by grazing livestock and wild animals.

According to figures from the UNCCD, drylands occupy 41.3% of the world's land surface whilst supporting 44% of its 'cultivated systems', 50% of livestock (mostly grazing on rangelands) and 33% of human population.[67] It has been estimated by the UN that some 12 million hectares of biodiverse or productive land, or approaching 0.1% of the total land surface area, is lost each year to drought and desertification.

The influence of humans on ecosystems is evident across much of the planet. According to Chapter 4 of IPCC WG2 AR5 some 25% of the planet's land surface has been 'highly modified' by human activity.[68] Apart from the urban environment and agricultural land this includes plantation forests and land managed for recreation, hunting, conservation and amenity. Even intrinsically 'natural' landscapes frequently have been impacted by human activity, sometimes over thousands of years.

3.9 Pollination and Pests

We saw in Chap. 2 how key pollinators, such as bees, are under threat from pesticides, such as neonicitinoids. Data held by the Department of Agriculture in the United States show that bee colonies declined from nearly six million in the 1940s to 2.3 million in 2008. The reasons for this collapse are complex and multi-factorial. However, according to Chapter 4 of the IPCC WG2 5th Assessment Report: 'after land use changes, climate change is regarded as the second most relevant factor responsible for the decline of pollinators'.[69] It goes on to state that '… the generally observed decline in honeybees is a clear indication of an increasing susceptibility to global change phenomena, with pesticide application, new diseases, and stress (and a combination of these) as the most relevant causes. Climate change may contribute by modifying the balance between honeybees and their environment (including exposure or susceptibility to diseases)'.

One study by Robbirt et al., published since the IPCC report, found that an increase in temperature of 2 °C led honeybees to leave their hives in search of pollen before the flowers that they would normally target had bloomed. The researchers concluded that 'our results demonstrate a significant potential for coevolved plant-pollinator relationships to be disrupted by climatic warming'.[70]

---

[67] http://www.un.org/en/events/desertification_decade/whynow.shtml

[68] https://iucn.org/sites/default/files/import/downloads/report_seagrass_and_climate_change.pdf

[69] https://archive.ipcc.ch/pdf/assessment-report/ar5/wg2/WGIIAR5-Chap4_FINAL.pdf

[70] http://www.cell.com/current-biology/abstract/S0960-9822(14)01342-6?_returnURL=http%3A%2F%2Flinkinghub.elsevier.com%2Fretrieve%2Fpii%2FS0960982214013426%3Fshowall%3Dtrue

It is interesting to note that despite all of the above there has been a steady increase in bee keeping as the demand for honey has increased across much of the developed world. The increase in cities in particular is problematic since the area of suitable landscape for pollination is far smaller than the that needed by the increase in honeybee populations. In the UK, it has been estimated that honeybees are only responsible for 30% of pollination services, whilst globally it is wild bees that are under threat.[71]

Warming may also interfere with the predatory and climatic controls over pest numbers: either by removing a natural predator or reducing the controlling influence of frosts, for example. A pest might be defined as a plant or animal that is detrimental to humans or human concerns. Whilst climate changes may improve the quality and quantity of certain foods, it may also extend the breeding season of pests. This may impact pests that are vectors for disease—transmitted to either human or animal—as well as those that reduce the productivity of crops. 'Vector-borne diseases of livestock such as African horse sickness and bluetongue may expand their range northward to the Northern Hemisphere because rising temperatures increase the development rate and winter survival of vectors and pathogens'.[72] Predictions of human exposure to the risk of contracting malaria and dengue fever indicate a potential increase in risk based on gross domestic product (GDP) per capita, depending on population growth, climate change and increased poverty. Increased temperature, precipitation, humidity and pooling create ideal conditions for mosquito larvae (see Sect. 3.11.9).

Some are predicting that if GDP remains at 2010 levels, some 5.2 billion people could be exposed to the risk of malaria, out of a population of 8.5 billion (61%), whilst ignoring the effects of climate change. World Health Organisation (WHO) figures published in April 2015 indicated that 3.2 billion, or around 50% of world population, were at risk of contracting malaria, with some 198 million cases reported in 2013, resulting in 584,000 deaths: a fall of 47% since the year 2000.[73] However, with strong economic growth the estimate for malaria cases for 2050 falls to 1.74 billion without climate change and 1.95bn with climate change. However the risk of exposure to dengue fever is increasing, and it is likely that both of these mosquito-borne diseases will spread to higher latitudes as climates warm.[74]

[71] https://news.un.org/en/story/2021/03/1087592#:~:text=Coral%20reefs%20are%20in%20grave,a%20result%20of%20climate%20change

[72] https://archive.ipcc.ch/pdf/assessment-report/ar5/wg2/WGIIAR5-Chap7_FINAL.pdf

[73] http://www.who.int/features/factfiles/malaria/en/

[74] https://archive.ipcc.ch/pdf/assessment-report/ar5/wg3/ipcc_wg3_ar5_chapter11.pdf

3.10 Impact on Animals

Like the plants on which many of them depend, some animals are having to move poleward or to higher altitudes as part of the shifting biomes referred to above. Animals that live in the Arctic are experiencing perhaps the greatest impact since it is not vegetation that is shifting but the very ice under their feet. For example, studies of polar bear sub-populations show declining numbers, leading the IPCC to conclude that: 'Polar bear populations are projected to decline greatly in response to continued Arctic warming'.[75] One collaborative international study published in 2015 that measured '*Polar bear dynamics in the Southern Beaufort Sea during a period of sea ice decline*' found that numbers decreased by between 25% and 50% between the years 2004 and 2006.[76] If this rate of decline occurs across all sub-populations, then there appears to be agreement that the global polar bear population will drop by two-thirds by the middle of the twenty-first century, with some local populations being lost completely.[77]

It is not only the polar bear that is at risk, although in most cases there are multiple factors involved, including climate change. The IPCC warned in 2014 that:

> A large fraction of the species that have been assessed are vulnerable to extinction as a result of climate change, often in interaction with other threats. Species with an intrinsically low dispersal rate, especially when occupying flat landscapes where the projected climate velocity is high, and species in isolated habitats such as mountain tops, islands, or small protected areas are especially at risk.[78]

Climate change is already forcing animals to move large distances to find more favourable habitats, whilst 'models of future climate change-induced shifts in the distribution of ecosystems suggest that many species could be outside of their preferred habitats within the next few decades'.[79]

As far back as 2005 researchers were raising the alarm that animals and plants were experiencing shifts in phenology—'the timing of seasonal activities such as timing of flowering or breeding'—due to climate change. Hence, 'many species are becoming mistimed due to climate change'.[80]

Increased temperature and fluctuations in rainfall patterns will also impact livestock animals, potentially through negative impacts on forage quality, increased heat stress, reduced feeding and hence inhibited growth rates. For example modelling quoted in Chapter 7 of the IPCC WG2 5th Assessment Report predicts that for cattle farmed in the UK, 'in some regions, milk yields will be reduced and mortality increased because of heat stress throughout the current (21st) century, with annual

---

[75] https://archive.ipcc.ch/pdf/assessment-report/ar5/wg2/WGIIAR5-Chap4_FINAL.pdf

[76] https://esajournals.onlinelibrary.wiley.com/doi/abs/10.1890/14-1129.1

[77] https://archive.ipcc.ch/pdf/assessment-report/ar5/wg2/WGIIAR5-Chap28_FINAL.pdf

[78] https://archive.ipcc.ch/pdf/assessment-report/ar5/wg2/WGIIAR5-Chap4_FINAL.pdf

[79] https://archive.ipcc.ch/pdf/assessment-report/ar5/wg2/WGIIAR5-Chap4_FINAL.pdf

[80] http://rspb.royalsocietypublishing.org/content/272/1581/2561

production and mortality losses amounting to some £40 million by the 2080s under a medium-high GHG (greenhouse gas) emission scenario'.[81]

We have already discussed the threats to sea creatures from climate change, but many of the same issues apply to animals and plants that occupy freshwater habitats. 'Freshwater ecosystems are considered to be among the most threatened on the planet', and recent surveys have found that 'freshwaters have some of the highest rates of extinction of any ecosystem for those species groups assessed for the IUCN (International Union for Conservation of Nature and Natural Resources) Red List (estimated as much as 4% per decade for some groups, such as crayfish, mussels, fishes and amphibians in North America)'.[82] Freshwater habitats support around 46% of fish species and 33% of invertebrates. Of the 32,500 known species of fish, some 15,000 are found in freshwater habitats and around 3000 of these are threatened with extinction, 17% of which is due to climate change.[83]

As temperatures increase and precipitation patterns change, it has been predicted that river flows will alter, with some rivers experiencing more frequent floods, whilst flows reduce significantly in others. For example, as glaciers melt, there will initially be an increase in outfall river flows, but these will gradually lessen as the ice mass in the glaciers reduces. Lower flows will impact water quality, nutrient loads and the physical habitats for flora and fauna, whilst, when combined with rising sea levels, increasing the salinisation of coastal wetlands. Higher than normal flow can increase flooding (see Sect. 3.11), surface run-off from fields and overflow from sewers, resulting in polluted rivers, lakes and coastal waters, the destruction of habitats and deposition of silt: in some instances changing the course of rivers and reshaping the landscape through which they pass.

Similarly, according to the IPCC, 'there is growing evidence that climate-induced changes in precipitation will significantly alter ecologically important attributes of hydrologic regimes in rivers and wetlands, and exacerbate impacts from human water use in developed river basins'.[84]

3.11 Extreme Weather

In Chap. 2 above we examined some of the potential causes of extreme weather events and examples from history that have had particular geopolitical impact. In Chapter 2 of the IPCC Working Group 1 Assessment Report 5 the authors attempt to identify trends in extremes of temperature, hydrological events, tropical storms

---

[81] https://archive.ipcc.ch/pdf/assessment-report/ar5/wg2/WGIIAR5-Chap7_FINAL.pdf

[82] https://archive.ipcc.ch/pdf/assessment-report/ar5/wg2/WGIIAR5-Chap4_FINAL.pdf

[83] https://www.iucn.org/press-release/202312/freshwater-fish-highlight-escalating-climate-impacts-species-iucn-red-list

[84] https://archive.ipcc.ch/pdf/assessment-report/ar5/wg2/WGIIAR5-Chap4_FINAL.pdf

and extratropical storms since the middle of the twentieth century. Despite problems with the identification of trends from intermittent events, it is concluded that:

> There is strong evidence that warming has led to changes in temperature extremes—including heat waves—since the mid-20th century. Increases in heavy precipitation have probably also occurred over this time, but vary by region. However, for other extremes, such as tropical cyclone frequency, we are less certain, except in some limited regions, that there have been discernible changes over the observed record.[85]

For example, most global land areas with available data have experienced a larger number of heat waves since the middle of the twentieth century. Trends are more easily identifiable for regions such as Europe, across most of which records are available for the last several centuries. Here it is clear that there has been a disproportionate number of heat waves since the 1990s.

Trends in extremes of precipitation are inconsistent: with North America, Central America and most of Europe experiencing greater extremes, whilst Southern Australia and West Asia have experienced fewer extreme precipitation events. Drought events have increased in the Mediterranean, but decreased in North America. Storm activity has increased in the North Atlantic since the 1970s, along with a poleward shift in extratropical cyclones. An extratropical cyclone is a synoptic (large) scale storm in middle or high latitudes, with very low central pressure, typically measuring more than 1000 km across and associated with very high wind speeds and precipitation.

In 2012, prior to the publication of its Fifth Assessment Report, the IPCC pulled together all of the evidence relating to extreme weather events from across its Working Groups in a report on *Managing the risks of extreme events and disasters to advance climate change adaptation* (SREX). In the executive summary of the chapter on the changes in climate extremes and their impact on the natural physical environment the evidence for anthropogenic influences on extreme events is graded in terms of their likelihood from 'likely' through to 'low confidence' and 'challenging'[86]:

> There is evidence that some extremes have changed as a result of anthropogenic influences, including increases in atmospheric concentrations of greenhouse gases. It is likely that anthropogenic influences have led to warming of extreme daily minimum and maximum temperatures at the global scale. There is medium confidence that anthropogenic influences have contributed to intensification of extreme precipitation at the global scale. It is likely that there has been an anthropogenic influence on increasing extreme coastal high water due to an increase in mean sea level. The uncertainties in the historical tropical cyclone records, the incomplete understanding of the physical mechanisms linking tropical cyclone metrics to climate change, and the degree of tropical cyclone variability provide only low confidence for the attribution of any detectable changes in tropical cyclone activity to anthropo-

[85] https://archive.ipcc.ch/pdf/assessment-report/ar5/wg1/WG1AR5_Chapter02_FINAL.pdf

[86] AR5 adopts a slightly different probabilistic scale whereby 'likely' corresponds with a probability ranging from 66 to 100%, 'about as likely as not' to 33-66% and 'unlikely' 0-33%. The latter two would appear to be similar to medium and low confidence respectively. See: https://archive.ipcc.ch/pdf/assessment-report/ar5/wg1/WG1AR5_Chapter01_FINAL.pdf

> genic influences. Attribution of single extreme events to anthropogenic climate change is challenging.[87]

It is important to note that although the number of tropical cyclones and hurricanes is likely to remain the same or even decrease with global warming, the intensity of individual storms is likely to increase, resulting in greater wind speeds and rainfall, whilst rising sea levels are likely to exacerbate the associated surge impacts (see Table SPM.1 in SREX[88]).

As the atmosphere warms it can hold more water and both temperature and moisture content gradients and associated energy generation increase. This results in significant changes to the forces that drive and control the global weather systems. For example, changes in the temperature gradients between the poles and equator create atmospheric instabilities which impact on cyclone creation and growth (baroclinicity), whilst increasing precipitation associated with higher moisture content in the atmosphere produces latent heat release, which in turn can increase the intensity of the cyclone. Tropical cyclone and hurricane genesis requires a combination of weak vertical wind sheer force and high sea surface temperature.

Differences in air temperature rise between the poles and equator are also significant in changing global weather patterns. The change in surface albedo in the Arctic due to melting of both sea ice and snow and the 'dark snow' phenomenon referred to earlier in this chapter is both weakening the polar vortex and reducing the temperature gradient in the atmosphere over the equator and polar region. The resultant instability is producing 'significantly amplified quasi-stationary mid-tropospheric planetary waves' (Rossby waves) which it is suggested in a paper by James Screen and Ian Simmonds published in 2014, are likely to 'preferentially increase the probabilities of heat waves in western North America, and central Asia, cold outbreaks in eastern North America, droughts in Central North America, Europe and central Asia and wet spells in western Asia'.[89]

In Chap. 2, we referred to historic droughts in the Great Plains of North America as well as the more recent prolonged drought in California. Long-term projections that indicate a propensity for drought or otherwise are based on some combination of predicted changes in precipitation, soil water content or the difference between precipitation and demand: such as the Palmer Drought Severity Index (PDSI).

According to a paper published in 2015 based on recent modelling carried out at the Lamont-Doherty Earth Observatory at Columbia University it is likely that most of The United States west of the Mississippi will be subject to drought through to the end of the twenty-first century.[90] The paper by Cook, Ault and Smerdon: *Unprecedented 21st century drought risk in the American Southwest and Central*

---

[87] https://www.ipcc.ch/site/assets/uploads/2018/03/SREX_Full_Report-1.pdf

[88] https://www.ipcc.ch/site/assets/uploads/2018/03/SREX_Full_Report-1.pdf

[89] http://www.nature.com/articles/nclimate2271.epdf?referrer_access_token=7_8mSBW4796BZZ9Is-xCptRgN0jAjWel9jnR3ZoTv0MfExFWAZHo033Wh5_2beillVHlO8nAU6on6lEMHZ0NK6cNTaERMH0Ad3O05wHauUYh8LOsuiF70vd1-dY_Z7-e

[90] http://advances.sciencemag.org/content/1/1/e1400082

Plains, discusses the results from the CMIP5 model, adopted for the IPCC 5th Assessment Report (AR5), to predict the soil moisture availability for the RCP8.5 and RCP4.5 scenarios. These plot the distribution of PDSI based on the moisture balance from the differences between precipitation and evapotranspiration (evaporation and moisture uptake by crops and vegetation) and suggest that the Southwest of the USA will fare worse than the Central Plains under both scenarios, and that drought is likely, although more modest, under the RCP4.5 scenario in all areas for which modelling was carried out.

Chapter 14 of the IPCC Working Group 1 AR5, which reports on future regional climate change, presents the results of modelling precipitation for the RCP4.5 scenario in map form. Although these do not take account of evapotranspiration they do, however, indicate a reduction in precipitation for much of central and Southwestern USA, stretching also down into Central America. Whilst the regions east of the Mississippi and north into Canada and Alaska are indicated as likely to have greater mean precipitation for the years 2080 to 2099 compared with 1986 to 2005.

These maps also indicate lower rainfall for southern Europe and North Africa, whilst Northern Europe, including the UK should experience greater precipitation in winter months. They also show that precipitation is likely to increase significantly in much of Asia, including China and India, whilst western Australia is likely to receive less rainfall.

On a global scale IPCC Working Group 2 has published maps that show projected changes in moisture content for the top 0.1 m of soil. According to the associated text in Chapter 12 of this report 'the large-scale drying in the Mediterranean, southwest USA, and southern Africa appear across generations of projections and climate models and is deemed likely as global temperatures rise and will increase the risk of agricultural drought'.

At the other end of the spectrum, although global average rainfall is not predicted to increase in proportion to temperature rise, IPCC report that 'consistently, climate models project future episodes of more intense precipitation in the wet seasons for most of the land areas, especially in the NH (northern hemisphere) and its higher latitudes, and the monsoon regions of the world, and at a global average scale'.[91] Of course the potential for flooding is not only a function of precipitation, but also of storm surges, storm damage, snow and ice melt and human activity and interventions. However, according to SREX 'there is medium confidence (based on physical reasoning) that projected increases in heavy rainfall would contribute to increases in local flooding in some catchments or regions'.[92]

A multi-model assessment of global flood risk is referred to in Chapter 3 of Assessment Report 5 by IPCC Working Group 2. This assessment provides projections for where the return period (interval) between flood events towards the end of the twenty-first century for scenario RCP8.5 is predicted to decrease, that is, the

[91] https://archive.ipcc.ch/pdf/assessment-report/ar5/wg1/WG1AR5_Chapter12_FINAL.pdf

[92] https://www.ipcc.ch/site/assets/uploads/2018/03/SREX_Full_Report-1.pdf

frequency of floods will increase: in particular Southeast Asia, China, peninsula India and Bangladesh, the central belt of Africa and the north-western and south-eastern areas of South America. It is interesting to note that the frequency of flood events for large parts of southern and eastern Europe is projected to decrease.

The assessment also shows projections for the number of people globally who are likely to be exposed to flood for the four RCP scenarios. 'Under a scenario of moderate population growth (as defined in the UN's World Population Prospects), the global number of exposed people is projected to increase by a factor of 7–25, depending on the RCP, with strong increases in Asia and Africa due to high population growth'.[93]

3.11.1 Social and Economic Impacts

> Impacts from recent climate-related extremes, such as heat waves, droughts, floods, cyclones, and wildfires, reveal significant vulnerability and exposure of some ecosystems and many human systems to current climate variability (very high confidence).
>
> Technical Summary to IPCC Working Group 2 Assessment Report 5: *Impacts, Adaptation & Vulnerability*.[94]

It is clear from this chapter so far that there are very few aspects of human life and natural ecosystems that have not already and will not in the future be impacted upon by climate change.

Increasingly the social and economic implications of the *status quo* compared with mitigating against and adapting to these impacts has been exercising climate scientists, politicians and many of those potentially impacted. We will be discussing these issues in the final chapter of this book, but here we will focus on the potential and predicted impacts on society of climate change, based primarily on the work that has been reported in the IPCC's 5th and 6th Assessment Reports.

An important concept in this context is the 'social cost of carbon' (SCC) which evolved in the USA as a primary tool for legislators in order to translate 'the effects of climate change into economic terms to help policymakers and other decision makers understand the economic impacts of decisions that would increase or decrease emissions'.[95] Using an Integrated Assessment Model (IAM) that links global climate and economic models, it takes into account the annual cost per tonne of carbon dioxide of lost agricultural productivity, human health impact and flood damage, using a predicted discount rate to estimate present value (PV). However, it does not take into account the impact of climate change on the growth rate of

---

[93] https://archive.ipcc.ch/pdf/assessment-report/ar5/wg1/WG1AR5_Chapter03_FINAL.pdf

[94] https://archive.ipcc.ch/pdf/assessment-report/ar5/wg2/WGIIAR5-TS_FINAL.pdf

[95] https://www.rff.org/publications/explainers/social-cost-carbon-101/#:~:text=The%20social%20cost%20of%20carbon%20(SCC)%20is%20an%20estimate%2C,carbon%20dioxide%20into%20the%20atmosphere

economies nor does it separate the likely greater impact on the more vulnerable developing economies from the robust developed ones. A study published by researchers from Stanford University in the USA in 2015 concluded that the median value for SCC used by legislators in the USA is a factor of six below what it should be, and that 'many more mitigation measures will pass a cost-benefit analysis' if the higher SCC were adopted.[96] Changes in the political landscape in the USA in 2017 and 2025 created an environment in which the use of SCC in Federal decision making was disrupted.

In evaluating the economic impact of climate change on ecosystems, it is useful to define the various 'ecosystem services' that encompass the benefits that humans derive from ecosystems and biodiversity. The 2003 Millennium Ecosystem Assessment[97] defined ecosystem services in terms of provisioning, regulating, supporting and cultural services. Provisioning includes the products obtained from plants and animals, such as food, clothing, goods and energy. Regulating services include climate modification, pollination, pest and disease control. Supporting services include primary production (see above) and nutrient cycling—the processing of chemical and biological agents back into the production of living matter, such as through composting. Cultural services include recreation and aesthetic value, for example. The world's stock of natural resources is sometimes referred to as 'natural capital'.

> Primary production, the process of plant growth, is fundamental to the global carbon cycle and underpins provisioning ecosystem services such as food, timber and grazing. Trends in the amount, seasonal timing, variability, location and type of primary production are therefore important indicators of ecosystem function.[98]

The severity of the impact of extreme weather on humans, either individually or communally, depends partially on their vulnerability to the stressors created. Vulnerability is typically defined in terms of 'the propensity or predisposition to be adversely affected' and is commonly associated with poverty and exposure to ongoing conflict or political unrest (see Chap. 2). The chapter on *Human Security* in the IPCC WG2 5th Assessment Report states that 'climate change can exacerbate major political changes given certain social conditions, including a predominance of subsistence producers, conflict over territory, and autocratic systems of government with limited power in peripheral regions'.[99]

In the Executive Summary to this chapter, IPCC sets out the key threats to human populations as follows:

> Climate change is an important factor threatening human security through undermining livelihoods; compromising culture and identity; increasing migration that people would

[96] https://news.stanford.edu/2015/01/12/emissions-social-costs-011215/

[97] http://www.millenniumassessment.org/en/index.html

[98] https://archive.ipcc.ch/pdf/assessment-report/ar5/wg2/WGIIAR5-Chap4_FINAL.pdf

[99] https://archive.ipcc.ch/pdf/assessment-report/ar5/wg2/WGIIAR5-Chap12_FINAL.pdf

> rather have avoided; and challenging the ability of states to provide the conditions necessary for human security.[100]

The costs associated with climate change fall broadly into three categories: mitigation, adaptation and dealing with the outcomes of the impacts. The IPCC concludes in the Executive Summary of WG3 AR5 Chapter 3 that the 'impacts of extreme events may be more important economically than impacts of average climate change (high confidence)'.[101] Clearly as the number and severity of extreme weather events increase then their economic impact and ability to disrupt lives, destroy buildings and infrastructure and displace populations grows disproportionately.

Inevitably it is the more vulnerable populations in the developing world who will be most severely hit by both long-term impacts of global warming and sea level rise and the violence of extreme weather events. Many of these people live off the land and do not have the resources or protection from central government to cope with the impacts of climate change:

> Major extreme weather events have in the past led to significant population displacement, and changes in the incidence of extreme events will amplify the challenges and risks of such displacement. Many vulnerable groups do not have the resources to be able to migrate to avoid the impacts of floods, storms, and droughts.[102]

Despite high mortality rates most developing countries are projected to experience higher population growth rates than the developed world, thus putting agricultural land under greater stress. Extreme climate events have historically had a disproportionate impact on agricultural systems, and since 'developing countries rely heavily on climate-dependent agriculture and especially in conjunction with poverty and rapid increase in population they are vulnerable to climate change'.[103] This is especially true for those communities in low lying areas that are exposed to sea level rise and storm surges, for example, as well as those in drylands exposed to rising temperatures, unreliable precipitation, dust storms and the increasing incidence of wild fires.

Some of the most vulnerable people live in the so-called informal settlements, where homes are built on land to which the occupants have no legal claim. At particular risk of the impacts of climate change and extreme weather are the urban poor living in these settlements. Worldwide, about one billion people live in these settlements, which are growing faster than formal settlements. 'The vulnerability is high due to very low-quality housing and limited capacity to cope due to a lack of assets, insurance, and marginal livelihoods, with less state support and limited legal protection'.[104]

---

[100] https://archive.ipcc.ch/pdf/assessment-report/ar5/wg2/WGIIAR5-Chap12_FINAL.pdf

[101] https://archive.ipcc.ch/pdf/assessment-report/ar5/wg3/ipcc_wg3_ar5_chapter3.pdf

[102] https://archive.ipcc.ch/pdf/assessment-report/ar5/wg2/WGIIAR5-Chap12_FINAL.pdf

[103] https://archive.ipcc.ch/pdf/assessment-report/ar5/wg2/WGIIAR5-Chap7_FINAL.pdf

[104] https://www.ipcc.ch/site/assets/uploads/2018/03/SREX_Full_Report-1.pdf

3.11.2 *Cyclones, Flood Risk and Extreme Rainfall*

Extremely energetic storms in the form of tropical cyclones and Atlantic hurricanes are arguably the most dramatic and costly manifestation of extreme weather, becoming more aggressive as air and sea temperatures increase. The IPCC's report on extreme weather and climate change (SREX) concludes that 'exposure and vulnerability (to cyclones and hurricanes) are increasing due to growth in population and increase in property values, particularly along the Gulf and Atlantic coasts of the United States. Some of this increase has been offset by improved building codes'.[105] SREX also reports that exposure to tropical cyclones has increased from 73 million people in 1970 to 123 million in 2010 and is projected to increase to 137 million in 2030 based on population rise alone.

The United Nations' International Strategy for Disaster Reduction (UNISDR) publishes an annual Global Assessment Report on Disaster Risk Reduction (GAR). Chapter 2 of the 2011 report sets out the key risks to the most vulnerable populations worldwide. It reports that the 'economic loss risk for cyclones is increasing in all regions. It has almost quadrupled (increasing by 265%) since 1980 in the OECD (Organisation for Economic Co-operation and Development), almost tripled in sub-Saharan Africa (181%), and is more than two-and-a-half times greater in other regions (over 150% higher)'.[106]

In much of the world 'exposure and the estimated risk of economic loss are growing faster than GDP per capita. Thus the risk of losing wealth in disasters associated with tropical cyclones is increasing faster than wealth itself is increasing'.[107]

According to figures from the year 2000 quoted by the IPCC, the 'Low Elevation Coastal Zone' (LECZ) at that time comprised around 2% of global land area, whilst being home to 10 percent of the world's population and 13 percent of the world's urban population. Referring to a paper from 2007 IPCC states that some 65% of cities having populations greater than 5 million were found in this Zone. Also, between 1970 and 2010, the number of people exposed to a one in a hundred year 'extreme sea level' nearly doubled to around 270 million, exposing some 13 trillion US dollars' worth of assets to this level of risk, with the greatest exposure being in Asia. Estimates vary, but according to modelling reported in the same chapter[108] some 187 million people could be displaced by inundation and erosion in the LECZ by the year 2100 if global average sea level rise reaches 2 m. In the IPCC's 2018 Special Report on *Managing the Risks of Extreme Events and Disasters* (SREX) the value of assets exposed is estimated to reach $35 trillion by the 2070s, assuming no additional coastal protection.[109]

---

[105] http://advances.sciencemag.org/content/1/1/e1400082

[106] http://ec.europa.eu/health/ph_projects/2005/action1/docs/action1_2005_a2_15_en.pdf

[107] http://ec.europa.eu/health/ph_projects/2005/action1/docs/action1_2005_a2_15_en.pdf

[108] https://archive.ipcc.ch/pdf/assessment-report/ar5/wg2/WGIIAR5-Chap5_FINAL.pdf

[109] https://www.ipcc.ch/site/assets/uploads/2018/03/SREX_Full_Report-1.pdf

Major storms, cyclones, hurricanes and the associated storm surges and flooding frequently result in populations being displaced. For example, GAR 2011 reveals that 'Pakistan's 2010 floods have to date left an estimated 6 million people in need of shelter; India's 2008 floods uprooted roughly 6 million people; Hurricane Katrina displaced more than half a million people in the USA; and Cyclone Nargis uprooted eight hundred thousand people in Myanmar and South Asia'.[110]

Flood risk from the breaching of river banks is predicted to increase, such that: 'by the end of the 21st century, the number of people exposed annually to the equivalent of a 20th-century 100-year river flood is projected to be three times greater for very high emissions (RCP8.5) than for very low emissions (RCP2.6) (multi-model mean) for the fixed population distribution at the level in the year 2005'.[111] It is extraordinary to note, if perhaps not surprising, that of the thousands of fatalities that have occurred as a result of river floods since the 1970s some 95% have been in developing countries, whilst the greatest economic damage in total (although not per GDP) has been in developed countries.

As a growing population is accommodated through increased urbanisation, hard surfaces replace soft thus modifying surface water flows and increasing flood risk. It is evident that, as stated in the IPCC SREX, 'economic development and land use change can … lead to changes in natural systems. Land cover changes induce changes in rainfall-runoff patterns, which can impact on flood intensity and frequency'.[112]

Extreme rainfall not only contributes to flooding but also increases the incidence of landslides and mudslides. When this coincides with the activities of loggers and others in removing the trees and root systems that have historically stabilised slopes and prevented slippage, disaster can strike. The huge mudslide that killed 43 people and destroyed 49 buildings in Steelhead Haven, near the town of Oso in Washington State on 22 March 2014 is thought to have been due to a combination of factors. Although this area had been subject to significant mudslides previously and the slopes were known to be unstable, none had resulted in fatalities. In the month prior to this disaster precipitation had been two to three times higher than normal for March, whilst over the previous 4 year period the region had been subject to the highest rainfall quantities ever recorded.

Globally there have been a number of landslides that have resulted in tens of thousands of fatalities—many triggered by earthquakes or volcanic activity, but with some notable examples that have resulted primarily from a combination of excessive rainfall and human activity. One of the worst of these occurred on 15 December 1999 in the State of Vargas in Venezuela when the cumulative effects of relentless rainfall produced catastrophic landslides down the slopes of the Sierra de Avila burying the towns of Carmen de Uria and Cerro Grande resulting in the deaths of some 30,000 people, most of whose bodies were carried out to sea or buried too

---

[110] http://ec.europa.eu/health/ph_projects/2005/action1/docs/action1_2005_a2_15_en.pdf

[111] https://archive.ipcc.ch/pdf/assessment-report/ar5/wg2/WGIIAR5-Chap3_FINAL.pdf

[112] https://www.ipcc.ch/site/assets/uploads/2018/03/SREX_Full_Report-1.pdf

deep to recover. The coast of Venezuela bordering the Caribbean is extremely steep with little vegetation and with settlements perched on the coast directly in the path of potential landslides, which do occur periodically, although the one that occurred at the end of 1999 was particularly severe.

3.11.3 Heat Waves, Drought and Drylands

It is a fundamental tenet of climate change science that most regions of the planet will experience increasing temperatures in the coming years. Reporting on a range of climate models the IPCC SREX predicts that 'it is very likely that the length, frequency and/or intensity of warm spells or heat waves will increase over most land areas'.[113] The impact of temperature extremes is potentially dramatic, particularly for the populations of urban areas. 'Heat waves combined with urban heat islands can result in large death tolls: with the elderly, the unwell, the socially isolated and outdoor workers being especially vulnerable'.

Such a heat wave occurred in July and August 2003 across Europe, with average temperatures exceeding norms by up to 4 °C. Resulting fatalities were reported shortly after to be in excess of 35,000, although subsequent analysis commissioned by the European Union and published in 2008 has increased this estimate to between 70,000 and 77,000.[114] Excess deaths were particularly prevalent amongst the elderly and socially isolated, for example, on the 12th of August, the worst day of the heat wave in France, 46% more people over 95 died than expected, whilst overall some 80% of excess deaths were people who were 75 or over.[115]

The heat wave resulted in stresses on the health services, water supply, food storage, energy infrastructure and agriculture, including an estimated 13 billion US dollars of uninsured losses to the agriculture sector across Europe.[116] The resultant drought created an increased incidence of wildfires and lower river levels, which not only impacted water supply and agriculture, but also many of the nuclear power plants in France, which also had to cope with a massive increase in electricity demand from air conditioning. Power cuts followed, resulting in more distress and fatalities.

Although the heat wave of 2003 was reported to have been the worst in 500 years it is interesting to note that the temperatures that occurred are similar to those predicted for Europe towards the end of the twenty-first century under the 'business as usual' scenarios in the IPCC 5th Assessment Report.

Similarly in 2022 more than 61,000 heat-related deaths were recorded, with Italy being the worst affected. In England, where the highest temperatures on record

---

[113] https://www.ipcc.ch/site/assets/uploads/2018/03/SREX_Full_Report-1.pdf

[114] http://ec.europa.eu/health/ph_projects/2005/action1/docs/action1_2005_a2_15_en.pdf

[115] https://archive.ipcc.ch/pdf/assessment-report/ar5/wg3/ipcc_wg3_ar5_chapter11.pdf

[116] http://advances.sciencemag.org/content/1/1/e1400082

occurred, there were more than 4500 deaths attributed to this heat wave towards the end of July 2022.

According to Chapter 11 of the IPCC WG2 5th Assessment Report 'the increase in heat-related mortality by mid (21st) century will outweigh gains due to fewer cold periods, especially in tropical developing countries with limited adaptive capacities and large exposed populations'. In the same chapter, references are made to reports from 2010 and 2013 that indicate that 'worldwide, more than half of all non-household labor-hours occur outdoors, mainly in agriculture and construction. Individuals who are obliged to work outside in hot conditions, without access to shade, or sufficient water, are at heightened risk of heat strain and heat stroke'.[117]

As we saw earlier in this chapter, drought and desertification are a growing problem. As increasing numbers are exposed to the resultant stress on agriculture and food production: poverty, malnutrition, illness and shorter lives will be the inevitable outcomes. As the UN Educational, Scientific & Cultural Organisation (UNESCO) explains:

> Desertification leads to poverty with all the social, economical and cultural consequences that this entails. Poverty then drives populations to over-exploit the remaining natural resources triggering a vicious cycle, accelerating land degradation still further. Poverty is thus both a cause and a consequence of desertification. Desertification affects the lifestyles of nearly one billion of the world's population.[118]

With population growth, pressure on land, food price fluctuations and climate change this number is set to grow by the end of the twenty-first century. Many of the regions that are experiencing desertification or that are transitioning from rain-fed croplands to semi-arid subsistence drylands are also conflict zones, adding further pressures that in many cases drive people out of their homes and away from their lands (Fig. 3.15).

> Desertification can drive whole communities to migrate towards cities or regions where survival conditions are initially more promising but ultimately are very difficult: social stability and cultural identity are threatened and the makeshift dwellings, which are insanitary and illegal, are sometimes sources of ethnic or religious conflict.[119]

Higher temperatures in regions that already experience some of the highest on the planet will increase the heat stress experienced by people working the land, forcing some to abandon croplands altogether. For example projections for Mali, reported by the IPCC, anticipate that by 2025, whose population grew from 15.3 million in 2013 to 20.26 million in 2020, there will be a shift from rain-fed millet and sorghum based agriculture to semi-arid livestock subsistence farming, resulting in an additional 6 million people suffering from malnutrition.[120]

Although the IPCC reports a lack of data enabling it to conclude that this increasing desertification will result in more dust storms in the future, a North American

[117] https://archive.ipcc.ch/pdf/assessment-report/ar5/wg3/ipcc_wg3_ar5_chapter11.pdf

[118] https://unesdoc.unesco.org/ark:/48223/pf0000125816

[119] https://unesdoc.unesco.org/ark:/48223/pf0000125816

[120] https://archive.ipcc.ch/pdf/assessment-report/ar5/wg2/WGIIAR5-Chap13_FINAL.pdf

Fig. 3.15 The encroaching desert threatens communities in Ningxia, China (From: https://en.m.wikipedia.org/wiki/File:Desertification_Control_Project,_Ningxia_China_-_Planet_Labs_satellite_image.jpg)

study from 2011 by Munson et al. on the *Responses of wind erosion to climate-induced vegetation changes in the Colorado Plateau* concludes that 'the effects of increased temperature on perennial plant cover and the correlation of declining plant cover with increased aeolian flux strongly suggest that sustained drought conditions across the southwest (USA) will accelerate the likelihood of dust production in the future on disturbed soil surfaces'.[121] Extrapolating this to semi-arid zones globally it is reasonable to conclude that as some grasses and shrubs die back due to lack of precipitation and rising temperatures the risk of wind erosion and associated dust storms will increase.

It has been mooted that the prolonged drought in California referred to earlier was caused by climate change. A study by researchers from Stanford University concluded that an area of high pressure over the Pacific diverted storms northwards to Washington State, Alaska and beyond. Unusually this 'blocking system', which they colourfully refer to as the 'ridiculously resilient ridge' or 'Triple R', although fluctuating in strength from time to time, remained in place between 2011 and 2017. Using a climate model that compared the conditions in 2013 with those in the

[121] https://www.pnas.org/doi/10.1073/pnas.1014947108#:~:text=protect%20soil%20surfaces.-,Conclusions,emission%20on%20the%20Colorado%20Plateau

Fig. 3.16 Before and after images of the Hoover dam on the Colorado River showing impact of drought (From: https://commons.wikimedia.org/wiki/File:Drought-before-after.jpg)

pre-industrial era the team published findings that showed that the increasing occurrence of extremely high atmospheric pressure over the Northeastern Pacific is "very likely" linked to global warming'.[122] Not only has there been less precipitation, but higher air temperatures have resulted in a smaller 'snowpack' which has melted earlier in the year.

The impact of this drought on California, its economy and its inhabitants was dramatic. In March 2015 the Governor introduced the latest in ever tightening restrictions on water use. The cost of irrigating crops had risen significantly, increasing the price and reducing the volume of a wide range of food crops, especially those which are primarily grown in California, such as almonds, artichokes, olives and grapes. The lower snow cover not only impacted winter sports and water supplies but also resulted in greater glacier melt. For example, Mount Shasta, in the north of California, had around 30% of normal snowfall in the winter of 2013/14 so that the snow cover that would normally protect the Konwakiton Glacier from the summer sun was insufficient to prevent the ice from melting, resulting in a massive mudslide on the night of the 21st September 2014. Fortunately there were no injuries or property damage, although a local road was rendered impassable.[123]

Water supply for many US States was dramatically impacted as reservoirs were depleted, as can be seen in Fig. 3.16 that shows Lake Mead, which sits behind the Hoover Dam on the Colorado river on the border between Nevada and Arizona, seen here in 2001 and 2015 at around one-third of its full capacity.

---

[122] https://pangea.stanford.edu/d7-archive/sesd7/news/rise-ridiculously-resilient-ridge-california--drought-patterns-becoming-more-common/index.html

[123] http://www.livescience.com/47948-mount-shasta-mudslide-california-drought.html

3.11.4 Fire Risk

Although California has a long history of large wildfires, there has been perhaps a doubling or more of their incidence since the drought has taken hold. Wildfires globally occur as a result of lightning strikes, volcanic activity or human activity. Some 90% of uncontrolled large scale wildfires are caused by humans: due to either arson, illegal land management or carelessness. On the plus side wildfires can provide something akin to a heavy prune: improving reproduction and ecosystem health. Unfortunately most of the impacts of large scale fires are negative, including greenhouse gas and particle emissions to atmosphere, destruction of ancient forests and peatlands, air pollution, destruction of root systems that bind slopes and reduce soil erosion and loss of human lives and property. Fighting fires is costly and dangerous and can tie up precious firefighting resources for months in some cases.

In 2014 the US Cost of Carbon Project produced a paper *Flammable Planet: Wildfires and the Social Cost of Carbon* which drew together research indicating that globally some 3.5 million km^2 of land on average was being burnt annually or about 0.7% of total global land surface area.[124] Researchers predict that this figure could rise during the twenty-first century, with models indicating 'meteorological fire danger' increasing, particularly in the northern hemisphere. This was demonstrated by a proliferation of wildfires across the northern hemisphere in particular, during 2023. The global economic implications of this are considerable, predicted to be between 50 and 300 billion US dollars annually by 2050.

The relationship between climate change and anthropogenic fire setting is nicely illustrated by the recent history of transboundary pollution in Indonesia and Malaysia. Fires burning out of control in Sumatra regularly throw a pall of smoke over neighbouring Singapore and Kuala Lumpur (see Fig. 3.17). The hundreds of separate fires identified have been blamed primarily on the illegal activities of wood pulp and palm oil concessions, although they have worsened recently due to changes attributed to climate change. A blog by the World Resources Institute in March 2014 reported that: 'this February has …. been one of the driest on record in Indonesia and neighbouring countries, affecting crop yields and setting the stage for burning. These sorts of dry spells may become more frequent and severe as climate change worsens'.[125]

During the summer of 2023 wildfires in Canada created pollution events in New York, with smoke being carried across the Atlantic by the jet stream and briefly impacting atmospheric conditions in the UK and beyond.

[124] http://costofcarbon.org/files/Flammable_Planet__Wildfires_and_Social_Cost_of_Carbon.pdf

[125] http://www.wri.org/blog/2014/03/indonesian-fires-bring-more-haze-southeast-asia

Fig. 3.17 Image showing the haze generated over Singapore and southern Malaysia by fires in Sumatra (From: https://commons.wikimedia.org/wiki/File:Satellite_image_of_2013_Southeast_Asian_haze_-_20130619_%28annotated%29.jpg)

3.11.5 Permafrost Thaw

At the other end of the temperature scale, warming is impacting Arctic communities through the degradation of permafrost; through the timing and type of precipitation and increase in 'rain on snow' conditions; and through the extent and stability of ice, coastal erosion and salinisation of freshwater resources. Communities that have hitherto relied on hunting, fishing and gathering are finding food sources being displaced and terrain unsuitable for travel. This is in turn leading to the westernisation of eating habits and resultant health impacts: both physiological and psychological. The thawing of permafrost is not only releasing methane as discussed earlier in this chapter but also undermining structures such as landfill sites, reservoirs, sewers, roads, bridges, railway lines, airfields, pipelines and buildings (see Fig. 3.18).

In 1994 a pipeline serving the Vozei oilfield in Northern Russia fractured, as the permafrost on which it was laid sank unevenly, releasing some 160,000–200,000 tonnes of oil: the world's largest terrestrial oil spill. The spill contaminated some 76.5 ha of forest tundra in the Usinsk area of the Republic of Komi and reached as far as the Barents Sea. Although a number of different remedial measures have been attempted over the years, subsequent surveys have found that extremely high

Fig. 3.18 Building damaged by permafrost thawing in Dawson city, Canada (From: https://commons.wikimedia.org/wiki/File:Permafrost_Damage_in_Dawson_City_-_panoramio.jpg)

concentrations of hydrocarbons remain, particularly in areas where the primary remedial measure had been ploughing.[126]

'Rain on snow' conditions occur when rain falls on snow and then freezes, making travel difficult and blocking access to forage for grazing animals such as reindeer. These animals are also under threat from human activities such as gas exploration, which is occurring in many prime reindeer grazing grounds, such as the Yamal area of Western Siberia, which contains some 90% of Russia's gas reserves, whilst also hosting the largest reindeer herd on the planet.

Thawing permafrost is also impacting communities, structures and tourism in mountain regions such as the Alps, the Andes and the Himalayas. These regions have experienced a significant increase in rock falls and landslides in recent decades, corresponding with a warming climate:

> Thawing permafrost in steep mountain terrain increases the risk of rock falls and landslides. Talus (rocks) cemented together by ice in mountainous permafrost zones can form rock glaciers that creep downhill at velocities of centimetres to several metres per year. If temperatures increase, the ice softens and rock glacier flow velocities increase. If the permafrost thaws, the "cement" holding the talus together drains away, increasing the risk of sudden collapse and landslides.[127]

There is a high degree of certainty that the number of people impacted by permafrost thaw will steadily increase as surface temperatures rise, resulting in the displacement of populations into more stable regions. Coastal communities in the Arctic are likely to be the worst hit since the high salt content of coastal permafrost

[126] https://urgentcomm.com/2000/03/01/oil-on-the-tundra-a-spill-in-the-komi-republic/

[127] https://epic.awi.de/id/eprint/33086/1/permafrost.pdf

means that it is more vulnerable to a rapid thaw, particularly when combined with sea level rise.

3.11.6 Food and Agriculture

It should be clear to the reader that many of the impacts referred to so far will result in changes to agriculture and food production, whether it be crops, livestock or fish and seafood. According to the IPCC 'the largest known economic impact of climate change is on agriculture because of the size and sensitivity of the sector, particularly in the developing world and to a lesser extent in parts of the developed world'.[128] The ripple effect of these changes will impact people's livelihoods, health and survival. The developing world may see whole populations displaced or communities wiped out if the food chain fails. At the other end of the spectrum industrial scale agriculture supplies an international market which is inherently more robust and flexible to component failure.

There have been numerous analyses and predictions of the socioeconomic impact of climate change on agriculture, most of which have examined specific countries or regions. Studies reported by the IPCC include one that predicts that the Gross Domestic Product (GDP) of the largest countries in SE Asia will fall by 1.4% by 2080 due to a reduction in agricultural productivity, whilst welfare costs will have a negative impact on GDP to the tune of 1.7%. Another study estimates that the average annual spend on dealing with 'adverse agricultural impacts' across the whole of Asia will be 4.2–5 billion US dollars between 2010 and 2050.[129]

Although rising temperatures and carbon dioxide concentrations may improve the productivity of some grasslands used for livestock grazing, this will depend on the availability of a reliable supply of water. Growing seasons may lengthen, but forage quality may be impacted. Although the vulnerability of livestock to high temperatures varies with animal and breed, in general the traits that make animals more productive, such as milk yield in cattle or high growth rates and leanness in pigs and poultry, tend to make them more prone to heat stress.[130]

IPCC report on predictions from a 2009 study that by 2050 the number of undernourished children aged 5 or younger will increase by 20–25 million or between 17% and 22%. In a 2018 study by the same authors reported in Chapter 5 of IPCC AR6 WGII comparing the numbers experiencing food insecurity under the RCP8.5 scenario with zero climate impacts estimates an increase globally of 65 million or 10% of global population.[131]

In its 2013 chapter on food insecurity, the IPCC concludes that:

---

[128] https://archive.ipcc.ch/pdf/assessment-report/ar5/wg2/WGIIAR5-Chap9_FINAL.pdf

[129] https://archive.ipcc.ch/pdf/assessment-report/ar5/wg2/WGIIAR5-Chap9_FINAL.pdf

[130] https://archive.ipcc.ch/pdf/assessment-report/ar5/wg2/WGIIAR5-Chap7_FINAL.pdf

[131] https://www.ipcc.ch/report/ar6/wg2/chapter/chapter-5/

> … if global yields are negatively impacted by climate change, an increase in both international food prices and the global headcount of food-insecure people is expected (limited evidence, high agreement). However, it is only about as likely as not that the net effect of climate and CO_2 changes on global yields will be negative by 2050, but likely that such changes will occur later in the 21st century.[132]

The authors make it clear, however, that they foresee a degree of socioeconomic and technological adaptation that could provide a countervailing force to these impacts. There will be some countries which are net exporters of food and other products that will benefit from price increases, although these will not always trickle down to every link in the supply chain.

The problem of pollinator mortality referred to earlier could impact agriculture severely. 'Pollination is critical for the food supply as well as for overall environmental health. Its value has been estimated globally at US$350 billion for the year 2010 (range of estimates of US$200–500 billion)'.[133] One approach to estimating the future impact of pollinator decline has been adopted by Dana Bauer and Ian Wing of Boston University in their paper on the *Economic Consequences of Pollinator Decline: A Synthesis*, published in 2010, which uses a 'General Equilibrium Analysis' to estimate the cost to the global economy of a catastrophic loss of pollination services. General equilibrium assumes a balance between supply and demand based on the economic system as a whole. They estimate that the 'total economy-wide losses' would be 334.1 billion US dollars, including 'the price effects on downstream sectors and households'. Their model divides the world into regions, concluding that West Africa would be particularly badly hit because of its reliance on pollinator-driven crops.[134]

3.11.7 Fisheries and Aquaculture

Fish, including shellfish and fish products, are an important food resource for the majority of the world's population. According to the 2022 review of *'The State of World Fisheries and Aquaculture'* from the Food and Agriculture Organisation (FAO) of the UN, some 178 million tonnes of fish and shellfish were processed worldwide in 2020 through 'capture fisheries and aquaculture production', resulting in a global market worth around 406 billion US dollars.[135] Asian countries dominate, producing some 70% of the entire global output in 2020. In that year approximately 51% of global production was from capture fisheries (fishing in open water) the remainder being from aquaculture (fish farming), Capture fisheries include trawling and line fishing, whilst aquaculture is the cultivation of fish and shellfish.

---

[132] https://archive.ipcc.ch/pdf/assessment-report/ar5/wg2/WGIIAR5-Chap7_FINAL.pdf

[133] https://archive.ipcc.ch/pdf/assessment-report/ar5/wg2/WGIIAR5-Chap4_FINAL.pdf

[134] https://people.bu.edu/isw/papers/Bauer%20and%20SueWing%20ARER%20Final.pdf

[135] http://www.fao.org/3/a-i3720e.pdf

Cultivation and harvesting of algae represented an additional 30 million tonnes in 2020.

Fifty percent of ocean fish are from small-scale fisheries in developing countries, generating some 47 million jobs, including 12.5 million in fishing. According to the TEEB report referred to above 'some 30 million people in coastal and island communities are totally reliant on reef-based resources as their primary means of food production, income and livelihood'.[136] In the box dealing with coral reefs in the IPCC 5th Assessment Report the authors state that 'coral reefs account for 10 to 12% of the fish caught in tropical countries, and 20 to 25% of the fish caught by developing nations',[137] although these only represent between 2% and 5% of global fisheries.

In their 2007 paper Newton et al. compare resource consumption with sustainable reef production for 49 islands, representing some 41% of coral reef areas.[138] More than half (55%) of these island countries were found to be exploiting their coral reef fisheries in an unsustainable way, whilst 'production of coral reef fish in the Pacific is projected to decrease 20% by 2050 under the Special Report on Emission Scenarios (SRES) A2 emissions scenario'.[139] A more recent analysis, published in 2023 by researchers from the Department of Nutrition at Harvard, used fish stock data from 2000 coral reefs worldwide and found that not only were 'fish stocks dangerously low and/or levels of fishing outmatch fish stocks' capacity to recover', but more than 50% were not meeting at least one sustainability criterion.[140]

Earlier predictions may turn out to overly optimistic since the Global Coral Reef Monitoring Network has reported that the area of coral reefs shrank by 14% between 2009 and 2022, whilst the UN has predicted that a further 90% could be lost by 2050, primarily due to climate change.[141]

The outlook for fisheries production is projected to vary with latitude, with the IPCC concluding in its chapter on *Food Security and Food Production Systems* that 'overall, there is high confidence that climate change will impact on fisheries production with significant negative impacts particularly for developing countries in tropical areas, whilst more northerly, developed countries may experience benefits'.[142] Whilst the chapter on *Coastal Systems and Low Lying Areas* reports that 'economic losses in landed catch value and the costs of adapting fisheries resulting

[136] http://www.teebweb.org/publication/mainstreaming-the-economics-of-nature-a-synthesis-of-the-approach-conclusions-and-recommendations-of-teeb/

[137] https://archive.ipcc.ch/pdf/assessment-report/ar5/wg2/WGIIAR5-CCboxes_FINAL.pdf

[138] http://www.dulvy.com/uploads/2/1/0/4/21048414/newton_et_al_2007_curbiol.pdf

[139] https://archive.ipcc.ch/pdf/assessment-report/ar5/wg2/WGIIAR5-CCboxes_FINAL.pdf

[140] https://www.hsph.harvard.edu/news/hsph-in-the-news/a-path-to-sustainability-for-coral-reef-fisheries/#:~:text=The%20study%20found%20that%20over,fish%20stocks%27%20capacity%20to%20recover

[141] https://news.un.org/en/story/2021/03/1087592#:~:text=Coral%20reefs%20are%20in%20grave,a%20result%20of%20climate%20change

[142] https://archive.ipcc.ch/pdf/assessment-report/ar5/wg2/WGIIAR5-Chap7_FINAL.pdf

from a 2°C global temperature increase by 2050 have been estimated at US$10 to 31 billion globally'.[143]

Modelling carried out by Narita et al., published in 2012 and referred to in the chapter in the IPCC report on *Key Economic Sectors and Services*, indicates that this figure could be dwarfed by the economic impact of ocean acidification on shellfish. 'Using a partial equilibrium model, this research estimates that 'by the turn of this century the aggregate cost could be greater than US$100 billion'.[144] Partial equilibrium models employ similar techniques as those used for the general equilibrium analysis referred to earlier but only requires the analysis of single markets.

Coastal fisheries are also impacted by 'dead zones' or anoxic regions caused by nutrients carried into the ocean in river water and sewage, for example (see Sect. 3.4). Faced with inadequate oxygen for survival, sea creatures will either move away from the coast or die, resulting in smaller catches for fishing fleets and longer trips to find populated fishing grounds. Similarly the pelagic zone contains oxygen minimum zones (OMZ), the ongoing expansion of which is directly linked with warmer sea temperatures. A review of the impacts of this expansion has been carried out by Gilly et al. in the 2013 *Annual Review of Marine Science* in which he discusses the '*Oceanographic and Biological Effects of Shoaling of the Oxygen Minimum Zone*', where 'shoaling' refers to expansion of the OMZ and a corresponding contraction of the layer of the water column that can support the majority of pelagic fish. Gilly et al. conclude that 'the responses of pelagic ecosystems and fisheries to OMZ shoaling will occur across all levels of biological organisation and will interact with other forms of environmental change, including ocean warming and acidification'.[145] 'Warm surface waters are stressful to some species, and expansion of this zone will restrict the depth to which organisms can retreat to burn off an oxygen debt accrued during night-time forays into OMZs'. The night-time habitat range for vertically migrating species are predicted to become compressed due to a combination of warming, ocean acidification and expanding OMZs.[146] The effects of OMZ expansion '… include altered microbial processes that produce and consume key nutrients and gases, changes in predator-prey dynamics, and shifts in the abundance and accessibility of commercially fished species. Although many species will be negatively affected by these effects, others may expand their range or exploit new niches. OMZ shoaling is thus likely to have major and far-reaching consequences'.[147]

Clearly, as the IPCC concludes in its chapter on *Oceans*, 'decreasing oxygen, consequently, is very likely to increase the vulnerability of fisheries and aquaculture (medium confidence), and consequently puts livelihoods at risk'.[148]

---

[143] https://archive.ipcc.ch/pdf/assessment-report/ar5/wg2/WGIIAR5-Chap5_FINAL.pdf

[144] https://archive.ipcc.ch/pdf/assessment-report/ar5/wg2/WGIIAR5-Chap10_FINAL.pdf

[145] https://pubmed.ncbi.nlm.nih.gov/22809177/

[146] http://ec.europa.eu/health/ph_projects/2005/action1/docs/action1_2005_a2_15_en.pdf

[147] https://pubmed.ncbi.nlm.nih.gov/22809177/

[148] https://archive.ipcc.ch/pdf/assessment-report/ar5/wg2/WGIIAR5-Chap30_FINAL.pdf

Similarly freshwater fisheries are impacted by eutrophication which, although primarily caused by discharges from fertilisers in surface water run-off and the nutrients in sewage, is exacerbated by warming. For example, the red green algal blooms, that are common in freshwater lakes and reservoirs during summer months, create toxic conditions for animals and humans alike. As summarised in the chapter on *Terrestrial and Inland Water Ecosystems* in IPCC WG2 AR5: 'primary production in freshwater ecosystems can be either nitrogen and phosphorus limited or both, but the biodiversity and capacity of freshwater ecosystems to deliver high quality water, recreational amenity and fisheries services is severely reduced by the addition of nutrients beyond their capacity to process them'.[149]

3.11.8 Forests

For millennia, trees have been harvested for fuel, manufacture and construction, such that large tranches of the ancient woodland have been plundered and replaced with farmland, urban sprawl or infrastructure. Deforestation continues apace whether it is due to commercial exploitation: such as timber production, wood pulp for paper and other products, or biomass for fuel, clearing for agriculture or development, destruction from disease, or from drought, flooding, landslides or wildfires. Some of these are offset by tree planting and natural regeneration.

The Economics of Ecosystems and Biodiversity (TEEB) Synthesis Report from 2010 states that 'forest ecosystems account for over two-thirds of net primary production on land, that is, the conversion of solar energy into biomass through photosynthesis—making them a key component of the global carbon cycle and climate'.[150] Although both IPCC and TEEB conclude that deforestation has slowed since 2000, partly due to the major planting project in China and partly due to policy and market changes in Brazil which slowed forest clearing in the Amazon for a time (although reversed during the Bolsenaro years) and resulting in a 1.5% reduction in global anthropogenic carbon emissions, deforestation remains a major challenge. It has been reported that some 2.3 million km^2 of forest were lost between 2000 and 2012.[151]

In the summary to the TEEB report referred to above it is concluded that 'halving deforestation rates by 2030 would reduce global greenhouse gas emissions by 1.5 to 2.7 $GtCO_{2eq}$ per year, thereby avoiding damages from climate change estimated at more than US$ 3.7 trillion in NPV (net present value) terms. This figure does not

[149] https://archive.ipcc.ch/pdf/assessment-report/ar5/wg2/WGIIAR5-Chap4_FINAL.pdf

[150] http://www.teebweb.org/publication/mainstreaming-the-economics-of-nature-a-synthesis-of-the-approach-conclusions-and-recommendations-of-teeb/

[151] http://www.teebweb.org/publication/mainstreaming-the-economics-of-nature-a-synthesis-of-the-approach-conclusions-and-recommendations-of-teeb/

include the many co-benefits of forest ecosystems', such as reduction in species degradation and extinction, and benefits to recreation and tourism.

The IPCC warns that despite the reduction in forest clearing and wildfires in the Amazon, some 50% of the rain forest could be lost by 2030. Whilst referring to a study that uses five climate model simulations for the 2040–2069 period under the 'business-as-usual scenario' (RCP8.5) it asserts that 'climate change, deforestation, fragmentation, fire or human pressure place virtually all (97%) of the remaining tropical dry forests at risk of replacement or degradation'.[152] Dry tropical forests are defined by strong seasonality in rainfall, whilst tropical rainforests remain moist throughout the year. Statistics are hard to come by, but the area of tropical rainforest in 2007 was reported to have been around 8 million km^2, compared with 1 million km^2 of dry tropical forest in 2006. In both cases more than 50% of the tropical forest was found in South America.

When examining projections through to the end of the twenty-first century the IPCC reports that 'most models suggest that rising temperatures, drought, and fires will lead to forests becoming a weaker sink or a net carbon source before the end of the century'.[153]

3.11.9 *Human Health*

Scattered through this and previous chapters has been mention of the health impacts of climate change, particularly on those who are most vulnerable. There will inevitably be an increase in trauma associated with extreme weather events such as violent storms and storm surges, but there will also be increased heat stress and disease associated with a warmer and wetter climate, along with a potential rise in the number of people suffering from malnutrition and concomitant illnesses.

Modelling reported in Chapter 11 of IPCC WG2 AR5 predicts that by 2050 more than half of afternoon work hours will be lost because temperatures will be too high for productive work; whilst another study from 2013 indicates that under the RCP4.5 scenario productivity globally will fall by around 20 percent because of heat stress.[154] One study, reported in the WHO publication which provides a '*Quantitative risk assessment of the effects of climate change on selected causes of death* (in the) *2030s and 2050s*', predicts that there will be over a quarter of a million additional deaths due to 'heat-related mortality' in 2050.[155]

In its checklist providing '10 facts on climate change and health', dated April 2015, the WHO concludes that 'more variable rainfall patterns are likely to

[152] https://archive.ipcc.ch/pdf/assessment-report/ar5/wg2/WGIIAR5-Chap4_FINAL.pdf

[153] https://archive.ipcc.ch/pdf/assessment-report/ar5/wg2/WGIIAR5-Chap4_FINAL.pdf

[154] https://archive.ipcc.ch/pdf/assessment-report/ar5/wg3/ipcc_wg3_ar5_chapter11.pdf

[155] https://www.who.int/teams/global-malaria-programme/reports/world-malaria-report-2023

compromise the supply of fresh water. Globally, water scarcity already affects four out of every 10 people. A lack of water and poor water quality can compromise hygiene and health. This increases the risk of diarrhoea which kills approximately 2.2 million people every year, as well as trachoma (an eye infection that can lead to blindness) and other illnesses'.[156] In its publication on mortality and climate change (referred to above), the WHO predicts that deaths associated with diarrhoeal disease that can be attributed to climate change of children aged 15 and under, will reach between 35,300 and 86,300 in 2030, falling away to between 19,800 and 81,300 in 2050, depending on temperature rise and economic growth. This decreasing trend is the result of a reduction in the underlying burden of disease due to predicted improvements in child health. The impacts are likely to be particularly concentrated in southern Asia and sub-Saharan Africa.

The WHO Global Burden of Disease study as reported by Kristie Ebie in a paper on the '*Adaptation costs for climate change-related cases of diarrhoeal disease, malnutrition and malaria in 2030*' includes estimates of the excess cost of interventions, taking into account the increased numbers of cases due to climate change using a business as usual model—or 'unmitigated emissions'.[157] The maximum costs predicted in 2030 are 9 million US dollars for diarrhoea, US$ 166.2 million for malnutrition and US$ 8.78 million for malaria.

According to the 2023 WHO World Malaria Report fatalities from malaria in 2022 were 608,000, compared with 864,000 in 2000. We saw earlier how projections of human exposure to malaria risk are heavily influenced by both GDP per capita and climate change. This is further exemplified by research quoted by the WHO which concluded that 'climate factors have a substantial effect on malaria transmission in countries where GDP per capita is less than US$ 20,000'.[158]

There has been an unprecedented international effort to tackle malaria in the developing world and hence between 2000 and 2022, the scale-up of interventions helped to reduce malaria incidence cases by 2.1 billion, with the largest reduction being in the WHO African Region,[159] and with an associated reduction in mortality of 11.7 million deaths.

On the assumption that these interventions continue and that they remain effective an optimistic forecast might see the battle against malaria succeeding. According to an earlier WHO World Malaria Report if 'GDP changes as projected by economic models but the global climate remains constant, then much of the world, with the notable exception of Africa, is projected to become free from malaria by the 2050s'. Unfortunately for much of Africa 'the social burden of malaria has been proposed as an important factor retarding economic growth. In one study (albeit dating from 2001) countries with intensive malaria grew by 1.3% per year less than other

[156] No longer available on line

[157] http://www.ncbi.nlm.nih.gov/pmc/articles/PMC3335675/

[158] https://www.who.int/teams/global-malaria-programme/reports/world-malaria-report-2023

[159] WHO African Region is all of Africa apart from Egypt, Sudan, Morocco, Libya, Somalia, Tunisia and Western Sahara

countries'.[160] This same study reported the economic impact of malaria on Africa as being around 12 billion US dollars per year. It is this self-perpetuating brake on economic growth that is probably the main mitigating factor against wiping out malaria and other diseases linked to poverty.

Another important vector-driven disease is dengue fever which, like malaria, is transmitted by mosquitos, although the genus involved is more likely to be found in an urban environment, having adapted very well to breeding in water tanks. The mild form produces 'flu like symptoms, whilst the rarer severe form, known as dengue haemorrhagic fever, is more likely to be fatal. Although the WHO quotes figures from a paper published in 2014 estimating that some 3.9 billion people were exposed to the risk of contracting dengue fever, only around half a million were reported as having been hospitalised each year, with some 12,500 fatalities. Elsewhere WHO predicts that the number exposed to risk will increase to around 4.6 billion by 2050, with around two-thirds in southern Asia and sub-Saharan Africa. However, additional deaths attributable to dengue fever linked with climate change are predicted to be in the low hundreds worldwide.[161]

Despite the extraordinary number of people exposed to the risk of dengue fever worldwide, it has largely been overlooked as a public health issue. Hence there has been only limited investigation of its economic impact, apart from a study of dengue cases in Puerto Rico by Halasa et al. published in 2012.[162] Dengue fever is endemic in Puerto Rico which, with a population of only around 3.5 million had 22,648 confirmed cases between 2002 and 2010, of which less than 370 were haemorrhagic, but which drained 418 million US dollars from the Puerto Rican economy, nearly half of which was a direct burden on households.

Malnutrition is not strictly speaking a disease, but an outcome, and the IPCC prefers to use the term 'undernutrition', since malnutrition includes the effects of eating too much as well as too little. It is closely linked with diarrhoeal disease as well as socioeconomic and environmental factors. The IPCC reports on modelling which indicates that on average globally, without climate change, child undernutrition would fall by some 23.4% by 2050 compared with the year 2000.[163] This research, however, ignores the albeit countervailing impacts of reduced productivity as temperatures rise against the beneficial effects of rising carbon dioxide concentrations on crop productivity. Taking into account climate change, this same model predicts that child undernutrition will fall from 147.9 million in 2000 to 138.5 million in 2050: a reduction of only 6.4%. As with other health impacts referred to above the exception to this pattern is sub-Saharan Africa, which is predicted to face increased child undernutrition with or without climate change: rising by 60% and 27.5%, respectively.

---

[160] https://www.who.int/publications/i/item/9789241564694

[161] https://www.who.int/teams/global-malaria-programme/reports/world-malaria-report-2023

[162] http://www.who.int/mediacentre/factsheets/fs117/en/

[163] https://archive.ipcc.ch/pdf/assessment-report/ar5/wg3/ipcc_wg3_ar5_chapter11.pdf

3.11.10 Surface Air Pollution

We saw in Chap. 2 how anthropogenic sources of pollution can combine with weather conditions to create pollution events that may have serious impacts on human health. We have also seen how anthropogenic emissions of greenhouse gases can accumulate in the atmosphere and create warming at the planet surface. Some pollutants are accumulating in the atmosphere at the earth's surface through a combination of global warming, anthropogenic emissions and biogenic emissions.

3.11.10.1 Ozone

Although ozone is well known for its role in the atmosphere for absorbing ultraviolet light from the sun, its concentration in the air that we breathe is also important. Exposure at low concentrations can have severe health effects, particularly on lung function, resulting in inflammation, increased lung permeability, susceptibility to infection and asthma attacks in sensitised individuals. Prolonged or excessive exposure can lead to hospitalisation and even death.

According to data published in 2014 by the Institute for Health Metrics and Evaluation (IHME) at the University of Washington as part of the Global Burden of Disease (GBD) project, there were more than 150,000 deaths worldwide from exposure to ambient ozone in 2010, as well as nearly 2.5 million DALYs (disability-adjusted life-years) lost. 'The DALY is based on years of life lost from premature death and years of life lived in less than full health'.[164]

A retrospective study carried out by deDear et al. on the causes of death in France from the 2003 European heat wave that we referred to earlier concludes that:

> Elevated minimum and maximum temperature and peak ozone all increase mortality, with substantial interaction effects between temperature and ozone. Researchers expect extreme weather events, along with elevated ozone levels and temperatures, to become more common. Our results suggest that ozone will be an important contributor to the adverse health effects of such events.[165]

Ozone is a highly reactive gas created from precursor compounds and ultra-violet light through photochemical processes. A combination of higher temperatures and water vapour will tend to reduce its generation, whilst increased emissions of precursors, such as nitrogen oxides (NO_x), methane (CH_4), carbon monoxide (CO) and volatile organic compounds (VOCs), may increase its concentration near the earth's surface in a warming world. For example, modelling summarised in Chapter 11 of the IPCC Working Group 1 Fifth Assessment Report predicts that ozone concentrations under the RCP8.5 scenario could increase by an average globally of up to 5 parts per billion (ppb) towards the end of the twenty-first century, compared with

[164] http://www.tandfonline.com/doi/abs/10.3200/AEOH.60.4.205-212

[165] http://www.ncbi.nlm.nih.gov/pmc/articles/PMC2556651/

the 1986 to 2005 mean.[166] This predicted rise is due to a combination of emissions of methane and VOCs from vegetation and soils, the thermal decomposition of nitrates into NO_x and emissions from power stations, industry and wildfires.

Figures published in 2008 indicate that global mean annual concentrations of ozone ranged between approximately 30 and 55 ppb (58 and 108 μg/m^3), whilst the WHO Air Quality Guidelines, revised in 2021, recommended a maximum average exposure over 8 h of 100 μg/m^3, with an average monthly peak season value of 60 μg/m^3.[167] It is clear that exposure to potentially unhealthy levels of ozone could increase markedly as the impacts of climate change accrue.[168]

3.11.10.2 Particulate Matter

Another pollutant of global concern is fine particulate matter, with the greatest attention focussed on particles smaller than 2.5 microns, known as $PM_{2.5}$, which can penetrate into the depths of the human lung. Health effects overlap with those for ozone in that exposure can aggravate asthma, cause respiratory symptoms, cardiovascular diseases and even lung cancer. The 2019 Global Burden from Disease Study reported that 15% of deaths from lung disease were due to exposure to $PM_{2.5}$ (compared with 62% from tobacco smoking).[169] The WHO Air Quality Guidelines dating from 2021 recommend a limit of 10 micrograms per cubic metre (μg/m^3) for average annual exposure to $PM_{2.5}$ with no more than 3 days reaching a 24 h average of 25 μg/m^3 in any 1 year. For $PM_{10,}$ the Guidelines are 20 μg/m^3 and 50 μg/m^3, respectively, for the same exposure. The equivalent US Federal standard for $PM_{2.5}$, set by the Environmental Protection Agency, was tightened to 12 μg/m^3 (annual average) and 35 μg/m^3 (24 h average) at the end of 2012.

An important source of these fine particulates is wildfire, with one international study by Johnston et al. published in 2012 estimating that some 339,000 premature deaths per year were attributable to exposure to PM2.5 emanating from 'landscape fires' globally for the years 1997 to 2006. This included 157,000 in sub-Saharan Africa, whilst the greatest emissions occurred during El Niño years.[170]

Although, as we saw above, it is expected that the incidence of 'meteorological fire danger' will increase as the planet warms the models summarised in the relevant IPCC AR5 chapter indicate a steady reduction in $PM_{2.5}$ concentrations during the twenty-first century, even under the RCP8.5 scenario.[171]

---

[166] https://archive.ipcc.ch/pdf/assessment-report/ar5/wg1/WG1AR5_Chapter11_FINAL.pdf

[167] https://iris.who.int/bitstream/handle/10665/345329/9789240034228-eng.pdf?sequence=1

[168] http://www.who.int/mediacentre/factsheets/fs313/en/

[169] https://www.sciencedirect.com/science/article/abs/pii/S1556086423006019

[170] https://pubmed.ncbi.nlm.nih.gov/22456494/

[171] https://archive.ipcc.ch/pdf/assessment-report/ar5/wg1/WG1AR5_Chapter11_FINAL.pdf

3.11.11 Poverty and Living Standards

Poverty can be defined in a number of ways. The most commonly used definition refers to extreme poverty in absolute terms as absence of money needed to meet basic needs. The World Bank uses an adjusted definition from the UN Sustainable Development Goals, defining extreme poverty as an income of less than 1.15 US dollars per day. It reports that in 2023 some 700 million people were attempting to survive on an income below this figure. Remarkably this represented a significant improvement on 1981 when some 52% of those in the developing world were living in poverty, whereas in 2010 it was 21%, representing some 1.2 billion earning less than $1.25. However, this improvement was largely due to the 'economic miracle' in China which saw the number in extreme poverty dropping from 835 million to 156 million between 1982 and 2010, whilst in sub-Saharan Africa the number more than doubled from 205 million to 414 million, which represented a slight decrease per capita since total population increased from around 400 million to 853 million during that period.

As we saw in Chap. 2 Sustainable Development Goals replaced the Millennium Development Goals in 2015, the former having resulted from the UN Millennium Declaration made by the United Nations General Assembly in September 2000, which included the statement:

> We will spare no effort to free our fellow men, women and children from the abject and dehumanizing conditions of extreme poverty, to which more than a billion of them are currently subjected. We are committed to making the right to development a reality for everyone and to freeing the entire human race from want.[172]

Although the Millennium Development Goal to 'halve, between 1990 and 2015, the proportion of people whose income is less than $1.25 a day'[173] has been met overall, sub-Saharan Africa remains a challenge, particularly since it is projected that its population will reach around 1.5 billion by 2050. Hence the Sustainable Development Goal to 'eradicate extreme poverty for all people everywhere' by 2030 presents an enormous challenge.[174]

Poverty and hunger are closely associated, and together open the door to a range of diseases associated with undernutrition, limited access to health care, open sewers, poor housing, poor air quality, limited and contaminated drinking water and compromised immune systems.

When examining the link between climate change and poverty the IPCC perhaps unsurprisingly conclude that 'individuals and households most vulnerable to climate hazards tend to be those with relatively low socioeconomic status'.[175] Whilst

---

[172] https://www.ohchr.org/en/instruments-mechanisms/instruments/united-nations-millennium-declaration

[173] https://www.sciencedirect.com/science/article/abs/pii/S1556086423006019

[174] https://sdgs.un.org/goals/goal6

[175] https://archive.ipcc.ch/pdf/assessment-report/ar5/wg2/WGIIAR5-Chap11_FINAL.pdf

looking to the future, Chapter 13 of its Working Group 2 Assessment Report 5 predicts that:

> Poor countries will face greater poverty as a result of climate change and extreme events (medium confidence), owing to location and low latitude high temperatures, anticipated further decline in adaptive capacity combined with reductions in agricultural productivity, greater inequality … and lower levels of education and large numbers of young dependents.[176]

Bearing in mind that the lack of access to safe drinking water and sanitation are key indicators of poverty, one of the Sustainable Development Goals (SDGs) is to 'by 2030, achieve universal and equitable access to safe and affordable drinking water for all' and '… achieve access to adequate and equitable sanitation and hygiene for all and end open defecation, paying special attention to the needs of women and girls and those in vulnerable situations'.[177]

Although some 2.3 billion people gained access to 'improved sources of drinking water' between 1990 and 2012, according to the United Nations MDG monitoring programme, in 2012 there remained 2.5 billion people without access to 'improved sanitation'. In both cases the word 'improved' refers to avoidance of contact with contaminants, such as faeces:

> An improved drinking-water source is defined as one that, by nature of its construction or through active intervention, is protected from outside contamination, in particular from contamination with faecal matter.
>
> An improved sanitation facility is defined as one that hygienically separates human excreta from human contact.[178]

However, under a 'business as usual scenario' with mid-range population growth, the Organisation for Economic Co-operation and Development (OECD) has predicted a reduction to about 1.4 billion worldwide in the number of people not having access to basic sanitation by 2050, albeit with a growing population.[179] This gives some indication of the extent of the challenge in meeting this SDG.

In its Executive Summary to Chapter 11 of the 5th Assessment Report Working Group 2 the IPCC considers that 'climate change will create new poor between now and 2100, in developing and developed countries, and jeopardise sustainable development. The majority of severe impacts are projected for urban areas and some rural regions in sub-Saharan Africa and Southeast Asia (medium confidence, based on medium evidence, medium agreement)'.

---

[176] https://archive.ipcc.ch/pdf/assessment-report/ar5/wg2/WGIIAR5-Chap13_FINAL.pdf

[177] https://sdgs.un.org/goals/goal6

[178] https://www.who.int/data/nutrition/nlis/info/improved-sanitation-facilities-and-drinking-water-sources#:~:text=Improved%20sanitation%20facilities%20are%20defined,with%20slab%20or%20composting%20toilets

[179] http://www.oecd.org/env/indicators-modelling-outlooks/49846090.pdf

3.11.12 Impacts on Cities, the Built Environment and Infrastructure

We have already touched on the vulnerability of the built environment to extreme weather events and such ongoing climate-related phenomena as permafrost melt. We will also be examining the environmental impact of buildings and infrastructure in later chapters. Although the human and ecological cost of climate change is potentially devastating, the impact on human-made structures represents a massive economic burden, particularly for developing countries.

According to the 2014 report on *World Urbanization Prospects* from the UN's Department of Economic & Social Affairs (DESA) Population Division the percentage of people living in urban areas globally in 1950 was 30%, increasing to 54% by 2014 and potentially 66% by 2050.[180] It predicts that more than a third of this growth will occur in India, China and Nigeria. The world's urban settlements will have to accommodate around 2.5 billion more people in 2050 compared with 2014.

Most of the so-called mega-cities, having populations of 10 million or more, are located on the coast, and will therefore tend to suffer from the effects of storm surges and coastal flooding. Hurricanes and cyclones also tend to hold their greatest threat to human-made structures along coasts, since their energy is dissipated as they cross land.

In general terms buildings are likely to have to cope with changes in ambient conditions, such as higher temperatures and humidities, which may create discomfort in some passively ventilated buildings or render air conditioning systems incapable of meeting comfort conditions. Drainage systems may not be able to cope with the higher intensities of precipitation, whilst energy infrastructure may have to be reinforced to meet higher demand.

Transport infrastructure is likely to be impacted directly by extreme weather events creating flooding, landslips, sink holes, subsidence, buckling of rails and damage to bridges. Shipping and the operation of commercial ports are particularly vulnerable to extreme weather events. According to the IPCC 'more than US$3 trillion in port infrastructure assets in 136 of the world's largest port cities are vulnerable to weather events'.[181]

References

1. IPCC. (2021). Summary for policymakers. In V. Masson-Delmotte, P. Zhai, A. Pirani, S. L. Connors, C. Péan, S. Berger, N. Caud, Y. Chen, L. Goldfarb, M. I. Gomis, M. Huang, K. Leitzell, E. Lonnoy, J. B. R. Matthews, T. K. Maycock, T. Waterfield, O. Yelekçi, R. Yu, &

[180] https://population.un.org/wup/

[181] https://archive.ipcc.ch/pdf/assessment-report/ar5/wg2/WGIIAR5-Chap10_FINAL.pdf

B. Zhou (Eds.), *Climate change 2021: The physical science basis* (Contribution of Working Group I to the Sixth Assessment Report of the Intergovernmental Panel on Climate Change) (pp. 3–32). Cambridge University Press. https://doi.org/10.1017/9781009157896.001

2. IPCC. (2022). Summary for policymakers. In P. R. Shukla, J. Skea, A. Reisinger, R. Slade, R. Fradera, M. Pathak, A. Al Khourdajie, M. Belkacemi, R. van Diemen, A. Hasija, G. Lisboa, S. Luz, J. Malley, D. McCollum, S. Some, & P. Vyas (Eds.), *Climate change 2022: Mitigation of climate change* (Contribution of Working Group III to the Sixth Assessment Report of the Intergovernmental Panel on Climate Change). Cambridge University Press, Cambridge. https://doi.org/10.1017/9781009157926.001
3. Bassis, J. N., Berg, B., Crawford, A. J., & Benn, D. (2021). Transition to marine ice cliff instability controlled by ice thickness gradients and velocity. *Science, 372*(6548), 1342–1344. https://doi.org/10.1126/science.abf6271
4. Nicholls, R. J., & Cazenave, A. (2010). Sea-level rise and its impact on coastal zones. *Science, 328*(5985), 1517–1520.
5. Watkiss, P. (2015). The cost of climate change in Europe. In K. W. Steininger, M. Konig, B. Bednar-Friedl, L. Kranzl, W. Loibl, & F. Prettenthaler (Eds.), *Economic evaluation of climate change impacts* (pp. 9–29). Springer.

Chapter 4
Greenhouse Gases: Emitters, Consumers and Sinks

Few of those familiar with the natural heat exchanges of the atmosphere, which go into the making of climate and weather, would be prepared to admit that the activities of man could have an influence upon phenomena of so vast a scale .... Such influence is not only possible, but is actually happening at the present time.

From: Guy Callendar, 1938

4.1 Introduction

In this chapter, we are going to examine in more detail the sources of the greenhouse gases that were discussed in the early part of Chap. 3 and their relative contributions to global warming. Greenhouse gases find their way into the atmosphere directly from either anthropogenic sources or natural processes. Any process or product that does not emit greenhouse gases directly into the atmosphere but results in emissions at some point in its life cycle is known as an indirect source. For example, the manufacture of foam thermal insulation requires energy to be used in its manufacture: from extraction of its raw materials through various stages of processing, storage, packaging and eventual delivery to site, installation and disposal; including fuel used in transportation at various stages. Some types of foam insulation are blown using gases such as hydrofluorocarbons (HFCs) which typically have a global warming potential more than a thousand times greater than carbon dioxide. These gases find their way into the atmosphere both through fugitive emissions during manufacture and outgassing as the foam ages and are eventually disposed of. The total emissions from extraction of raw materials through to disposal are known as the lifecycle greenhouse gas (GHG) emissions. Virtually every product or process has an associated lifecycle GHG emission.

The main sources for the direct emission of GHG to the atmosphere are: combustion of fossil fuels for electricity and heat generation; burning of waste; leakage and flare offs associated with fossil fuel extraction and transmission; evaporation and

P. Appleby, *The Overburdened Planet*,
https://doi.org/10.1007/978-3-032-01510-5_4

outgassing associated with manufacturing; transport; emissions from biodegradation, animals and animal wastes; sewage and wild fires.

Many of these sources also provide energy for servicing buildings and industrial processes, including food processing; processing waste and sewage; treating and distributing water; agriculture, forestry and other land use (AFOLU). These are sometimes referred to as indirect sources of GHG.

Figure 3.4 in the previous Chapter shows the percentage contribution of each of the main sources of GHG in 2016. In 2022 the total GHG emissions (as represented by the full area of the outer ring of the pie chart) were 53.8 gigatonnes of carbon dioxide equivalent ($GtCO_{2eq}$), of which approximately 36.3 Gt was from the combustion of fossil fuels.[1] The contribution of energy supply to GHG emissions, including flaring, fugitive and other emissions associated with extraction and distribution of fossil fuels came to nearly 35% of the total.

In the Executive Summary to the International Energy Agency's *World Energy Outlook* for 2014, the authors concluded that 'in order to limit (global mean surface) temperature increase to 2 °C—the internationally agreed goal to avert the most severe and widespread implications of climate change–the world cannot emit more than around 1,000 gigatonnes of CO_2 from 2014 onwards'.[2] If GHG emissions continued at the rate for 2010 this budget would be used up by around 2034—or by 2040 in the IEA's 'central scenario'.[3] Examining the carbon content of fossil fuel reserves and resources the IPCC states in Chap. 7 of WG3 5th Assessment Report that 'the estimate of the total fossil fuel reserves and resources contains sufficient carbon, if released, to yield radiative forcing above that required to limit global mean temperature change to less than 2 °C'.[4] Indeed according a 2013 report by the IEA:

> Policies that have been implemented, or are now being pursued, suggest that the long-term average temperature increase is more likely to be between 3.6 °C and 5.3 °C (compared with pre-industrial levels), with most of the increase occurring this century.[5]

---

[1] https://www.iea.org/reports/global-energy-review-co2-emissions-in-2021-2

[2] https://unfccc.int/files/documentation/submissions_from_parties/adp/application/pdf/international_energy_agency._redrawing_the_energy-climate_map._world_energy_outlook_special_report._submitted_by_the_u.s.pdf

[3] https://www.ipcc.ch/site/assets/uploads/2018/02/ipcc_wg3_ar5_chapter1.pdf

[4] https://www.ipcc.ch/site/assets/uploads/2018/02/ipcc_wg3_ar5_chapter7.pdf

[5] https://unfccc.int/files/documentation/submissions_from_parties/adp/application/pdf/international_energy_agency._redrawing_the_energy-climate_map._world_energy_outlook_special_report._submitted_by_the_u.s.pdf

4.2 Energy, Fuel Reserves and Resources

When examining sources of energy that may be available for exploitation by humankind, it is important to understand the differences between energy resources and reserves. This is analogous to the money that a person might possibly earn in the future (resources) and money that they actually have in the bank (reserves). The definitions used in the Global Energy Assessment have been adopted by the IPCC as follows:

> Resources are detected quantities that cannot be profitably recovered with current technology, but might be recoverable in the future, as well as those quantities that are geologically possible, but yet to be found. Undiscovered resources are what remain and, by definition, one can only speculate on their existence.
>
> Reserves are generally those quantities that geological and engineering information indicate with reasonable certainty … can be recovered in the future from known reservoirs under existing economic and operating conditions.[6]

Renewable energy resources represent the estimated maximum technological potential, whilst reserves represent what would be economically viable at a given point in time.[7]

4.2.1 *Fossil Fuels*

The main types of coal are lignite, sub-bituminous coal and bituminous coal. Lignite is sometimes referred to as 'brown coal' and is coal's closest relation to peat, found closest to the surface, which means it is mostly strip mined (see Fig. 4.1), and has the lowest heat content, highest moisture content and greatest carbon emissions of the three.

In simple terms coal is graded or ranked from soft and crumbly lignite to hard and rock-like anthracite, depending on the amount of pressure the original material has experienced, resulting in metamorphosis from vegetation through peat that was buried and subsequently compressed and heated, losing moisture and volatile gases as it slowly hardened into coal. The greater the pressure the harder the material became and hence lignite tends to be the most recently formed coal, mostly during the Tertiary period (otherwise known as the Paleogene and Neogene periods) between 2.6 million and 65.5 million years ago. Most of the seams of bituminous, sub-bituminous and anthracite coal were formed during the Carboniferous period: 360–290 million years ago, although lesser amounts continued to be formed through into the Tertiary period. Anthracite tends to be found where there has been

[6] https://pure.iiasa.ac.at/id/eprint/10061/

[7] https://www.ipcc.ch/site/assets/uploads/2018/02/ipcc_wg3_ar5_annex-iii.pdf

Fig. 4.1 Lignite strip-mine in Garzweiler, Germany (From: https://commons.m.wikimedia.org/wiki/File:Open_pit_lignite_mining_in_Garzweiler,_Germany_03.jpg)

considerable pressure from earth movement and upheavals during the formation of mountains and hills.[8]

Because lignite has a low energy density, it is usually not cost effective to transport great distances, so historically strip mines have tended to serve nearby power stations. Sub-bituminous coal ranges in its properties between lignite and bituminous coal: that is from brown and crumbly to blacker and harder, with a corresponding variation in energy density, moisture content and CO_2 intensity. Despite being harder and having fewer impurities and a lower moisture content, anthracite has a significantly higher carbon intensity and lower moisture content than other coals. It only represents some 1% of total coal production globally and is not an important fuel for power generation, partly because it is up to three times more expensive per tonne to purchase.

According to the data published in Chap. 7 of the IPCC WG3 5th Assessment Report there was up to 11,230 billion tonnes of coal carbon content (Gt C) to be found as a potential resource globally in 2012, but with reserves at between 446 and 542 billion tonnes carbon (Gt C) estimated to be sufficient for 100 years, it is unlikely that this huge resource will ever be exploited. The Total Primary Energy Supply (TPES) in 2010 using coal and peat was 145.52 Exajoules (EJ = 10^{18} Joules),

[8] https://www.worldenergy.org/assets/images/imported/2013/10/WER_2013_1_Coal.pdf

whilst the energy content of reserves in 2012 were reported as being up to 21,000 EJ, or around 144 times the supply for the whole of 2010.

Gas has the lowest carbon intensity of all the fossil fuels and has the lowest concentration of pollutants associated with combustion. As we saw in Chap. 2, there is a huge international trade in gas served by a complex infrastructure of pipelines, along with a growing market in liquefied gas carried by ship. Most gas is derived from natural sources, with the so-called unconventional gas reserves exceeding conventional ones by a factor of ten, by some estimates.[9] Unconventional gas consists primarily of shale gas, along with deep formation and tight gas and coal-bed methane.

Before the surge in the development of unconventional gas sources such as shale fields, natural gas was primarily either drawn off as a by-product from oil fields, having previously been flared-off and lost to atmosphere, or drawn from gas fields. In the energy sector, these are known as associated and non-associated gas developments, respectively. The gas recovered from associated sources is linked to the life-cycle of the oil field, whilst non-associated fields usually require the costly application of compression gases and separation of water and contaminants before distribution.

According to the IEA publication *Resources to Reserves 2013: Oil, Gas and Coal Technologies for the Energy Markets of the Future* 'the IEA divides unconventional gas into four broad categories:

- Tight gas is natural gas trapped in extremely low-permeable and low-porous rock, sandstone or limestone formations. Such gas may contain condensates;
- Shale gas is natural gas contained in organic-rich strata dominated by shale. Because of the types of reservoir, it is sometimes considered a sub-category of tight gas;
- Coal-bed methane (CBM) is methane adsorbed on to the surface of coal within coal seams;
- Methane hydrates are made up of methane molecules trapped in a solid lattice of water molecules under specific conditions of temperature and pressure'.[10]

Natural gas is also re-injected into oil fields to increase production of oil: this practice is particularly common in the Middle East and especially in Iran, which has the world's second largest reserves of natural gas after Russia.

Oil and gas have steadily become more difficult, and more expensive, to access. The IEA reports in its 2013 book on *Resources to Reserves* referred to above, that in 2010 'more than half of all conventional oil discovered since 2000 was in deep-water offshore reservoirs and the Arctic is one of the world's largest remaining prospective areas for oil and gas. Developing resources in deepwater and ice-prone areas in the Arctic can be considered the most complex and expensive challenges

[9] https://www.ipcc.ch/site/assets/uploads/2018/02/ipcc_wg3_ar5_chapter7.pdf

[10] https://iea.blob.core.windows.net/assets/afc6bec5-22e8-4105-a1b3-ceb89eaec1e9/Resources2013.pdf

currently facing the oil industry'.[11] Since the disastrous Macondo well blowout, more popularly referred to as the Deepwater Horizon oil spill, in the Gulf of Mexico in April 2010, and the less well known Montara oil spill in the Timor Sea off the coast of Western Australia some 8 months previously, there has been a much greater focus on the safety and potential environmental impacts of oil exploration in sensitive and fragile environments. The Macondo disaster resulted in 11 fatalities from the initial explosion and some 4.9 million barrels (668,000 tonnes) of oil being discharged over a 3 month period, spreading over an area of more than 60,000 km^2. This resulted in both short- and long-term damage to wildlife and ecosystems, including some 3000 oil contaminated sea birds, stranding of turtles and dolphins, along with serious damage to deep coral reefs, fish and invertebrates, interference with the marine food web and contamination of coastal wetlands. The cost of the clean-up, the economic cost to businesses impacted, the loss of ecosystem services and the fines imposed on BP, are still accruing at the time of writing, but will ultimately be many billions of dollars.

In a similar manner to gas, oil that has to be extracted by means other than tapping into free flowing underground sources is known as 'unconventional oil'. This includes kerogen shale (or oil shale), oil sands (or tar sands), light tight oil (LTO) and oil derived from coal, gas or biomass. Kerogen refers to the organic matter found in sedimentary rocks either derived from rotting vegetation (humic), resulting in the formation of peat and coal or from decomposed aquatic organisms at the bottom of lakes or oceans (sapropelic), resulting in the creation of oil and natural gas. This is why most coal fields are found underground and a large proportion of oil and gas deposits under seabeds. Oil and gas bearing rock found below dry land are the result of the deposition of marine plankton, algae and other microorganisms, when these regions were covered in oceans. Where these deposits are found near to the earth's surface, such as the so-called tar sands in Alberta (Fig. 4.2) and Venezuela, the massive upheavals that formed the Rockies and Sierra Orientale forced the oil into sand deposits left behind by ancient river beds. The oil thus formed is mostly 'extra heavy crude' or bitumen, which is too thick to flow and thus has to be mined like coal or heated before extraction.

As can be seen in Fig. 4.2 the surface deposits in Alberta are essentially strip mined, involving the clearance of some 900 km^2 of boreal forest and peatlands up to 2015, but with a total mineable area of 4800 km^2. The amount of energy required to liquefy this solid oil results in greenhouse gas emissions between 1.75 and 2.75 times those for conventional oil and between 50% and 80% of the GHG emissions associated with combustion of the end product.[12] The deeper oil sands that constitute the greatest proportion of the 142,000 km^2 in Alberta, as well as the fields in Venezuela, require even more energy-intensive high-pressure steam injection to liquefy the oil and flush it to the surface.

[11] https://iea.blob.core.windows.net/assets/afc6bec5-22e8-4105-a1b3-ceb89eaec1e9/Resources2013.pdf

[12] https://www.ipcc.ch/site/assets/uploads/2018/02/ipcc_wg3_ar5_chapter7.pdf

Fig. 4.2 Oil tar sands excavations in Alberta, Canada (From: https://commons.wikimedia.org/wiki/File:Tar_sands_in_alberta_2008.jpg)

As we saw in the section on *Environmental impact* in Chap. 2, where oil or gas resources are trapped in porous rock similarly extreme methods are being used to open up the pores and propel the oil or gas to the surface, but using high pressure water rather than steam. Hydraulic fracturing or fracking has resulted in a boom in natural gas production in the USA and elsewhere. Shale gas production increased by a factor greater than six between 2007 and 2013 in the USA, surpassing conventional gas production for the first time. Global tight gas reserves were around 40% of those confirmed for shale gas in 2011.[13]

Tight oil which, like tight gas, is extracted by fracking sandstone, has also been steadily increasing its contribution to oil production in the USA, such that by 2035 it is expected to constitute up to two-thirds of total domestic production.[14] Shale was originally clay and sandstone sand: as the former settled and was heated during the formation of oil and gas, cracks formed, whilst sand settled more evenly with near spherical grains creating a lower porosity. Hence shale beds tend to release their load more easily and more rapidly than does sandstone.

---

[13] https://iea.blob.core.windows.net/assets/afc6bec5-22e8-4105-a1b3-ceb89eaec1e9/Resources2013.pdf

[14] http://www.eia.gov/todayinenergy/detail.cfm?id=15731

Based on 2010 figures for the whole planet, only around 21% of crude oil and its products were used for non-transport energy during that year, whilst 62% fuelled transport and 17% made its way into non-energy uses such as manufacture. Globally oil's share in electricity generation was around 4.62% and gas 22.25%.

4.2.2 *Nuclear Power*

Apart from military applications nuclear technologies are primarily used in the generation of electricity and, although no greenhouse gases are emitted as part of the generation process, there are lifecycle emissions associated with the extraction, processing and transport of uranium, the construction, operation and decommissioning of nuclear power stations and the processing and storage of waste. However, median values for these are similar to those associated with wind turbines and between 25% and 50% of the lifecycle GHG for solar and hydroelectricity installations.

On the other hand, uranium is not a limitless resource. The life expectancy of known reserves is likely to depend on the price of uranium, hence authorities disagree, but one estimate from 2024 suggests resources should last 90 years if the rate of consumption does not increase and at economic extraction costs.[15] This would be doubled if the cost of uranium increases to a level at which all spent fuel is reprocessed and uranium and plutonium is recovered from used fuel.[16] If all the 434 reactors operating in 30 countries worldwide in 2013 were replaced with fast breeder reactors (FBRs), the life expectancy of uranium reserves would increase by a factor of 50. However, because FBRs have so far proved to be some 25% more expensive than light water reactors (LWRs) most countries abandoned their development programmes once it became obvious that abundant sources of uranium were available.

The light water reactor is by far the most common reactor used for nuclear power generation worldwide, representing more than 80% of those operating in 2013, and with more than three quarters of these being of the pressurised water type (PWR). 'Light water' refers to the fact that ordinary water is used for cooling and moderation of the neutron bombardment, and to differentiate it from a heavy water reactor (HWR), which uses deuterium oxide as a coolant and which contains an additional neutron in its hydrogen nuclei (H_3O). However, LWRs require slightly enriched uranium, a controversial process since it represents one of the steps in the manufacture of nuclear weapons. HWRs do not require enriched uranium as a fuel, but they cost more to build. 'Pressurised water' differentiates those reactors that use heat exchangers to transfer the heat from fission to turbines for electricity generation from the 'boiling water' type (BWRs) that generate steam directly to drive the turbines.

---

[15] https://world-nuclear.org/information-library/nuclear-fuel-cycle/uranium-resources/supply-of-uranium

[16] https://www.ipcc.ch/site/assets/uploads/2018/02/ipcc_wg3_ar5_chapter7.pdf

Unfortunately PWRs employ a "once-through" process that actually captures only 1% of the energy potentially available from the uranium, the remainder is locked into the fuel as plutonium: the main waste product from the PWR. This has resulted in an accumulation of plutonium in nuclear countries across the world, creating a major storage challenge: the isotope of plutonium most commonly found in FBRs has a half-life of 24,000 years. FBRs can use recycled plutonium as a fuel by accelerating the bombardment of neutrons, which requires the use of a liquid metal such as sodium as a moderator. At the time of writing a number of countries are considering a programme of FBR construction, primarily to resolve the plutonium storage problem.

4.2.3 Renewables

As a resource renewable energy is virtually limitless, although each individual technology is limited both in practical and economic terms. According to the 2012 Special Report of the Intergovernmental Panel on Climate Change: *Renewable Energy Sources and Climate Change Mitigation* (SRREN), 'studies have consistently found that the total global technical potential for renewable energy is substantially higher than global energy demand'.[17]

The extent to which each renewable technology will contribute to primary energy and replace fossil fuels in the future will depend on a combination of technical development, economics, political momentum and consequent policy implementation. It has become increasingly clear that the IPCC SRREN's conclusion that 'under most conditions, increasing the share of renewable energy in the energy mix will require policies to stimulate changes in the energy system' was prescient.[18]

The investment required in renewable energy as part of a strategy to keep atmospheric carbon dioxide equivalent concentrations below 450 ppm were predicted in 2009 as 5.1 trillion US dollars between 2011 and 2020, with a further investment between 2021 and 2030 of around $US 7.18 trillion.[19] These estimates are less than 1% of predicted world gross domestic product for the same periods. According to tracking reported on the Statista website, the actual amount spent from 2011 to 2020 inclusive was $2.85 trillion.[20] This is around half of that required to keep CO_2 levels below 450 ppm, according to the International Energy Agency (IEA) '450 Scenario', corresponding with a long-term peak mean surface temperature rise of 2°, approximating to the IPCC's RCP2.6 scenario (see section on *Temperature change* in Chap.

---

[17] http://srren.ipcc-wg3.de/report/IPCC_SRREN_Full_Report.pdf

[18] http://srren.ipcc-wg3.de/report/IPCC_SRREN_Full_Report.pdf

[19] http://srren.ipcc-wg3.de/report/IPCC_SRREN_Full_Report.pdf

[20] https://www.statista.com/statistics/186807/worldwide-investment-in-sustainable-energy-since-2004/

3). The Statista tracking indicates that in 2022 $495 billion was spent on renewables globally, representing a 36% increase compared to 2020.

As referred to in Chap. 3, the IEA has recognised this shortfall in its 2013 report *Redrawing the Energy-Climate Map*[21] which describes the 'New Policies Scenario' consistent 'with a 50% probability of an average global temperature increase of 3 °C by 2100 (compared with pre-industrial levels) and of 3.6 °C in the longer term'. In order to mitigate against this scenario the IEA used this report to introduce its "4-to-2 °C Scenario" which proposed a two stage approach to meeting the 2° target which it is suggested will have minimum impact on economic growth in the short term. We will be discussing the relationship between economic growth and climate change in our final Chapter. It is interesting to note that the IEA report pre-dates the COP21 Paris Agreement which committed the Parties to holding 'the increase in the global average temperature to well below 2 °C above pre-industrial levels' and pursue efforts 'to limit the temperature increase to 1.5°C above pre-industrial levels'.

4.3 Electricity Generation

The GHG emissions associated with electricity generation depend primarily on the carbon intensity of the fossil fuel used, if any, and the proportion of coal, gas, nuclear and renewable electricity generation that feeds into the supply, which varies from country to country and at any one time (see Fig. 4.3). As well as the direct

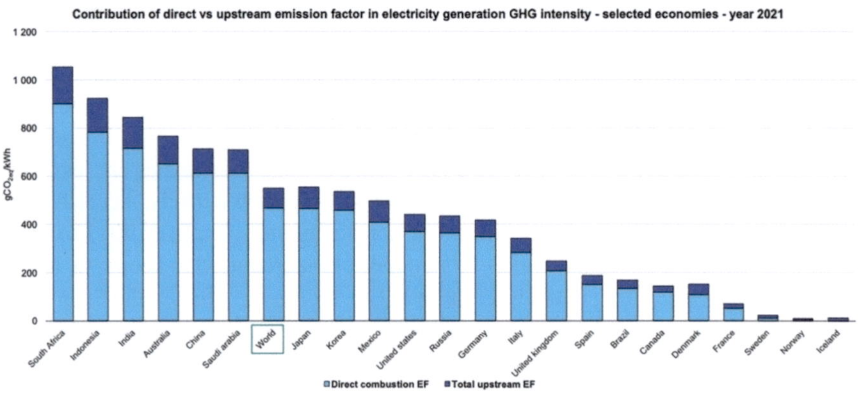

Fig. 4.3 Direct and upstream emissions from electricity generation for specific countries (Image: IEA: https://iea.blob.core.windows.net/assets/69b838f4-12ad-4f51-9155-9da6435b5d53/IEA_UpstreamLifeCycleEmissionFactors_Documentation.pdf)

greenhouse gas emissions associated with fuel combustion there are 'upstream'

[21] https://www.iea.org/reports/redrawing-the-energy-climate-map

emissions—whole life GHG emissions—associated with the construction, operation, maintenance and deconstruction of power stations and generation equipment (see Fig. 4.4). In 2021 the carbon intensity or direct emission factor in the UK electricity grid varied between 66 and 275 g of carbon dioxide per kWh of power generated (gCO_2/kWh)[22]; whilst in the same year, the average direct emission factor for electricity generated ranged from close to zero for Norway to around 880 gCO_2/kWh for South Africa (see Fig. 4.3). Electricity generation using fossil fuels in the United Kingdom fell by 67% between 2008 and 2023, including a 97% reduction in kWh associated with coal-fired power stations, whilst output from renewable electricity sources increased sixfold during the same period. Average direct GHG emissions in 2023 were 162 gCO_2/kWh, compared with 441 in 2011.

The reason for massive variations in emissions between individual countries is the different balance of energy generation. For example, most of Norway's electricity is obtained from hydroelectric sources, whilst South Africa derives around 80% of its electricity from carbon-intensive coal.

In general there has been a downward trend in GHG emissions associated with electricity generation globally: due to some combination of an increase in

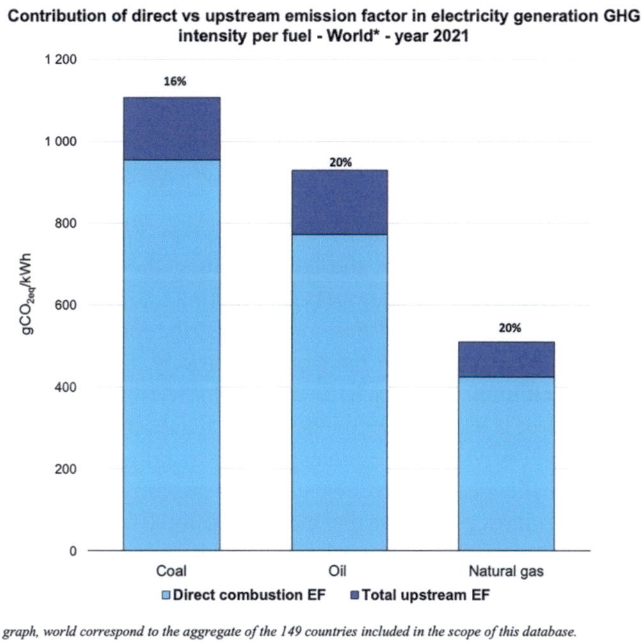

Fig. 4.4 World direct and upstream emissions from fossil fuel fired electricity generation (https://iea.blob.core.windows.net/assets/69b838f4-12ad-4f51-9155-9da6435b5d53/IEA_UpstreamLifeCycleEmissionFactors_Documentation.pdf)

[22] https://www.carbonbrief.org/analysis-uk-electricity-from-fossil-fuels-drops-to-lowest-level-since-1957/#:~:text=Overall%2C%20the%20electricity%20generated%20in,hour%20(gCO2%2FkWh)

renewable electricity generation; new nuclear power stations being constructed; old coal-fired power stations being decommissioned, typically being replaced by gas-fired plant; coal being replaced with biomass; more efficient (larger) coal-fired power stations; and the exploitation of non-conventional sources of lower carbon fossil fuels, such as shale gas. Much of this has been driven by the Kyoto protocol (see section on *The debate* in Chap. 2) and the subsequent CoP process, with nations having to meet international and national commitments to reduce greenhouse gas emissions.

Around 36% of global electricity production in 2023 used coal-firing, whilst gas had the second largest share, followed by hydropower and nuclear. Across most of the developed world electricity is generated in large power stations that have been constructed either close to fuel sources, near convenient bodies of water for heat rejection or in proximity to clusters of high demand intensity. For example power stations in the UK were originally built to meet a local demand: the first being constructed in 1881 in Godalming, Surrey, powered by a water wheel, followed shortly by installations in Brighton, Chesterfield and Norwich. Thomas Edison launched the first steam powered generator the following year 1882, at Holborn Viaduct in London. All of these small-scale experimental schemes provided direct current (DC) to street lighting. This led to a boom in relatively small generating plants either for private use or street lighting. The first power station in the world designed to supply three-phase alternating current (AC) for a wide range of uses was the Neptune Bank coal-fired power station near Newcastle upon Tyne, which opened in 1901, but in 1915, there were still some 600 separate undertakers in the UK, supplying a wide range of competing voltages, with no broad agreement on whether DC or AC should be the standard supply.

Electricity supply was gradually harmonised in the UK through Electricity Supply Acts in 1919 and 1925, the latter creating the Central Electricity Board (CEB) which established the National Grid, linking all the larger power stations to consumers via high voltage transmission and distribution lines, although many big industries carried on generating their own electricity. However nationalisation of electricity generation and supply by the Labour government following the Second World War resulted in a truly interconnected and standardised network that, although subsequently semi-privatised, continues into modern times.

One of the challenges of electricity supply in general has been matching generation to demand, through system balancing and load following. Most developed countries have some form of national grid or grids, sometimes interconnected with neighbouring countries. These grids are fed by various power stations ranging in size from individual grid-connected photovoltaic arrays and wind turbines having outputs that are weather dependant through to massive nuclear power stations such as Kashiwazaki-Kariwa in Japan rated at 7965 MW, or the Three Gorges Dam in China, the largest power generation facility in the world, rated at around 22,500 MW (see Chap. 2). As we saw above, as of 2010 coal-firing dominated electricity generation globally, with the largest coal-fired power station being in Taiwan and rated at 5500 MW, although this is steadily changing. Most countries have to match their generation capacity, including perhaps a variable component from renewable energy, with a continually varying demand. Historically larger coal-fired and nuclear plants provided baseload operation, whilst load following or 'peaking' has been provided by cycling more responsive gas or oil-fired plant and hydropower. Some hydroelectric installations have included the capability of using surplus electricity from the grid to pump water between lower and upper reservoirs, at night, for example, to be reversed later in the day to cater for peaks in demand.

However, where national policy has been to rely on an array of large single-fuel power stations, such as nuclear power in France, for example, which represents around 68% of total annual electricity production in 2021, at least some of the plants have to include an element of load following. PWR reactors typically allow a reduction to 30% of maximum output mostly through the use of control rods. However there is evidence that load following reduces the overall performance of the reactor, nearly doubles the rate of outages and is likely to reduce life expectancy.[23] Similarly coal-fired power plant not adapted for cycling can suffer fatigue and corrosion of boiler components; resulting in failure and down time after a year or two for older plant and between 5 and 7 years for new plant.

As can be seen in Fig. 4.5, as a means of converting fuel to electricity most power stations are extremely inefficient. The efficiency of a conventional coal-fired plant is typically around 32%, with a global average in 2015 of 33%.[24] However, efficiencies of up to 48% have been achieved with modern "ultra-supercritical" technology

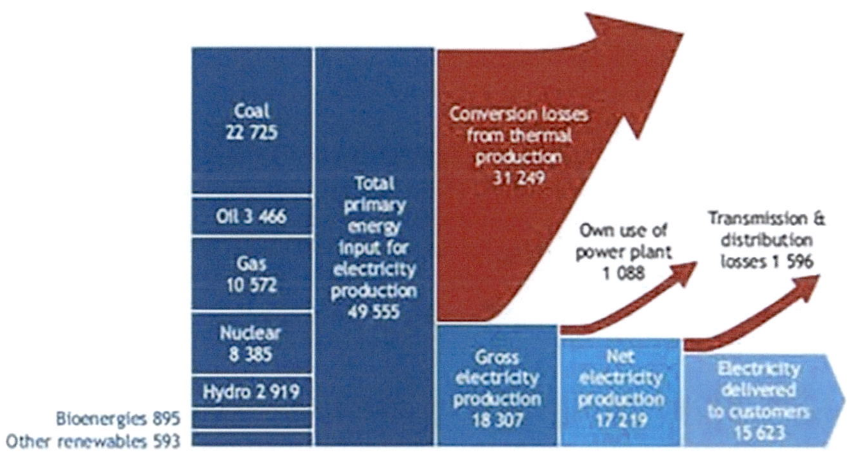

Fig. 4.5 Energy flows in the global electricity system for 2006 (TWh) (http://www.iea.org/media/files/chp/Energy_Flows_s.jpg). Note: Although the quantum of energy flows has changed year on year since 2006, this figure illustrates in general terms the large inefficiencies in electricity generation globally and the main routes of energy loss

[23] http://www.templar.co.uk/downloads/0203_Pouret_Nuttall.pdf

[24] https://www.worldcoal.com/coal/28082013/improving_coal_fired-power_plant_efficiency_a/

which employs temperatures and pressures of greater than 600 °C and 300 bar, respectively, to generate superheated steam. Similarly a basic gas turbine power station also has an efficiency of around 32%, although efficiencies as high as 60% are achievable with a combined cycle gas turbine, which incorporates a heat recovery steam generator in combination with a conventional turbine.

It has been estimated that if the average efficiency of coal-fired power stations existing globally in 2015 were increased from 33% to 40% then 2 $GtCO_2$ (billion tonnes of carbon dioxide) would be saved each year, roughly equivalent to the total emissions from all sources in India in 2013.[25] It is interesting to note that whilst embarking on a massive programme of constructing coal-fired power stations China were scrapping some of its older inefficient plant whilst installing large ultra-supercritical power stations, some of which have been achieving efficiencies exceeding 40%. However, since committing to achieving carbon neutrality by 2060, the Chinese authorities have instigated a programme to close down all of the more than 1000 coal-fired power stations by 2045.[26]

Figure 4.5 represents a snapshot of energy flows in 2006, since when the contribution to electricity generation from renewables has increased considerably, such that by 2023 it was 13.6%, or 30.2% including hydropower and bioenergy: reducing the conversion losses overall accordingly.[27] By 2023 gross electricity production had increased from around 18,000 TWh to nearly 29,500 TWh.

There are other inefficiencies that manifest themselves between the power station and the consumer, such as those associated with connected plant and equipment, whilst there are also significant losses within the transmission and distribution infrastructure. Losses occur through three independent effects: thermal, magnetic and dielectric. When electrical current has to combat a resistance, such as when passing along a cable or through the windings of a transformer or motor, heating occurs and energy is lost to the surroundings: known as the 'Joule effect' or I^2R losses (current squared x resistance). Heat is also generated in the magnetic core of transformers and motors due to an alternating magnetic field creating localised eddy currents and hysteresis in the magnetic field. Dielectric losses are caused by the dissipation of energy through the movement of charges in an alternating electromagnetic field as polarisation switches direction from positive to negative.

Transmission and distribution losses vary considerably globally depending on the age and efficiency of infrastructure, the distance between power stations and consumers, and the extent of pilfering from the network. Over the last 20 years in the UK, for example, losses have varied from 7.1% to 9.6% of total power generated, whereas in India losses have typically been around 25%. For some countries, such as Iraq and Haiti, transmission and distribution losses are reported as being in

[25] https://www.worldcoal.com/coal/28082013/improving_coal_fired-power_plant_efficiency_a/

[26] https://www.carbonbrief.org/china-should-rapidly-close-186-coal-plants-to-help-meet-its-climate-goals-study-says/

[27] https://ourworldindata.org/electricity-mix

excess of 50%.[28] Remarkably in 2010 the losses for Botswana were reported to be 62.59%, in part due to the rudimentary condition of the transmission lines, and widespread theft from the system.[29]

In India it has been suggested that around a third of the population were obtaining their electricity illegally, costing the Indian energy suppliers an estimated $US 16.2 billion per year, or nearly 20% of the total value of electricity theft globally as reported in research published in 2014.[30] The risks associated with stealing electricity are significant. The resultant tangle of wires creates a risk of fire and electric shock to both those attempting the theft and to anyone trying to rectify the faults that inevitably result.

Another impact of electricity theft is that lines are frequently having to cater for greater demand than their size permits, resulting in circuit breakers tripping and consequent power outage. In July 2012 this combined with a lack of resilience in the system such that 20 out of India's 28 states were without power for 2 days, impacting some 600 million people and costing the Indian economy more than $US 100 million:

> The July 30th blackout appears to have begun in Agra, Uttar Pradesh. The transmission lines were apparently carrying twice the permitted load. When the Agra-Gwalior line went down, the effect was a cascade, with lines tripping one after the other.[31]

4.4 Cogeneration

The efficiency of power stations can be vastly improved by capturing waste heat and using it for industrial or space heating. Remarkably the first cogeneration plant was built by Thomas Edison in New York in the same year that he built the first power station in London in 1882. The Pearl Street Station plant not only generated electricity but also provided heat to neighbouring buildings, achieving an efficiency of 50%, which is comparable with the global average for cogeneration power stations of 58% in 2010.[32]

The principle of cogeneration has been applied extensively at sizes ranging from residential scale combined heat and power units (micro-CHP) through district energy plants (see below) to some very large power stations, such as the 1365 MW gas-fired CHP plant on the Isle of Grain in Kent which supplies 350 MW of heat to the nearby National Grid liquid natural gas (LNG) terminal in order to vaporise liquid gas that has been shipped in from overseas at a temperature of −162 °C. This

[28] http://data.worldbank.org/indicator/EG.ELC.LOSS.ZS

[29] https://www.indexmundi.com/facts/botswana/electric-power-transmission-and-distribution-losses

[30] http://www.energylivenews.com/2014/12/10/electricity-theft-costing-90bn-a-year-globally/

[31] http://www.theautomaticearth.com/2012/08/india-power-outage-the-shape-of-things-to-come/

[32] https://www.ipcc.ch/site/assets/uploads/2018/02/ipcc_wg3_ar5_chapter7.pdf

plant was opened in 2010 and, according to its operator e.on, has subsequently achieved an efficiency of 72%.[33]

Essentially any plant that generates electricity rejects waste or surplus heat which can be used for a variety of purposes. There are many examples of power stations powered by gas, oil, coal, biomass and nuclear fission that have incorporated heat recovery. As well as LNG vaporisation, the heat from cogeneration has been used for space heating, domestic hot water, industrial processes, greenhouse heating, desalination, steam for recovery of unconventional oil, and the manufacture of heavy water (deuterium oxide) for nuclear reactor cooling. When the heat is used in the generator of an absorption chiller to provide cooling the process is referred to as trigeneration or combined cooling, heat and power (CCHP). Trigeneration may be used to provide both district cooling and chilled water for individual buildings (see below).

Cogeneration and trigeneration are employed in the generation of approximately 8% of the world's electricity supply, whilst in some countries, such as Denmark, the Netherlands and Finland it is significantly more than this. The field is led by Denmark, where more than 68% of electricity was generated from CHP plants in 2014, along with nearly 63% of private houses heated from district heating.[34] According to the same source, around 40% of fuel used for district heating in 2014 was biomass.

4.5 Community and District Energy

The first recorded example of supplying energy to multiple consumers from a single source appears to date from 1332 when hot water from a geothermal source was channelled via wooden pipes to some of the residents of Chaudes-Aigues in France. Consumers were able to draw off as much hot water as they required by adjusting a valve. The water was supplied direct from the geothermal source by gravity. Over the centuries this was extended across the whole town, heating public baths and private swimming pools, and Chaudes-Aigues has remained a spa town through to modern times.

The forerunner to the modern district heating system was installed by American hydraulic engineer and inventor Birdshill Holly who, in 1877 connected a steam boiler to pipework serving heating in a number of buildings in Lockport, New York, charging for the amount of heat used by metering the condensate return from each building. Holly patented his system and went on to oversee similar installations in Denver, Detroit, Springfield and elsewhere in the USA. By 1882 the New York Steam Company had installed 48 steam boilers and embarked on a programme of

[33] https://www.power-technology.com/projects/isleofgrain/?cf-view&cf-closed

[34] https://c2e2.unepccc.org/wp-content/uploads/sites/3/2023/02/regulation-and-planning-of-district-heating-in-denmark.pdf

laying pipework under 5th Avenue, supplying heat to the wealthy and powerful of New York.

Eventually Consolidated Edison (conEdison) took over the New York Steam Company, extending the network to some 160 km of underground steam pipes, connecting to more than 100,000 commercial and residential establishments across Manhattan. Most of the pipework that remains to this day dates from the first half of the twentieth century, much of it wrapped in the original asbestos insulation. From time to time a component has failed, with disastrous consequences; including 12 major steam explosions between 1987 and 2007. The 2007 explosion occurred at the junction of Lexington Avenue and 41st Street, propelling a tow truck 4 m into the air and profoundly injuring the driver and passenger. Others were hospitalised following the blast, whilst one woman died of a heart attack when fleeing the disaster area. The resultant steam jet was seen to rise well above the nearby 319 m tall Chrysler building and continued for some 2 h (see Fig. 4.6). The culprit was eventually found to be a condensate trap that was clogged by excess sealant emanating from a recently patched leak.

Because of the high pressures and temperatures normally associated with district heating, older installations are prone to leaks and inefficiency. Many of the earlier systems were poorly controlled and metered, resulting in considerable waste. Radiators fed directly with steam offer only limited controllability: either operating at full capacity or not at all, whilst exposed metal surfaces at 100 °C pose a significate scalding risk. Hence modern district heating systems generally incorporate heat

Fig. 4.6 New York steam explosion 2007—with the pinnacle of the Chrysler building to the right (From: http://news.bbc.co.uk/1/hi/in_pictures/6906069.stm)

exchangers at the building interface so that heat emitters can be supplied with low temperature hot water (LTHW), and with heat output controlled thermostatically.

The generation and supply of energy to multiple consumers takes many forms: for example, district energy schemes comprising some combination of heating, cooling and electricity generation (decentralised generation); whilst community energy schemes include some community involvement in funding and operation. The type of energy generation may include heating-only boilers, refrigeration plant supplying chilled water, or electricity generation from renewable sources, such as wind turbines or solar photovoltaics. Heating and/or cooling may also be produced from geothermal sources or from CHP/CCHP (see above), which also generates electricity, whilst cooling may be provided by the incorporation of absorption refrigeration (CCHP). Some neighbourhood schemes bundle together a number of other utility services with energy, such as waste management and recycling, sewage treatment and even composting.

As we saw above the earliest district energy schemes were built and operated by private companies and variations of this model have continued into modern times. District and community schemes are usually owned privately, sometimes with the involvement of local government. Across Europe, since the late 1980s, the Energy Service Company (ESCO) financial model has been widely adopted for the funding and operation of district and community energy schemes. More recently the multi-utility service company (MUSCO) model has been employed, where a bundle of services is provided under a single contract.

On its website the European Commission's Institute for Energy and Transport states that: 'what characterises (ESCOs) …. from the traditional energy consultants or equipment suppliers is the fact that they can also finance or arrange financing for the operation and their remuneration is directly tied to the energy achieved'.[35] This is commonly referred to as energy performance contracting (EPC), which is also used as a vehicle for energy efficiency improvements to existing buildings.

When applied to district or community energy projects the ESCO model can cover various services including design, financing, installation, commissioning, operation, maintenance, monitoring and energy management. The energy performance contract typically incorporates a guarantee of energy savings or carbon reduction, the financial savings from which can be used to repay monthly debt service costs, for example.

In the UK, the Government set up the Heat Networks Delivery Unit in 2013 to provide funding for the development of heat networks in England and Wales, including feasibility studies through heat mapping and project development. Up to the end of 2023, there were 13 funding rounds with bids awarded almost entirely to local authorities. In parallel with this the Heat Networks Investment Project was established in 2018 primarily to support the construction phase of a project, feeding into the Green Heat Networks Fund which ran between 2022 and 2024. To improve the efficiency of existing heat networks and communal heating schemes the Heat

[35] https://e3p.jrc.ec.europa.eu/node/190

Network Efficiency Scheme provided funding between 2022 and 2025, whilst the Heat Networks Zoning Pilot identified zoning opportunities in cities throughout England with six cities implementing a City Decarbonisation Delivery Programme drawing funding from all of the above sources.

It is not unusual for ESCOs to be owned by community partnerships or cooperatives: for example, in 2024, there were around 2250 energy cooperatives within the European Union's REScoop federation of citizen's energy cooperatives.[36] Much of the focus is on cooperative development of renewable energy schemes, such as wind and solar farms. For example in Germany in 2022 there were 914 community energy schemes reported by the DGRV (*Deutscher Genossenschafts und Raiffeisenverband e.V.*)—the federation for German cooperatives,[37] including wind and solar farms generating around 8 TWh of electricity, or 1.7% of the total electricity generated in that year. Germany has a history of locally operated energy supply, with some 1420 *Stadtwerke* (municipal utilities) across the country, many of which provide integrated energy, water, sewage and waste services. In 2013 only 43.8% of electricity supplied in Germany was from major national suppliers compared with 93.5% in the UK.

According to a report prepared by former British MP Alan Simpson for the consultation by DECC on its 2013 Community Energy Strategy, 'community ownership and community "first use" of the energy generated (in Germany) are central to both a culture change (and market change) in which communities take increasing responsibility for their own energy security and carbon emissions'.[38] Unlike Germany, Denmark and the USA, and with the honourable exception of Scotland, community energy has been slow to take off in the UK. However, in concert with DECC's Community Energy Strategy, Community Energy England was established in 2014 to join its long-standing cousin in Scotland to '…represent and support those committed to creating the conditions within which the community energy sector can rapidly scale and grow across the country'.[39] By 2024, its membership grew to over 300 community energy organisations.

One of the biggest challenges for electricity generated from grid-connected community energy schemes, or distributed energy generation, is the ability of the electricity network to handle 'reverse power flow'; that is when an infrastructure intended for electricity supply to the consumer receives electricity generated by the consumer, or some other generator connected to the distribution network. One pinch point is the primary substation which, for example, in the UK may step down from 33 kV or 132-kV incoming voltage to 11-kV distribution lines. The reverse power flow in these transformers may be as low as 15% of the capacity in the normal direction. Other limiting factors include fault levels that exceed equipment ratings, and

[36] https://www.rescoop.eu

[37] https://www.dgrv.de/wp-content/uploads/2023/07/DGRV_Survey_EnergyCooperatives_2023.pdf

[38] http://www.communityenvironment.org.uk/cms/wp-content/uploads/2013/08/Community-owned-energy-generation-Alan-Simpson.pdf

[39] http://communityenergyengland.org/about/

thermal or voltage constraints on cables and equipment. The importance of each of these factors varies with location: for example, in urban areas such as London, having a high density of demand and typically substations being used to step down from 132 to 11 kV, fault levels are responsible for 95% of constraints on distributed generation, whereas in rural areas having long circuits with high impedance, variations in voltage are more likely to create the constraint. With an ageing infrastructure such as that which serves much of the UK, facing both growth in the demand for electricity and in distributed generation, there are many networks that are constrained in the size of generating equipment that can be connected without reinforcement, such as an increase in cable size, new overhead pylons or additional substations.

The network in the UK is divided into 14 areas operated by six distributed network operators (DNOs). These companies administer 'Point of Connection' applications for distributed generation and typically publish network maps that indicate where circuits are approaching tolerable limits and where network reinforcement may be required. In recognition of the limitations, these constraints are placing on prospective community/district energy schemes the concept of a 'smart grid' with Active Network Management (ANM) has been developed under the auspices of the Energy Networks Association; which has coined the definition for ANM as: 'using flexible network customers autonomously and in real-time to increase the utilisation of network assets without breaching operational limits, thereby reducing the need for reinforcement, speeding up connections and reducing costs'.[40] This includes provisions for curtailing output from distributed generation equipment during constrained periods to avoid the need for reinforcement.

The Renewable Grid Initiative (RGI) is an EU wide pressure group, launched in Germany in 2009, that promotes 100% integration of electricity from renewable energy sources on the basis that strategic and collaborative action is required to create new grid infrastructure to meet future electricity needs. New grid architecture should enable the integration of all renewable energy sources hence expanding distributed generation and demand-side management. According to its mission statement, 'RGI is a unique collaboration of NGOs and TSOs from across Europe engaging in an 'energy transition ecosystem-of-actors'. We promote fair, transparent, sustainable grid development to enable the growth of renewables to achieve full decarbonisation in line with the Paris Agreement'.[41] One of the strategic objectives of this Initiative is that it 'collects and facilitates the development of Best Practice solutions for fair and environmentally sensitive grid development with a special emphasis on nature conservation, transparency and participation across Europe both in the development of new grids and modernising or maintaining the existing infrastructure.' This is unsurprising perhaps with partners that include the Royal Society for the Protection of Birds, WWF International, Friends of the Earth Ireland, as well as the main

[40] https://www.energynetworks.org/assets/images/Resource%20library/ANM%20Good%20Practice%20Guide%202015.pdf

[41] https://renewables-grid.eu/fileadmin/user_upload/Files_RGI/About_RGI/2019_11_15__New_RGI_Mission_Statement_and_Strategic_Objectives_final.pdf

Transmission System Operators (TSOs) in Germany, France, Belgium, Italy, Spain, Portugal, Ireland, Croatia, Switzerland, The Netherlands and Norway. As of 2024 none of the UK TSOs, such as the National Grid for England, were involved.

A crucial ingredient in the growth of community energy is funding. In a competitive investment market, this requires that funders can be guaranteed a return on investment that exceeds the low risk end of the market. In the 2010s crowdfunding of energy projects gained in popularity, through the growth in social media via peer to peer lending and cooperative share offers, for example, raising an estimated 2.7 billion dollars in 2012.

In many countries, Government support for renewable and community energy projects are available, usually through some combination of feed-in tariffs, tax relief and low interest loans. Feed-in tariffs (FiTs) are not defined as subsidies as such since they are financed by a surcharge on energy bills in general. Through its 1990 'Law on Feeding Electricity into the Grid' (*Stromeinspeisungsgesetz*) Germany was the first country to introduce a legal framework for guaranteeing distributed electricity generators access to the grid, which at the time related the price paid for electricity to its retail price. Spain and Denmark soon followed suit, heralding a boom in wind power in these countries. It was not until 2000, however, when Germany introduced its 'Act on Granting Priority to Renewable Sources' (*Erneuerbare Energien Gesetz—EEG*), that a mechanism was introduced relating the feed-in tariff to the cost of generation. This model was subsequently copied in many countries worldwide, including the UK which introduced its own scheme in 2010.

In the UK, the feed-in tariff system was initially applied to specified renewable technologies, running in parallel with a Renewables Obligation Certification scheme, introduced in 2002, that obliged electricity suppliers to source an increasing proportion of electricity from renewable sources: this was 15.4% during the financial year 2014/15 climbing to 41% by 2024. The value of these tradable certificates approaches zero as the cost of generating renewable energy approaches that of non-renewable sources. This scheme closed to photovoltaic installations and onshore wind farms in 2015 and all other new generators in 2017, but will carry on trading until 2037. Following the German FiT model, the amount that a generator is paid for electricity generated steadily decreases in concert with the cost of the technology, whilst the amount paid for electricity exported to the grid increases with the retail price index.

In 2015 the UK Government introduced Electricity Market Reform with the intention of controlling costs through a Levy Control Framework and a rationalised system of Feed-in Tariffs and Contracts for Difference (CfD) where:

> A Contract for Difference (CFD) is a private law contract between a low carbon electricity generator and the Low Carbon Contracts Company (LCCC), a government-owned company. A generator party to a CFD is paid the difference between the 'strike price'—a price for electricity reflecting the cost of investing in a particular low carbon technology—and the 'reference price'—a measure of the average market price for electricity in the GB market.[42]

---

[42] https://www.gov.uk/government/collections/electricity-market-reform-contracts-for-difference

CFDs were available for installations of 5 MW and larger and the scheme was intended to run in parallel with ROCs during its phase-out. Perhaps unsurprisingly the initial tranche of CFDs were for solar PV and onshore wind farms. However, in response to a degree of local antipathy to onshore wind farms the UK Government introduced measures in 2015 to require planning applications to incorporate evidence of local support. However, this requirement was removed in 2023.[43] All renewable energy projects of 50 MW or larger required 'development consent' as defined in the 2008 Planning Act meaning that, after a lengthy consultation process and examination by the Planning Inspectorate, applications would have to be referred to the Energy and Climate Change Minister for approval, although this requirement was removed in 2016, allowing planning decisions to be made by the Local Planning Authority (LPA).

In a number of countries the tax system has been used to provide incentives for renewable energy generation. According to Pollalis et al. [1], in their 2012 book *Infrastructure Sustainability and Design* 'at least 83 countries have some type of tax-related policy to promote renewable power generation. Investment Tax Credits, reductions in VAT and import duty reductions, and other tax-related incentives provide financial support at the national level, and also at the provincial level in Canada, the USA, and Australia'.

In the USA either Production Tax Credits (PTC) or Investment Tax Credits (ITC) have been available for installations that generate electricity with zero greenhouse gas emissions for some years. These schemes were extended through the 2022 Inflation Reduction Act and became known as the Green Electricity PTC and ITC from January 2025. PTCs are essentially FiTs by another name, with credits being awarded based on kWh generated by a number of renewable technologies, whereas ITC offers a subsidy for the installation of prescribed renewable technologies.

China has a number of separate tax incentives for those generating renewable energy: through rebates in corporate income tax, reduced value added tax on the sale of electricity from renewable sources and ESCOs engaged in energy performance contracting. Some Indian states have also offered VAT savings on specific technologies, whilst exemptions are available for customs and excise duties on specific goods required to set up renewable energy projects. For a comprehensive country by country guide to 'Tax and Incentives for Renewable Energy' refer to the 2023 report by the World Economic Forum.[44]

In recent years, energy cooperatives have also been able to benefit from tax relief on investment. For example, the Enterprise Investment Schemes have been available in the UK alongside such schemes as Social Investment Tax Relief (SITR) and the Seed Enterprise Investment Scheme (SEIS). SITR closed in 2023 and was only made available to the so-called community benefit societies which, unlike cooperatives, cannot distribute surpluses to their members in the form of dividends and therefore offer a lower return on investment.

[43] http://researchbriefings.parliament.uk/ResearchBriefing/Summary/SN04370

[44] https://www.weforum.org/agenda/2023/05/renewable-energy-incentives-households-countries/

Since 2001 UK businesses have been able to obtain tax relief on the up-front costs of community energy schemes, whilst elsewhere in Europe a number of countries offer tax incentives to businesses investing in community energy projects. In Denmark, for example, community foundations have historically been subject to Business Tax Settlement Rules which allowed them to obtain tax relief on 'grants' allocated to community energy schemes.[45]

Also some countries have provided preferential or soft loans to support renewable and community power projects. In Germany renewable energy projects have been able to access low interest loans from Germany's State-owned development bank, *Kreditanstalt für Wiederaufbau* (KfW). Whilst Denmark's Act on *Kommune Kredit* offered 20 year low interest loans to municipalities which were used to finance community energy projects, but with strict rules on government oversight. Poland's 'Stork Programme', running between 2014 and 2022, provided financial support for 'distributed renewable energy sources' including preferential loans for between 30% and 70% of eligible costs.

4.6 Carbon Capture and Storage

So far in this chapter we have examined the principal modes of energy generation as sources of greenhouse gases, whilst also reviewing the renewable alternatives. In this context it is worth discussing integrated carbon capture and storage (CCS) as a means of reducing emissions associated with the combustion of fossil fuels.

CCS typically comprises apparatus to separate carbon dioxide from the fuel delivered to a process (pre-combustion) or from the products of combustion (post-combustion and oxyfuel), along with a pipeline to transport the CO_2 to a suitable store: usually either depleted oil and gas fields or deep saline aquifer formations. According to *An overview of current status of carbon dioxide capture and storage technologies* by Leung et al. dating from the end of 2014, based on US prices, between 70% and 80% of the cost of CCS is accounted for by the capture equipment.[46] Typically this uses a wet process, the most common of which employs an aqueous amine solution, using mono-ethanol amine (MEA) dissolved in water, to chemically absorb the CO_2 from combustion gases (post-combustion solvent scrubbing). The combustion gases are cooled to between 40 °C and 65 °C and passed through an absorption tower with the MEA and water. The carbon dioxide is subsequently recovered from the amine solution by reheating to between 100 °C and 150 °C and passed through a 'stripper' in which the MEA is regenerated for recycling to the absorption tower and the resultant CO_2 and water vapour mixture is

[45] https://friendsoftheearth.eu/wp-content/uploads/2014/11/community_power_model_legal_frameworks_june2014.pdf

[46] http://www.sciencedirect.com/science/article/pii/S1364032114005450

passed through a dryer and the essentially pure carbon dioxide compressed for transporting to storage underground.

Unfortunately MEA is both corrosive and environmentally toxic, necessitating costly corrosion-resistant materials and creating potential health and environmental damage in the vicinity of the plant. Typical emission rates to the atmosphere are between 40 and 160 kg of amines for each 1000 tonnes of carbon dioxide captured. Amines degrade into other compounds, such as nitrosamine and nitramine, which are potential carcinogens. Concern about exposure to these substances was one factor that led to the postponement of a large-scale post-combustion CCS plant at Mongstad in Norway in 2013. A study by the Norwegian Institute of Public Health had recommended that total long-term concentrations of nitrosamine and notramine together should not exceed 0.3 ng (10^{-9} g) per cubic metre (ng/m^3), whilst they also estimated that average concentrations within 1 km of the plant at Mongstad could be as high as 1000 ng/m^3, dropping to 400 ng/m^3 at a distance of 10 km. For this reason research has been focussing on alternatives to MEA as an absorption medium, such as phenoxides.[47]

Following the first patent for enhanced oil recovery (EOR) using carbon dioxide in the USA in 1952, there has been a gradual rise in the use of carbon capture and injection at high pressure underground to increase production from depleted oil formations. The first commercial application was in Scurry County, West Texas, in 1972, which became one of the world's largest EOR oil fields, using carbon dioxide captured from a nearby ammonia plant. Conventional oil production has historically comprised primary and secondary phases: the latter requiring water to pressurise the formation. Once production starts to tail off it was found that yield could be increased by injecting carbon dioxide at high pressure (see Fig. 4.7). 'Under high pressure and temperature in the reservoir, CO_2 mixes with the oil to generate a low surface tension and a low viscosity fluid that can be displaced easily. Moreover, CO_2 is able to invade zones that were not invaded by water flooding, resulting in reducing and releasing the trapped oil'.[48]

The application of CCS to remove carbon dioxide from power station gases has taken longer to evolve, however, primarily because of the lack of economic or legislative drivers. According to an IEA report from 2013, for OECD countries, CCS can increase the cost of new plants by between 33% and 64%, and even more for retrofits, whilst reducing the efficiency of electricity generation by between 7.5% and 10.5%.[49] These cost implications render CCS uncompetitive with most low or zero carbon electricity generation technologies, particularly when the cost of carbon is driven down by market forces. In the Executive Summary to its 2014 Energy Technology Perspectives the IEA states that 'the future of CCS is uncertain; at

---

[47] https://www.ncbi.nlm.nih.gov/pmc/articles/PMC7774268/#:~:text=Under%20high%20pressure%20and%20temperature,and%20releasing%20the%20trapped%20oil

[48] http://www.api.org/~/media/Files/EHS/climate-change/Summary-carbon-dioxide-enhanced-oil-recovery-well-tech.pdf

[49] https://www.iea.org/reports/technology-roadmap-carbon-capture-and-storage-2013

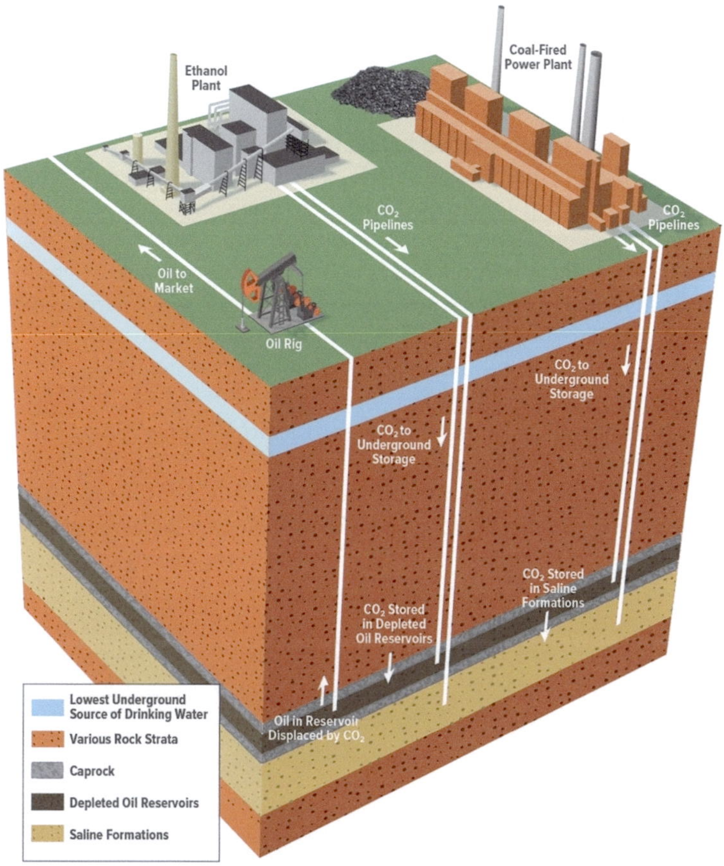

Fig. 4.7 Congressional Budget Office, U.S. Federal Government

present, the technology is advancing slowly, due to high costs and lack of political and financial commitment'.[50]

When CO_2 absorbed into water comes into contact with basalt over long periods of time it eventually solidifies and metamorphoses into rock. This phenomenon has been exploited in the Project Silverstone and Orca CCS installations developed in Iceland by a company called Carbfix to deal with the CO_2 in the volcanic gases used to generate heat and electricity at the Hellesheidi geothermal power plant in South Iceland.[51] These plants exploit the presence of veins of basalt within volcanic rocks in Iceland, but basalt is found in pockets globally and hence the potential for the

[50] https://www.ourenergypolicy.org/wp-content/uploads/2014/05/energy-tech.pdf

[51] http://www.theguardian.com/environment/2016/jun/09/co2-turned-into-stone-in-iceland-in-climate-change-breakthrough; https://www.carbfix.com/currentoperations

development of future CCS plants that convert CO_2 into a stable solid form is being explored, despite the high costs and significant energy demand.

According to the Global CCS Institute 41 complete end-to-end CCS systems were operational by July 2023, with 392 in the pipeline and having a total potential CO_2 capture capacity of 361 million tonnes per annum (Mtpa).[52] The 1.3 billion US dollar Boulder Dam carbon dioxide EOR plant, which opened in Saskatchewan, Canada, at the end of 2014, was the first large-scale commercial CCS linked to a coal-fired power station. It captures carbon dioxide from a local 110 MW coal-fired power station and injects it into a nearby oil field. The economic viability of this plant relies on the state-owned electricity provider, SaskPower International, selling the carbon captured to the Cenovus Oil Company to enhance oil extraction from the nearby field. Figure 4.8 shows the first CCS plant associated with cement manufacture, which started operation in 2025.

In Chap. 7 of the IPCC's 5th Assessment Report Working Group 3 concluded that 'the risks associated with a large-scale deployment of CCS technologies include concerns about the lifecycle toxicity of some capture solvents, the operational safety and long-term integrity of CO_2 storage sites, as well as risks associated with CO_2 transport via dedicated pipelines'.[53] Each site has specific geological characteristics

Fig. 4.8 Heidelberg carbon capture plant in Brevic, Norway (From: https://commons.wikimedia.org/wiki/File:Heidelberg_Brevik_Carbon_Capture.jpg)

[52] https://www.globalccsinstitute.com/wp-content/uploads/2024/01/Global-Status-of-CCS-Report-1.pdf

[53] https://www.ipcc.ch/site/assets/uploads/2018/02/ipcc_wg3_ar5_chapter7.pdf

which govern how the injected gas behaves over the long term, and it is the permeability of the so-called caprock layer between the storage region and the surface that is critical. This places a great reliance on the accuracy of site evaluation, characterisation and modelling of flow pathways, along with subsequent measurement, monitoring and verification (MMV) of potential leakage pathways.

Despite the ongoing concerns expressed by the IPCC and others, the International Energy Agency, which refers to CCS as carbon capture, utilisation and storage, has warned that despite the rapid increase in projects in the pipeline in 2024 the uptake is nowhere near that required by its Net Zero Scenario.[54] The IEA has historically assumed that coal-fired power stations will continue to be built, and that they will all be fitted with CCS, whilst half of all new gas-fired power stations and 20% of biomass plant will have CCS.[55] However, a 2023 paper emanating from the Smith School of Enterprise and the Environment at Oxford University assessing the relative costs of alternative pathways to keep global warming below 1.5° and achieve zero carbon globally by 2050 predicted that heavy dependence on CCS would cost 30 times more than investing in renewables and limited use of CCS [2]. With an average cost differential between the two options of around $1 trillion per annum, clearly the investment requirements for a carbon reduction strategy based on using CCS to mitigate emissions due to electricity generation using some combination of coal, biomass, gas and oil are likely to be astronomical. A lot of the investment would have to be front-loaded and require considerable intervention from governments since, as of 2023, the typical cost of CCS for a coal-fired power station for example is more than twice the carbon price in both the USA and Europe.[56] CCS is more efficient and hence more cost effective at removing CO_2 from a gas mixture having a high concentration of the gas. Furthermore, because of the high energy demand from CCS plant the electrical power generated from a coal-fired power station typically drops by 30–35%.

4.7 Buildings

Buildings contribute to greenhouse gas emissions directly: through the combustion products from heating boilers and leakage of refrigerant gases; and indirectly: through electricity consumed and embodied carbon of materials. The design and planning of buildings and the communities in which they sit can also impact on transport emissions: through the provision of cycle parking and the opportunities enabled for people to choose alternatives to private car use, for example. The total contribution of buildings to global greenhouse gas emissions, according to data

[54] https://www.iea.org/energy-system/carbon-capture-utilisation-and-storage#tracking

[55] https://www.iea.org/reports/technology-roadmap-carbon-capture-and-storage-2013

[56] https://sustainability.crugroup.com/article/carbon-capture-economics-why-usd-200-per-tco2-is-the-crucial-figure

assembled by the IPCC for its 2022 AR6 WG3 report, is approximately 21% with buildings responsible for 31% of final energy use (based on 2019 data).[57] According to Chap. 9 of its AR5 WG3 report:

> This energy use and related emissions may double or potentially even triple by mid-century due to several key trends. A very important trend is the increased access for billions of people in developing countries to adequate housing, electricity, and improved cooking facilities. The ways in which these energy-related needs will be provided will significantly determine trends in building energy use and related emissions. In addition, population growth, migration to cities, household size changes, and increasing levels of wealth and lifestyle changes globally will all contribute to significant increases in building energy use.[58]

Analysis reported by Working Group 2 for Chap. 10 of Assessment Report 5 indicates that air conditioning and refrigeration will play an ever increasing role in the impact buildings have on global warming during the twenty-first century.[59] This is reinforced in Chap. 9 of the AR6 WG3 report which states that 'All scenarios assessed project an increase of ownership of cooling appliances in all regions over the period 2020–2050'.[60]

Apart from the increased GHG emissions associated with the manufacture, use and disposal of refrigerants, and without mitigation beyond that in place in 2008, 'the increase (in energy consumption from the use of air conditioning) is from nearly 300 TWh (terawatt hours, or 10^{12} watt hours) in 2000 to about 4000 TWh in 2050 and more than 10,000 TWh in 2100, about 75% of which is due to increasing income in emerging market countries and 25% is due to climate change'.[61] Furthermore, according to research published in 2009 by Velders et al., hydrofluorocarbon (HFC) emissions weighted by their relative global warming potential will be nine times greater in developing countries than in the developed world by 2050.[62] This is primarily due to the increased use of air conditioning in buildings and road vehicles in those countries. This contributes to a predicted increase in radiative forcing due to HFC emissions by a factor of up to 30 by 2050 or according to projections based on Velders' work given in the UNEP Synthesis Report: *HFCs A Critical Link in Protecting Climate and the Ozone Layer* published in 2011, radiative forcing that corresponds to between 21% and 31% of the total value predicted for CO_2 in the atmosphere in 2050 by the IPCC-SRES scenarios used as the basis for global warming prediction in the IPCC 4th Assessment Report.[63]

We will be taking a closer look at the impact of buildings, along with strategies for low energy design and sustainable refurbishment in Chap. 7.

---

[57] https://www.ipcc.ch/report/ar6/wg3/chapter/chapter-9/

[58] https://www.ipcc.ch/site/assets/uploads/2018/02/ipcc_wg3_ar5_chapter9.pdf

[59] https://www.ipcc.ch/site/assets/uploads/2018/02/ipcc_wg3_ar5_chapter10.pdf

[60] https://www.ipcc.ch/report/ar6/wg3/chapter/chapter-9/

[61] https://www.ipcc.ch/site/assets/uploads/2018/02/ipcc_wg3_ar5_chapter10.pdf

[62] http://www.pnas.org/content/106/27/10949.full

[63] https://wedocs.unep.org/bitstream/handle/20.500.11822/8014/-HFCs_%20A%20Critical%20Link%20in%20Protecting%20%20Climate%20and%20the%20Ozone%20Layer-20111072.pdf?sequence=3&amp%3BisAllowed=

4.8 Industry

According to the IEA, industry was responsible for around 37% of global energy use in 2022, compared with 34% in 2002, and around a quarter of CO_2 emissions associated with energy use, or about 9 Gt of carbon dioxide.[64] A significant proportion of greenhouse gas emissions from industrial processes arise from energy-intensive industries, such as cement, iron, steel, aluminium, ammonia and paper manufacture. For example, an analysis of energy use statistics for the USA from 2010 by its Energy Information Administration (EIA) showed that energy-intensive industries were responsible for just over half of industrial energy consumption, or about 16% of the total, but only 7.6% of total US economic output in dollars.[65] Of the main energy-intensive industries, statistics reported by IEA from 2012 indicate that global cement production had increased by a factor of 6.5 since 1970, iron ore by 3.9 and ammonia by 3.5. Perhaps unsurprisingly much of this growth had been fuelled by the so-called Chinese economic miracle, with annual increases in iron ore production in China of 18%, for example, compared with a world average of 10%, whilst its steel and cement production increased by 10% and 11% per annum, respectively, compared with global averages of 4% and 6% (see also section on *Embodied carbon and other environmental impacts* in Chap. 7).

Most industries require buildings, typically to house the processes and provide a suitable environment for back-office staff, administration, catering, storage and laboratories. The energy consumption and greenhouse gas emissions associated with servicing these are grouped with the data for other buildings, whilst process energy and GHG emissions fall under the heading of industrial energy use and emissions. These may include heating, cooling, ventilation, lighting, direct combustion and electricity supply, where these are part of the process. Some chemical processes, such as solvent degreasing, may involve the release of vapours or gases having a high global warming potential directly to the atmosphere.

Refer to Chap. 9 for a review of the carbon footprint and cradle-to-grave environmental impact of manufactured products and consumables, including a look at the conflict between depleting natural resources and increasing demand.

4.9 Transport

Greenhouse gas emissions associated with transport include those from private road vehicles, delivery vehicles, buses, trains, aircraft, and commercial and leisure boats. It can be seen from Fig. 3.4 that the percentage of overall GHG in 2016 was 16.5%.

Emissions from transport are growing at a faster rate than for other sources. There has been a steady growth since 1970, when emissions were 2.8 $GtCO_{2eq}$,

[64] https://www.iea.org/energy-system/industry

[65] http://www.eia.gov/todayinenergy/detail.cfm?id=11911

largely attributable to road transport: which increased from just short of 60% of transport emissions in 1970 to 72% in 2010. At that time some 80% of passenger kilometres were attributable to 10% of the world population, dominated by OECD countries, although this is being challenged by Asian countries in particular where, for example, between 2000 and 2010, two and three wheel vehicle use doubled.

In Chap. 8 of its 2013 5th Assessment Report Working Group 3 of the IPCC predicts that global transport emissions will increase to 12 $GtCO_{2eq}$ by 2050, much of which will be driven by increased incomes and improved infrastructure in the developing and emerging economies.[66]

Please also refer to section on transportation greenhouse gas emissions in Chap. 8: *Sustainable Transport.*

4.10 Waste

Greenhouse gas (GHG) emissions associated with waste have historically been mainly from decomposing landfill, emitting primarily methane (CH_4), which is also generated from wastewater and sewage, along with a certain amount of nitrous oxide (N_2O). According to data published by the US Environmental Protection Agency, in 2010 16.2% of anthropogenic methane emitted in the USA arose from landfill sites.

Estimates differ for global GHG emissions associated with waste, but a 2011 report from the UN Food and Agriculture Organization (FAO): *Global Food Loss and Food Waste*, suggests that around 3300 million tonnes of CO_2 equivalent ($MtCO_{2eq}$) was being produced by food waste alone each year at that time.[67] This comprises emissions associated with agriculture, post-harvest activities, processing, distribution and consumption.

It is interesting to note that data for 2005 from the IPCC's 4th Assessment Report indicates waste was responsible for only some 1300 $MtCO_{2eq}$ per annum of GHG emissions, although this was based on post-consumer waste only, comprising landfill methane, wastewater methane and nitrous oxide, as well as incineration (mostly CO_2 and N_2O).[68] This figure had risen to 1435 $MtCO_{2eq}$ by 2010 (2.9% of 49.5 $GtCO_{2eq}$). It appears that this rise was mostly due to economic growth in the industrialised Asian countries which, according to FAO data from 2011, generate nearly one third of total global GHG emissions associated with food waste, for example.

The difference between the IPCC and FAO figures can be accounted for by the proportion of GHG emissions associated with food waste during growing, harvesting and processing within the data for AFOLU (Agriculture, Forests and other Land Use) and Industry.

[66] https://www.ipcc.ch/site/assets/uploads/2018/02/ipcc_wg3_ar5_chapter8.pdf

[67] http://www.fao.org/docrep/014/mb060e/mb060e00.pdf

[68] https://archive.ipcc.ch/publications_and_data/ar4/wg3/en/ch10.html

Research carried out by Christine Wiedinmyer of the US-based National Center for Atmospheric Research and published after the IPCC 5th Assessment Report in 2014, suggests that the GHG emissions reported previously underestimated the contribution from the open burning of domestic waste.[69] Her report concluded that some 1 billion tonnes of waste were being burnt globally, mostly in uncontrolled and unregulated open sites in the developing world. This represented an extraordinary 41% of the world's garbage. Although the estimated GHG emissions from open waste burning represent just 5% of total emissions, this was not incorporated into existing GHG inventories. It is interesting to note that according to Wiedinmyer's estimates for some countries 'such as Lesotho, Burundi, Mali, Somalia, and Sri Lanka—the trash burning produces more carbon dioxide than is tallied in official inventories' for all sources.[70]

Although not relevant to this chapter, it is worth mentioning that this study also found that open waste burning contributed 29% of total particulates smaller than 2.5 micron ($PM_{2.5}$) globally, 10% of airborne mercury and 64% of the carcinogenic polycyclic aromatic hydrocarbons (PAH).

4.11 Agriculture, Forests and Other Land Use (AFOLU)

As we saw in Chap. 3 in 2011 some 31% of the planet's land surface was covered in forest, whilst 38.4% was agricultural, with nearly 70% of this being permanent pasture, the remainder being planted with either permanent or rotation crops. In the Executive Summary to Chap. 7 of the IPCC's WGIII 6th Assessment Report it is stated that 'the AFOLU (managed land) sector, on average, accounted for 13–21% of global total anthropogenic greenhouse gas (GHG) emissions in the period 2010–2019 (medium confidence). At the same time managed and natural terrestrial ecosystems were a carbon sink, absorbing around one-third of anthropogenic CO_2 emissions (medium confidence)'.[71]

A number of processes associated with AFOLU result not only in the emission of carbon dioxide, but methane (CH_4) and nitrous oxide (N_2O), both of which have global warming potentials (GWPs) significantly higher than CO_2. CH_4 has a GWP of 28 over 100 years and N_2O 264 over the same time horizon.

According to the UN's Food and Agriculture Organization (FAO) greenhouse gas emissions from crops and livestock nearly doubled in 50 years: from 2.7 billion tonnes of CO_2 equivalent in 1961 to 5.3 $GtCO_{2eq}$ in 2011.[72] By 2019 this had risen to 7.89 $GtCO_{2eq}$.[73] Of the 555 petagrams of carbon (PgC) (555 billion tonnes), that

[69] https://news.ucar.edu/12239/trash-burning-worldwide-significantly-worsens-air-pollution

[70] https://news.ucar.edu/12239/trash-burning-worldwide-significantly-worsens-air-pollution

[71] https://www.ipcc.ch/report/ar6/wg3/chapter/chapter-7/

[72] http://www.fao.org/resources/infographics/infographics-details/en/c/218650/

[73] https://ourworldindata.org/food-ghg-emissions

were emitted into the atmosphere due to anthropogenic activities between 1750 and 2011, approximately 180 PgC or 32.4% evolved from changes in land use. However, less than half of this has remained in the atmosphere: the remainder being absorbed by the oceans and terrestrial ecosystems (155 and 160 PgC, respectively). The total mass of carbon in the atmosphere in 2011 was estimated to be 828 PgC, corresponding to an average concentration of 390.5 ppm.[74] By November 2015, the concentration recorded by the Mauna Loa laboratory in Hawaii had risen to 400.16 ppm, rising to around 418 ppm by 2022.[75]

Historically a significant proportion of the methane emissions associated with AFOLU globally have been from enteric fermentation from ruminant animals, with contributions also from rice cultivation, and manure management.[76] Manure management and fertiliser application also contribute to N_2O emissions.

Plants manufacture their own food and convert it to energy through a combination of photosynthesis, respiration and transpiration. Photosynthesis relies on the absorption of light by the plant surface and carbon dioxide through pores (stomata) in the underside of leaves, along with the transpiration of water through the root system, to produce food in the form of photosynthates (sugars, starches, carbohydrates and proteins), with oxygen and water vapour released as a by-product through the stomata. The process of respiration uses oxygen from the atmosphere to convert these photosynthates to energy for growth and other metabolic processes, releasing carbon dioxide and water as waste products. The difference between CO_2 absorbed and that which is emitted by plants globally averages at between 45% and 50% of the total dry weight of biomass. Hence undisturbed plants act as a sink for carbon dioxide, in a process referred to as biological carbon sequestration. However this carbon can be released to the atmosphere either through harvesting, processing and digestion as a food crop, through combustion or through decomposition, composting or biodigestion. In these latter cases other gases such as methane and nitrous oxide can be formed which, as we have seen, have a much higher GWP than carbon dioxide.

Methane is generated from the so-called methanogens, which are anaerobic microorganisms in the archaea domain, sometimes inaccurately referred to as methanogenic bacteria. Although both are single cell microorganisms there are differences in their cell wall material, whilst archaea can survive in more extreme environments than bacteria. Methanogens are responsible for enteric fermentation in the digestive tracts of ruminants (cattle, sheep, goats, and buffalo) and humans, resulting in gaseous emissions containing methane. They are also involved in emissions of methane from manure, marshes and melting tundra (see Chap. 2), which all contribute to the global 'methane cycle'.

[74] https://www.ipcc.ch/site/assets/uploads/2018/02/WG1AR5_Chapter06_FINAL.pdf

[75] https://www.metoffice.gov.uk/research/climate/seasonal-to-decadal/long-range/forecasts/co2-forecast

[76] https://www.ipcc.ch/report/ar6/wg3/chapter/chapter-7/

Nitrous oxide emissions occur also through anaerobic microbial processes, along with chemical processes in soils acting on nitrite ions to form N_2O. These occur through the use of nitrogen based fertilisers, the biological fixation of nitrogen by crops, the ploughing in of crop residues, cultivation of histosols (organic soils, typically high in peat content), spreading animal wastes on land and from manures dropped by animals grazing in the field. Also indirect emission sources include N_2O from atmospheric deposition of agricultural oxides of nitrogen (NO_x) and ammonia (NH_3), along with the emission of N_2O from the leaching of agricultural nitrates and runoff from fields, contributing to the global 'nitrogen cycle' and 'nitrogen cascade'. Nitrogen (N_2) becomes reactive (N_r) through processes such as combustion and food production, creating NO_x and NH_3 respectively, for example. The conversion of N_2 into ammonia is referred to as 'biological nitrogen fixation' (BNF) which is a microbiologically mediated process also referred to as the Haber-Bosch process.

Chapter 6 of the IPCC 5th Assessment Report from WG1 provides a useful example of the various components of the nitrogen cascade:

> In addition to its contributions to climate change and stratospheric ozone depletion, N_r contributes to the formation of smog; increases the haziness of the troposphere; contributes to the acidification of soils and freshwaters; and increases the productivity in forests, grasslands, open and coastal waters and open ocean, which can lead to eutrophication and reduction in biodiversity in terrestrial and aquatic ecosystems.[77]

Because forests and woodland have historically covered a large, if shrinking, percentage of the planet's land surface, they have always played an important role in balancing the greenhouse gas concentration in the atmosphere (see sections on *Forests* and *Terrestrial ecosystems and the biosphere* in Chap. 3). Changes in land use have frequently been associated with the destruction of ancient forests through burning or logging, exposure of peat substrates and the planting of crops, such as rice, or oil palm plantations. In Chap. 11 of its report WG3 also states that: 'Undisturbed waterlogged peatlands (organic soils) store a large amount of carbon and act as small net sinks. Drainage of peatlands for agriculture and forestry results in a rapid increase in decomposition rates, leading to increased emissions of CO_2, and N_2O, and vulnerability to further GHG emissions through fire.' Whilst 'burning vegetation releases CO_2, CH_4, N_2O, ozone precursors and aerosols (including black carbon) to the atmosphere. When vegetation regrows after a fire, it takes up CO_2 and nitrogen'.[78] Ozone precursors include NO_x, volatile organic compounds (VOCs) and carbon monoxide (CO), all of which, in the presence of solar radiation, are associated with the formation of both ground level and tropospheric ozone. It is interesting to note that rising carbon dioxide concentrations in the surface atmosphere increases plant growth through what has become known as the 'CO_2 fertilisation effect'.[79]

[77] https://www.ipcc.ch/site/assets/uploads/2018/02/WG1AR5_Chapter06_FINAL.pdf

[78] https://www.ipcc.ch/site/assets/uploads/2018/02/ipcc_wg3_ar5_chapter11.pdf

[79] https://www.ipcc.ch/site/assets/uploads/2018/02/WG1AR5_Chapter06_FINAL.pdf

4.12 Greenhouse Gas Sinks and Stores

It is clear from the above that as well as there being both natural and anthropogenic sources of greenhouse gases on earth there is a continuous absorption of these gases by land and ocean surfaces:

> On average, CO_2 molecules are exchanged between the atmosphere and the Earth surface every few years. This fast CO_2 cycling through the atmosphere is coupled to a slower cycling of carbon through land vegetation, litter and soils and the upper ocean (decades to centuries); deeper soils and the deep sea (centuries to millennia); and geological reservoirs, such as deep-sea carbonate sediments and the upper mantle (up to millions of years)[80]

Reservoirs, such as oceans and permafrost, contain vast quantities of carbon, nearly 50 times the mass that has accumulated in the atmosphere since 1750. As we saw in the section on *Ocean acidification and oxygen content* in Chap. 3 rising carbon uptake has increased concentrations of carbonic acid and associated acidification which, along with warming, pollution and deoxygenation, is causing changes to aquatic ecosystems worldwide.

We also saw in Chap. 3 that soils can be a sink for methane through an oxidation process facilitated by methanotrophic bacteria, which are found in relatively dry soils, absorbing on average 28 million tonnes of methane annually (i.e., between 9 and 47 $TgCH_4yr^{-1}$). There are also vast stocks of methane bound to water molecules as ocean hydrates and permafrost hydrates as well as in natural gas reserves, estimated to be between 3 and 10 million tonnes in total.

References

1. Pollalis, S. N., Georgoulias, A., Ramos, S. J., & Schodek, D. (Eds.). (2012). *Infrastructure sustainability and design*. Routledge.
2. Bacilieri, A., Black, R., & Way, R. (2023, December 4). *Assessing the relative costs of high-CCS and low-CCS pathways to 1.5 degrees* (Working Paper No. 23-08). Oxford Smith School of Enterprise and the Environment. https://www.smithschool.ox.ac.uk/sites/default/files/2023-12/Assessing-the-relative-costs-of-high-CCS-and-low-CCS-pathways-to-1-5-degrees.pdf

[80] https://www.ipcc.ch/site/assets/uploads/2018/02/ipcc_wg3_ar5_chapter11.pdf

Chapter 5
Exploitation of Finite Resources: Reaching the Peak?

> *If the subsistence for man that the earth affords was to be increased every twenty-five years by a quantity equal to what the whole world at present produces, this would allow the power of production in the earth to be absolutely unlimited, and its ratio of increase much greater than we can conceive that any possible exertions of mankind could make it.*
>
> *Thomas Malthus [1]. An essay on the Principle of Population*

5.1 Introduction

Toward the end of the eighteenth century, Thomas Malthus was living in a pre-industrial England where he was wrestling with the relationship between an increasing population and its ability to produce sufficient food to meet its needs. In the twenty-first century Malthus' concerns still resonate, not only with regard to 'feeding the world' but also the depletion of finite resources, such as fossil fuels, minerals, ancient forests and peat.

The nineteenth century English economist William Jevons identified a phenomenon that became known as the 'Jevons Paradox' observing that technological progress resulted in the increased efficiency with which a resource is used, whilst also increasing demand. In his landmark book *The Coal Question* he summarised this point most elegantly with the comment that 'it is a confusion of ideas to suppose that the economical use of fuel is equivalent to diminished consumption. The very contrary is the truth' [2]. It is illuminating to note that the Jevons' Inquiry was carried out at a time when there was concern that England's coal mines would soon be exhausted. Nearly a century later, in his 1981 book *The Ultimate Resource* (revised in 1998) the American economist and business professor Julian Simon saw this relationship between technological innovation and increased production as a positive, suggesting that:

> The most important elements in raw-material price trends have been (1) the rate of movement from richer to poorer ores and mining locations—that is, the phenomenon of

P. Appleby, *The Overburdened Planet*,
https://doi.org/10.1007/978-3-032-01510-5_5

> "exhaustion", and (2) the continued development of technology, which has more than made up for exhaustion. Is the rate of development of such new technology slowing up? To the contrary: the pace of development of new technology seems to be increasing. Hence if the past differs from the future, the bias is likely to be in the direction of understating the rate at which technology will develop, and therefore underestimating the rate at which costs will fall ([3], Chap. 1).

Controversially Simon took this as evidence that natural resources could last indefinitely, on the premise that humans would always find the technology to extract them. In reality, however, fluctuations in the market for resources have a direct impact on the viability of hard to access fields, whilst over-production and falling demand drive prices down, as evidenced by the downturn in the Chinese economy in 2015/16, the repercussions from which included catastrophic contractions in Chilean copper and Australian iron ore production, for example.

The concept of a potentially exponential growth in demand for non-renewable resources was taken up by the Club of Rome in its 1972 report on the *Limits for Growth* (Meadows et al., 1972) referred to in Chap. 1 above. The authors carried out an analysis of the main 'non-renewable natural resources', including fossil fuels and a range of mineral commodities ranging from aluminium to zinc. Table 4 of this report offers the authors' predictions for what they refer to as the 'Exponential Index', indicating the likely life expectancy for each resource based on an exponential increase in demand and a fivefold increase in reserves. Exponential growth was assumed by the authors because the trend in demand increase historically was found to be exponential. Interestingly this table shows predictions that indicate that oil (petroleum) reserves would be used up by 2010, natural gas by 2009 and coal by 2110, with many other mineral commodities disappearing in the first two decades of the twenty-first century. Although these are clearly illustrative and somewhat simplistic (as well as being inaccurate), these predictions sounded a warning that continues to resonate in modern times.

Although recognising that 'we are not running out of any mineral', the thirty-eighth report from the Club of Rome, published in 2014, opined that 'the production of many mineral commodities appears to be on the verge of decline… we may be going through a century-long cycle that will lead to the disappearance of mining as we know it' [4]. Employing the concept of Energy Return on Energy Invested the authors conclude that 'extraction is becoming more and more difficult as the easy ores are depleted. More energy is needed to maintain past production rates, and even more is needed to increase them'.

In this context, and for some years, commentators have been referring to 'peaks' in the production of resources such as oil, gas, coal and other minerals. More recently, however, as we saw in *The debate* section in Chap. 2, the Carbon Tracker Initiative (CTI) started a movement culminating in a call to keep fossil fuels in the ground, based on its prediction that, if burnt, they would emit some five times as much carbon dioxide required for average global surface temperatures to reach 2 °C above pre-industrial levels.

5.2 Fossil Fuels

The CTI formed the basis for the 2013 book by Mike Berners-Lee and Duncan Clark which asked '*The Burning Question*': 'We can't burn half the world's oil, coal and gas, so how do we quit?' [5].

Historically it was assumed that we would eventually run out of fossil fuels. A number of 'peak oil' panics that have occurred since the latter part of the nineteenth century, when world production of oil was synonymous with North American oil production. In the 1880s, it was thought that resources would be confined to the Pennsylvania fields and that no oil would be found west of the Mississippi river. Concern that the world was about to run out of oil continued through to the early twenty-first century, perpetuated by the influential earth scientist Marion King Hubbert who, in the 1940s predicted that US oil production would reach its peak between 1965 and 1970, which latterly became known as 'Hubbert's Peak'.[1]

Bearing in mind that up until the late 1960s, the USA had been responsible for 90% of the world's oil production this prediction had global implications. However, along with a massive increase in demand for oil post World War II, a combination of politics and conflict created the oil crisis of the early 1970s, during a time when the USA became dependent upon oil imports; and conflict in the Middle East was being fuelled by Russia and driven by anti-Israel sentiment in the Arab world.

US support for Israel led to an oil embargo by Libya, Saudi Arabia and the other Arab states. This soon escalated to draw in other countries, ultimately impacting on both the price and availability of oil in much of the world. This led to massive inflation in some countries, the formation of the International Energy Agency, a burgeoning environmental movement and a focus in many countries on energy efficiency and demand reduction. The embargo petered out following a peace agreement between Israel and Egypt, although subsequent shocks occurred with Iranian oil being embargoed by the USA after the Shah was deposed and the American embassy hostage crisis of 1979, again causing a hike in oil prices.[2]

However, these proved to be aberrations in the ever increasing exploitation of fossil fuels and indeed, as oil prices increased and as Julian Simon had predicted in 1981, it became economically viable to utilise ever more costly methods of extraction, thus increasing the accessible reserves, such as those below the ocean floor, in oil shale and in light tight oil (LTO) formations. Forecasts of peak oil being reached in 2007 or soon after were confounded by the exploitation of such unconventional oil reserves, particularly in North America.

According to the International Energy Agency's (IEA) 2023 *Oil Analysis and Forecast to 2028*: 'The global energy crisis has moved energy security to the fore of the international policy agenda and boosted the momentum behind the deployment

[1] https://www.chemengevolution.org/energy/oil-gas/the-story-of-peak-oil

[2] http://www.freerepublic.com/focus/news/918168/posts

of clean energy technologies. Investment in clean energy is accelerating at a faster rate than for fossil fuels, helping bring peak oil demand into view'.[3]

As can be seen from Figs. 5.1 and 5.2 taken from this report, with the energy efficiency measures and carbon emission limits introduced by signatories following COP21 in 2015, and reinforced in subsequent COPs, the IEA has predicted that from 2024 supply of oil globally would overtake demand and the gap would be maintained thereafter.

Although natural gas does not currently compete greatly with oil in the petroleum marketplace for fuelling road vehicles, the slump in oil prices in 2015/16 had a marked effect on gas production, driving prices down and necessitating large capital expenditure cuts upstream. Unconventional gas resources, such as shale gas fields, have combined with demand reduction to extend the date for 'peak gas' well into the future. Just how far is the subject of competing predictions: for example, a recent high resolution study by the University of Texas, which examines only the main shale gas formations or plays in the USA, indicates that peak production for the four main plays could be reached in the early 2020's.[4]

As we saw in the section dealing with the geopolitics of environmental impact in Chap. 2, globally there are massive theoretically 'technically recoverable resources' (TRR) of shale gas (and oil) (see Fig. 2.8), but up until 2016 only those in North America had been exploited extensively. Data provided in one study carried out for the Energy Security Unit of the European Commission in 2012 indicates that the total of best estimates of TRR in known shale beds outside of North America were

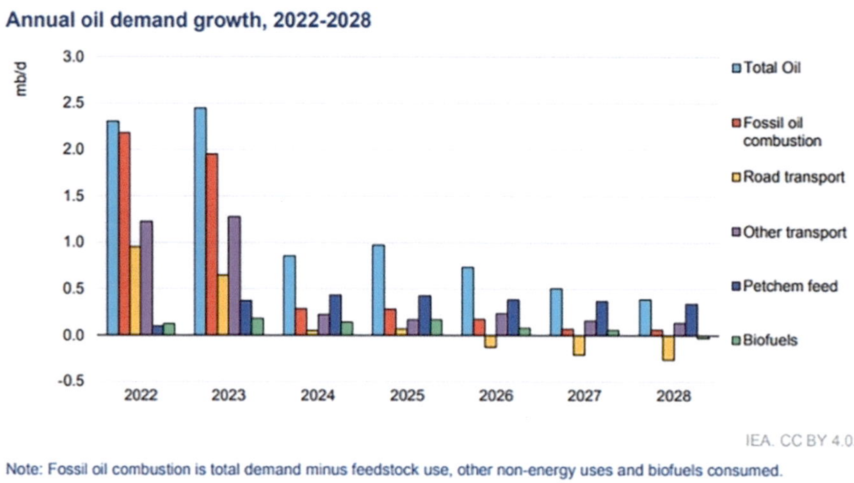

Fig. 5.1 Predicted Oil demand growth 2022–28 (From IEA Oil Analysis and Forecast to 2028: https://iea.blob.core.windows.net/assets/cc7fd38f-3d68-4796-a958-8dfa3f3ef4a6/Oil2023.pdf). Petchem feed is petrochemical feedstock refined from fossil fuels

[3] https://iea.blob.core.windows.net/assets/cc7fd38f-3d68-4796-a958-8dfa3f3ef4a6/Oil2023.pdf

[4] http://www.iea.org/Textbase/npsum/MTCMR2015SUM.pdf

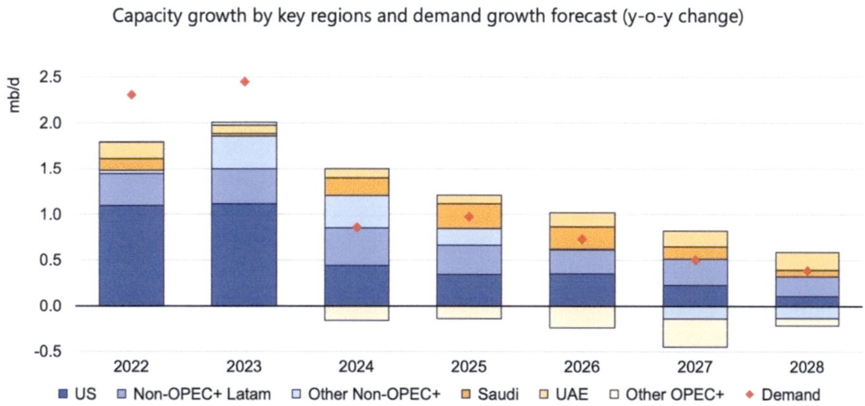

Fig. 5.2 Predicted oil supply 2022–28 (From IEA Oil Analysis and Forecast to 2028: https://iea.blob.core.windows.net/assets/cc7fd38f-3d68-4796-a958-8dfa3f3ef4a6/Oil2023.pdf). Non-OPEC+ Latam refers to Latin American countries not in OPEC (https://www.worldbank.org/en/publication/worldwide-governance-indicators)

between 4 and 7 times greater than for the US and Canada combined [6]. However, although these were considered to be 'technically recoverable' using currently available methods of extraction, most had not been exploited due to either economic reasons or resistance from the communities impacted.

In the 2023 edition of its *Coal Analysis and Forecast to 2026,* 'The International Energy Agency's latest projections see coal demand peaking within this decade under today's policy settings, primarily as a result of the structural decline in coal use in developed economies and a weaker economic outlook for China, which has also pledged to reach a peak in CO_2 emissions before 2030'.[5] However, the IEA predicts also that 'India becomes the driving force behind the upward pressure on global coal demand through to 2026, even if the global trend is decided in China. For the forecast period until 2026, we estimate Indian coal demand to rise by 3.5% annually to 1,397 Mt, with growth in every coal grade'. This is due to the Indian government's ambitious plans to provide access to electricity for the approximately 240 million people still without it, although they have a major renewable power generation programme, including conversion of existing coal-fired plant to co-firing with biomass. Indeed India, Indonesia, Vietnam and the Philippines are the only countries predicted to continue increasing coal use in the future.

In 2023 coal remained the dominant fuel for electricity generation globally, with more than 35% of power generated from coal-fired power stations.

Whilst the *IEA Coal Analysis and Forecast* looks only to 2026, a longer term has been examined by the MILES Project Consortium[6] under an EU funded project to examine the implications of intended nationally determined contributions (INDCs)

[5] http://www.iea.org/Textbase/npsum/MTCMR2015SUM.pdf

[6] The MILES Project is the 'Modelling and Informing Low-Emissions Strategies' project.

for greenhouse gas emissions beyond 2020 [7]. Each party to the UNFCCC CoP 21 in December 2015 committed to an INDC with a view to limiting global average surface temperature rise to a maximum of 2 °C or ideally 1.5 °C (See Chap. 3). In an article for the *energy post* blog the principal author of this report, Thomas Spencer, summarises the impact that these commitments are likely to have on global coal demand as an 'absolute drop … of 5% by 2030 from 2012 levels'.[7] However, projections made in the 2023 *Energy Outlook* report from BP suggest that in order to achieve 'net zero' CO_2 emissions by 2050 coal use in power generation will have to reduce to near zero, with a massive increase in renewables and nuclear power.[8]

Whichever prediction proves correct, there is little doubt that coal production will be slowing down during the first half of the twenty-first century, despite growing energy demand globally. Not because of a depletion in resources, but partly due to a worldwide move away from carbon intensive energy sources resulting in a boom in renewable energy, including hydropower, along with the exploitation of previously difficult to access gas resources and a renewed confidence in nuclear power in China and Russia.

In the drive to a zero carbon future natural gas is considered as a transition fossil fuel. As well as being a major fuel for electricity generation globally (about 22% in 2023), it has an extensive piped distribution network for both heating and industrial applications. There is a global market for both piped natural gas and liquid natural gas (LNG), the latter market for which grew significantly following the sanctions introduced after Russia invaded Ukraine in 2020.

According to the IEA's *Medium-Term Gas Report 2023,* the demand for gas grew globally by on average 2.5% each year between 2017 and 2021 and by 3% annually in North America. However, it projected a slowing to 1.6% annually through to 2026.[9] The IEA anticipated that globally industry would dominate demand with around 60% of the total, with much of this in Asia Pacific and the Middle East. Although there would still be a growth in the demand for gas for electricity generation this would slow to 0.5% annually, with reductions in demand in Europe, North America and Northeast Asia primarily because of the rise in renewables and energy efficiency programmes. Countries that have historically had a heavy dependence on coal, such as China and India, have ongoing programmes of transitioning from coal to gas and biomass as well as accelerating renewables programmes.

There is, however, a countervailing pressure that works against nations reducing fossil fuel exploitation within their ambit. It is the Investor State Dispute Settlement (ISDS) system that enables companies that have a project cancelled by a

[7] https://reneweconomy.com.au/the-future-of-coal-the-long-comedown-40923/#:~:text=It%20estimates%20an%20absolute%20drop,a%20plateau%20by%20around%202020

[8] https://www.bp.com/content/dam/bp/business-sites/en/global/corporate/pdfs/energy-economics/energy-outlook/bp-energy-outlook-2023.pdf

[9] https://iea.blob.core.windows.net/assets/f2cf36a9-fd9b-44e6-8659-c342027ff9ac/Medium-TermGasReport2023-IncludingtheGasMarketReportQ4-2023.pdf

government to sue the government for what can be huge sums.[10] In parallel with this a similar facility has been available to investors in the fossil fuel industries since 1991 through the Energy Charter Treaty.[11] It has been suggested that these two devices have had a chilling effect on divestment from fossil fuels and other environmental protection strategies. It is interesting to note that the UK left the Treaty in 2024 citing its failure to align with the Government's 'net zero' strategies, whilst the EU left in the same year due to the Treaty being incompatible with the 2015 CoP 21 Paris Agreement.

5.3 Minerals

Beyond fossil fuels, 'non-fuel' mineral commodities in global trade include the metallic minerals: iron ore, copper, aluminium, nickel, tin and zinc; whilst non-metallic minerals include crushed stone, cement, sand and gravel, phosphates, lime, salt, soda ash and clays. The US Geological Survey 2024 report on Mineral Commodities stated that in 2023 the mineral commodities worth more than $1 billion to the US economy were, in the descending order: crushed stone, construction sand and gravel, cement, gold, copper, iron ore, industrial sand and gravel, salt, lime, zinc, phosphate rock, soda ash, molybdenum and helium.[12]

The European Commission (EC) produces a list of 'Critical Raw Materials' (CRM) which is updated every 3 years. Criticality is assessed on the basis of economic importance and supply risk, defined by the EC [8] as follows:

> Economic importance looks in detail at the allocation of raw materials to end-uses based on industrial applications. Supply risk looks at the country-level concentration of global production of primary raw materials and sourcing to the EU, the governance of supplier countries, including environmental aspects, the contribution of recycling (i.e. secondary raw materials), substitution, EU import reliance and trade restrictions in third countries.[13]

The 2020 EC list of CRMs includes both heavy and light rare earth elements (REE) which comprise 16 materials that occur together in the periodic table, having similar properties, and which are used extensively in the manufacture of batteries, lighting, electronics, nuclear reactors and in petrol refining through fluid catalytic cracking. Between 87% and 99% of REEs are mined in China. The CRM list includes another 18 materials ranging from antimony, used in the manufacture of flame retardants for example; platinum group metals (PGM), used primarily for auto catalysts; through to tungsten, used in alloys and cemented carbides.

---

[10] https://www.europarl.europa.eu/RegData/etudes/BRIE/2015/545736/EPRS_BRI%282015%29545736_EN.pdf

[11] https://www.iisd.org/articles/explainer/energy-charter-treaty

[12] https://pubs.usgs.gov/periodicals/mcs2024/mcs2024.pdf

[13] https://eur-lex.europa.eu/legal-content/EN/TXT/PDF/?uri=CELEX:52020DC0474

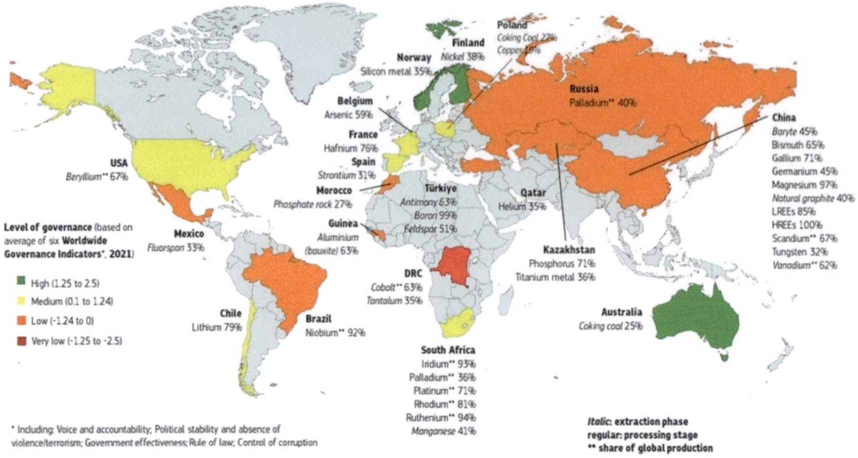

Fig. 5.3 Critical raw materials—main suppliers to EU (From: https://rmis.jrc.ec.europa.eu/images/CRMs_2023.jpg)

These CRMs are ingredients for a range of industries, including the manufacture of mobile phones, computers, flat screen televisions, batteries, optics, specialist steels and alloys; whilst many of them have only limited alternatives and/or recyclability. As can be seen from Fig. 5.3 that the dominant supplier for most CRMs is China. The Worldwide Governance Indicators are based on the 'six dimensions of governance' referred to in Fig. 5.3 and have been developed using an extensive survey of a wide variety of organisations worldwide.[14]

Minerals that are critical to battery manufacture globally are lithium, cobalt and nickel: the key minerals used in the manufacture of lithium ion batteries. As we will see in Chap. 8 these batteries have evolved as the key power source for electric vehicles, as well as many electronic devices, relying on manufacture in China which imports lithium from the likes of Chile and Australia to supplement its own lithium mining industry, whilst importing cobalt from the Democratic Republic of Congo and nickel from Indonesia and the Philippines.

The risks to supply associated with one country dominating production are unrelated to the peak materials issue discussed above for fossil fuels. Although there have been some that have suggested that 'peak metal' is possible, in fact, according to the British Geological Survey (BGS), this argument is flawed since its advocates have based their calculations on depleting reserves, even when resources are known to be abundant. Its view is that 'metal production is highly cyclical, with intermittent peaks and troughs which are closely linked to economic cycles. Declining production is generally driven by falling demand and prices, not by scarcity'.[15] One

[14] https://www.worldbank.org/en/publication/worldwide-governance-indicators

[15] BGS reference no longer available.

recent example of this comes from the contraction of the Chinese economy during 2015 and 2016 which resulted in a reduction in demand for numerous minerals including copper and iron ore.

An article by Chris Lo in the mining-technology.com website from September 2015 states that 'economic turbulence in China has been laying bare the mining industry's reliance on the country's insatiable demand for resources. With Chinese construction and manufacturing experiencing a sharp downturn that looks set to last, the knock-on effect for commodity prices and the health of the mining sector has been severe'.[16]

In 2022/3 around 22% of reserves and 23% of production of copper globally were in Chile, 57% of which was exported to China. Copper prices on world markets have experienced large fluctuations, falling by some 60% in the aftermath of the 2008 global financial collapse and experiencing a steady decline by 50% between 2011 and 2016. Due to a combination of external forces, poor management and aging infrastructure, Chile's economy has been badly hit by the closure of copper mines and thousands of miners being laid off. Similarly world iron ore prices fell to 30% of their 2011 peak in mid-2016: an 8 year low, resulting in serious damage to ore mining globally, and especially in Australia which held 26% of global reserves (based on iron content) in 2014 and was responsible for 20% of global production. According to an article by Vladimir Basov on the Mining.com website from September 2015 'a number of high-cost iron ore mines have been closed and suspended throughout the world in 2014, with up to 30% of low-grade iron mines shut down in China in 2014 alone'.[17]

Not all mineral resources have suffered from this "China syndrome"; indeed, despite aluminium prices having fallen by 40% between 2011 and 2016 China increased its smelting capacity and, with its own mines nearing exhaustion, invested heavily in bauxite mines in Australia.

As discussed in the section dealing with *Nuclear power* in Chap. 4 uranium is also a finite mineral resource, although its depletion rate depends on a variety of somewhat unpredictable factors. The price of uranium peaked in 2007 before plummeting during the world financial crisis, recovering slightly just prior to the tsunami which damaged the Fukushima nuclear plants in Japan on March 11, 2011 and the subsequent decisions to phase out nuclear power made by Governments in Germany, Switzerland, Spain and Italy. In June 2016, the price of uranium had fallen by 80% compared with that in 2007. This has resulted in many mines becoming economically unsustainable, despite demand growing from the construction of power stations in China, India and Russia, with Kazakhstan producing 39% of the world's uranium in 2015.[18] However it is interesting to note that a report by the World Nuclear Association published in 2015 predicted that nuclear power generation

---

[16] http://www.mining-technology.com/features/featurechasing-the-dragons-tail-global-mining-and-chinas-economic-slowdown-4678039/

[17] http://www.mining.com/true-giants-of-mining-worlds-top-10-iron-ore-mines/

[18] http://www.mining.com/uranium-market-getting-crushed/

capacity would grow by 45% between 2015 and 2035, resulting in a 20% increase in the demand for uranium, with Chinese uranium consumption alone increasing five-fold.[19]

5.4 Ancient Woodland and Peat

In Chap. 3, the role of forests is examined in the context of climate change, whilst Chap. 4 discusses their contribution to the emission and absorption of greenhouse gases. So far in this chapter we have considered the exploitation of resources that were formed through geological metamorphosis many millions of years ago. Peat, as the initial stage in the creation of coal, has been around for at least 360 million years and is still forming at a rate of around 1 mm/year. Most modern peatlands however were formed either in high latitudes after the retreat of glaciers at the end of the last ice age some 9000 years ago and in tropical regions between 3500 and 6000 years ago.

The world's oldest trees are the last outposts of forests that grew across most of the earth's surface not covered by frozen or liquid water once the first trees evolved some 385 million years ago, towards the end of the Devonian period. As trees died and decayed these early forests formed the basis for the creation of peat and coal seams, in particular during the Carboniferous period that followed the Devonian when a phenomenon occurred caused by rapid cooling that has become known as the 'Carboniferous Rainforest Collapse'. This brought about the near extinction of trees that grew in swampland along the low lying margins of continents (Coal Forests) causing the rapid diversification of amphibians and reptiles which a 2010 study has suggested led the way to the evolution of dinosaurs [9].

Trees evolved from the original plant life on earth, initially as ferns, steadily getting taller as they competed for light, developing stronger stems which evolved over a period of roughly 100 million years into the bark-covered wooden trunks that characterise the modern tree. The first of these was the now extinct *Archaeopteris*, which eventually covered most of the land mass, successfully dominating other tree species until it represented some 90% of forest trees. Over the millennia that followed other species appeared, with some reaching perhaps 50 m in height in competition for light. The first conifers (gymnosperms) appeared around 300–360 million years ago whilst the hardwoods (angiosperms) began to proliferate some 95 million years ago, dominated by maple, sycamore, oak, magnolia and laurel.[20]

The oldest trees on the planet are thought to be the Great Basin Bristlecomb pines (*Pinus longaeva*) in the White Mountains of California, with one having been dated in 2010 at 5062 years old, although this had not been verified as of 2017, although another given the name of Methuselah was carbon dated to 4850 years of

[19] http://www.mining.com/how-the-global-nuclear-power-industry-is-transforming/

[20] https://www.britannica.com/plant/gymnosperm/Appearance-of-gymnosperm-divisions

age in the late 2010s.[21] Also in California, the giant redwood or sequoia not only grows to heights exceeding 80 m with girths at its base of 30 m or more but has been shown to live for more than 3000 years in some instances.

A major international survey published in 2015, based on satellite imagery and surface-based observations, estimated that there were approximately 3.04 trillion trees on the planet, some 46% lower than before humans started harvesting or clearing them for their own purposes [10]. The authors also estimated that around '15 billion trees are cut down each year'. The study does not differentiate between short-rotation harvesting, land clearing for agriculture and grazing, forest fires due to lightning strikes or human activity, or the destruction of slow-growing ancient trees however. Another satellite-based study published by Global Forest Watch in the same year estimated that around 18 million ha of forest was lost in 2014, with 9.9 million ha of this being in tropical countries.[22]

The Amazon catchment hosts around one half of the world's rainforest, the majority of which comprises trees that grow extremely slowly. A study by Vieira et al. published in 2005 reported that in some of the sites surveyed 50% of trees over 10 cm diameter were more than 300 years old. Age was estimated by carbon dating because the growth rings were too close together to count [11], with mean average tree ages between 153 and 422 years, depending on location, and individuals of 1000 years and older. Hence deforestation of the Amazon rainforest involves the destruction of many thousands of ancient trees. Data for Brazil, which contains around 60% of the Amazonian forest, indicates that although between 1970 and 2013 19% of forest had been lost, the rate of deforestation dropped by 80% between 2004 and 2012, with between 65% and 70% of deforestation due to clearing for cattle ranching.[23] However, with changes in government rates increased dramatically from 2012, such that by 2019 deforestation of the Brazilian forest reached levels not seen since 2007. However since a change of government in 2022 deforestation rates have slowed dramatically.[24]

Indonesia, with a primary forest area at around 40% of that in Brazil, has been heading in the same direction as Brazil did prior to 2022, with an authoritative study published in 2014 indicating that the area of forest destruction in 2012 had overtaken that in Brazil, when the annual loss was 460,000 ha compared with 840,000 ha in Indonesia [12]. However, the Global Forest Watch study referred to above reported that forest loss in 2014 from countries other than Brazil and Indonesia amounted to 65% more than the two added together. This includes countries such as Cambodia, where tree cover loss has been increasing at a faster rate than anywhere on the planet as trees are felled to make way for rubber plantations, whilst West Africa and the Congo basin, for example, has seen a surge in palm oil cultivation displacing primary forest.

---

[21] http://www.rmtrr.org/oldlist.htm

[22] http://www.wri.org/blog/2015/09/satellites-uncover-5-surprising-hotspots-tree-cover-loss

[23] http://rainforests.mongabay.com/20brazil.htm

[24] https://www.sciencemag.org/news/2019/11/brazil-s-deforestation-exploding-and-2020-will-be-worse

Peat is formed in waterlogged land when decomposition is slower than deposition of organic matter and requires either anaerobic or acidic conditions, with generally few nutrients. Peatlands make up some 3% of global land area and contain around 30% of global soil carbon stocks, which is approximately twice that of the forest biomass. In low temperature regions such as tundra, both plant growth and decomposition rates are low hence peat deposits tend to be thin but widespread, whilst in temperate zones where there has been prolific tree growth peat deposits tend to be locally thick where appropriate conditions of geology, geomorphology and soil chemistry allow. Peat formation in tropical rainforests tends to be very strong because of the sheer volume of organic material deposited.

Some of the world's most rapidly diminishing forests, such as those in Indonesia and Malaysia, grow in peat swamps. For example, 62% of the world's peat swamp forests are to be found in the Indomalaya region, where there has been widespread logging (both legal and illegal) and conversion to agricultural land, such as the Government sponsored Mega Rice Project in Indonesian Borneo (Kalimantan).[25] This involved the clearing of over a million hectares of primary peat swamp forest in the late 1990s and the construction of more than 4600 km of drainage channels which unintentionally dried out the peatlands entirely and rendered them both useless for growing rice and subject to periodic and devastating fires. Apart from the destruction of millions of slow-growing trees and associated ecosystems this disastrous project led to the death of an estimated 5000 orangutans and countless other animals, leaving the Bornean orangutan population in 2016 as low as 104,700 by some estimates[26] which is less than half of the number a century ago. Like ripples in a pond this project has had a number of other impacts such as the opening up of central Kalimantan to illegal loggers and hunters, with an increasing population coming into conflict with and threatening the existence of a wide range of wildlife, whilst 'the economy and quality of life of indigenous people had been irreparably disrupted'.[27]

> The combination of forest destruction, land clearance and an exceptionally severe El Nino climatic event in 1997 led to the severest forest and peatland fires ever known in this region. Between half a million and three million hectares of vegetation burned, much of it on peat. The fires penetrated into the dried-out surface peat to a depth of up to 1.5 metres.[28] (See also section on *Terrestrial ecosystems and biosphere* in Chap. 3).

In colder climes peatlands are frequently associated with sphagnum moss which works symbiotically with the peat formation process. The ongoing threats to peat in these regions are typically from some combination of: exploitation as a fuel, inundation with seawater, drainage for agriculture, fire and open-cast mining, such as the exploitation of tar sands in Canada.

[25] https://www.insideindonesia.org/editions/edition-6542/kalimantan-s-peatland-disaster

[26] http://www.worldwildlife.org/species/orangutan

[27] https://www.insideindonesia.org/editions/edition-6542/kalimantan-s-peatland-disaster

[28] https://www.insideindonesia.org/editions/edition-6542/kalimantan-s-peatland-disaster

According to the Wetlands International 2013 submission to the Subsidiary Body for Scientific and Technological Advice (SBSTA) of the UNFCCC: 'Drained peatlands lose their intrinsic properties in terms of water storage and regulation, and emit enormous quantities of CO_2. In addition, drained peatlands inevitably subside—exposing large tracts of land to permanent flooding and rendering them unsuitable for agricultural use, which in turn endangers livelihoods and food security'.[29] Natural and re-wetted peatlands emit methane but the net greenhouse emissions are close to zero because of their carbon absorption characteristics.

5.5 Water

We have already seen in Chap. 2 how important water resources are in geopolitical terms both in transboundary disputes and fuelling of conflicts. Despite the fact that 71% of the earth's surface is covered with water most of the world's population is dependent upon access to fresh water from rivers, reservoirs or sources underground. As the world population grows, fresh water is becoming an increasingly precious and finite resource.

Although some regions have access to plentiful supplies of fresh water many are impacted by drought and the resultant stress. According to a report from the UN Food & Agriculture Organization (FAO) published in 2013:

> People in drought-prone areas live under a continuous threat of hunger and malnutrition. Droughts in Africa during the 1980s forced 10 million farmers to abandon their land; over 1 million died. The impact of the drought was exaggerated in Ethiopia by government policies which reduced the reserves of food grains held on farms. The cost in lives would have been far greater but for the mobilization of food aid by the international community.[30]

Figure 5.4 shows areas that experienced drought and wet conditions during 2015, based on modelling of the Palmer Drought Severity Index (PDSI), which integrates data on precipitation and temperature to estimate evapotranspiration and relative soil dryness. We have seen in the section on *Terrestrial ecosystems and the biosphere* in Chap. 3 that desertification and drought had by 2013 occupied between 33% and 35% of the total land surface area, increasing by around 0.1% per year on average globally.

It follows that the water available for human populations varies enormously across the planet. This not only depends on precipitation but also on the infrastructure available to capture, store, treat and distribute water, which in turn depends on the finances and geopolitics of the country or region concerned. For example, the 7-year long drought in the US state of California which started in 2011 was partly mitigated by spending billions of dollars on increasing reservoir capacity and infrastructure. The average annual per capita water abstraction rate for the USA in 2016

[29] http://unfccc.int/resource/docs/2013/smsn/ngo/381.pdf

[30] http://www.fao.org/3/u8480e/U8480E05.htm

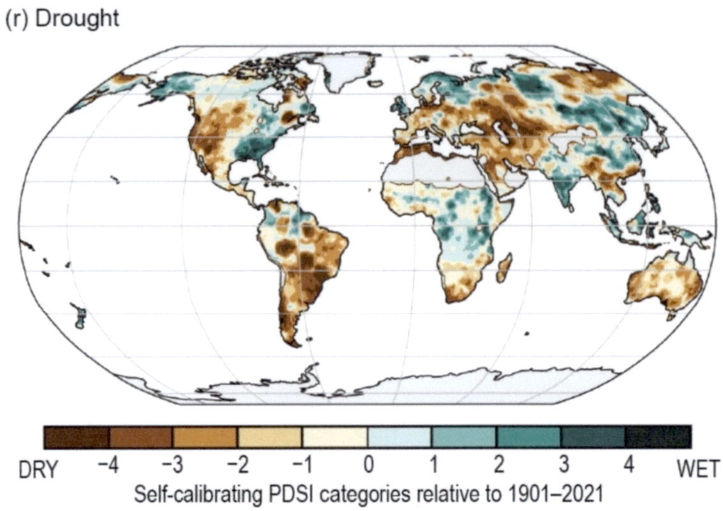

Fig. 5.4 Palmer drought severity index (From [13])

was 1583 m^3,[31] which compares with 213 m^3 for the UK (2007 data)[32] and 80.5 m^3/capita per annum for Ethiopia (2005 data).[33] These figures include water used by agriculture and industry as well as for domestic supplies. According to UN data, 844 million people worldwide were not able to source uncontaminated drinking water in 2015.[34] Over much of sub-Saharan Africa more than 50% of the population in that year were unable to access a basic drinking water service, that is water from protected wells or springs was not available within 30 min' journey time. In Eritrea, the figure was 81%. Between 1980 and 2015, an average of 780,000 people died globally from drinking contaminated water.

Comparing the world maps in Figs. 5.4 and 5.5, it is interesting to note that the countries that had the worst access to drinking water in 2015 were mostly in sub-Saharan Africa, for which there is an absence of PDSI data, whilst many other countries experienced extreme drought during that year. This indicates the importance of poverty, conflict and poor infrastructure in creating the access problems highlighted in Fig. 5.5.

Contaminated drinking water can spread diseases such as cholera, typhoid and hepatitis A, which are responsible for most of the deaths referred to above; whilst droughts can decimate agriculture and result in dehydration, famine and death. According to the UN FAO's AQUASTAT database water withdrawals for

---

[31] https://www.statista.com/statistics/263156/water-consumption-in-selected-countries/

[32] http://www.europarl.europa.eu/RegData/etudes/BRIE/2015/559470/EPRS_BRI(2015)559470_EN.pdf

[33] https://www.indexmundi.com/ethiopia/freshwater_withdrawal_(domestic_industrial_agricultural).html

[34] https://www.statista.com/chart/17445/global-access-to-safe-drinking-water/

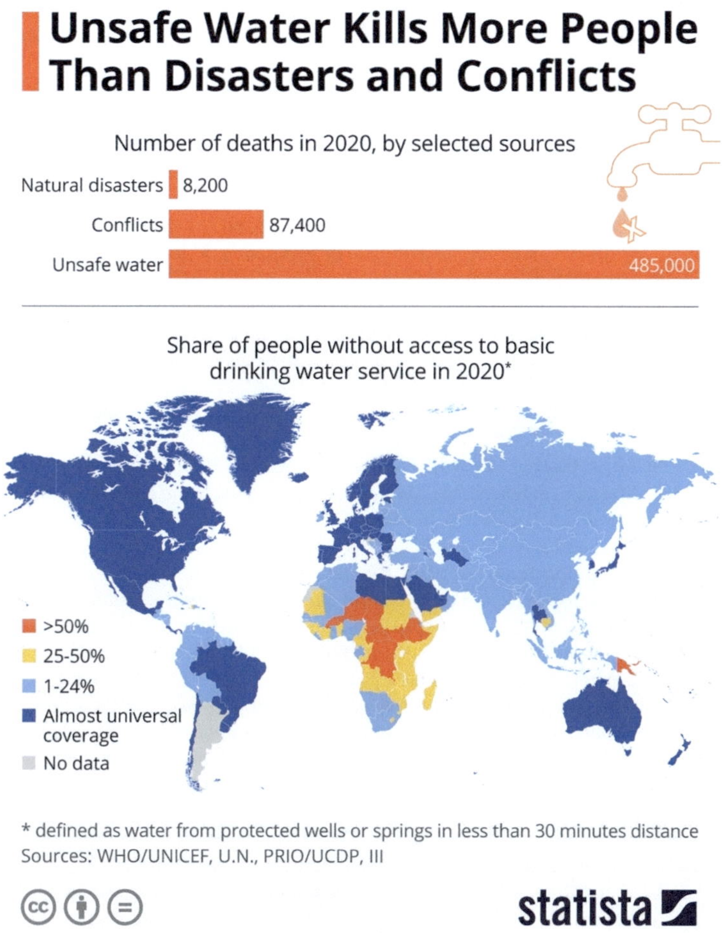

Fig. 5.5 Summary of statistics for access to drinking water (Image: Statista: https://www.statista.com/chart/17445/global-access-to-safe-drinking-water/)

agriculture make up some 69% of total water consumption globally, whereas industry demands 19% and 'municipal' use represents around 12%. The proportions vary dramatically between countries however: for example the percentages for the USA are 36%, 51.2% and 12.8% for agriculture, industry and municipal withdrawals respectively; for the UK 12.8%, 14.5% and 62.1%; and for Ethiopia 91.8%, 0.5% and 7.7%; which provides an illuminating snapshot of the socio-economic differences between these three countries.[35]

[35] http://www.fao.org/nr/water/aquastat/water_use/index.stm

We will be discussing the environmental needs and impacts of agriculture in Chap. 9 and the environmental impact, carbon and water footprints of construction in Chap. 7.

References

1. Malthus, T. (1798). *An essay on the principle of population*. J. Johnson. http://www.esp.org/books/malthus/population/malthus.pdf
2. Jevons, W. (1866). *The coal question: An inquiry concerning the progress of the nation, and the probable exhaustion of our coal-mines*. Macmillan. Republished by Forgotten Books, 2012.
3. Simon, J. (1998). *The ultimate resource 2*. Princeton University Press. http://www.juliansimon.com/writings/Ultimate_Resource/
4. Bardi, U. (2014). *Extracted, how the quest for mineral wealth is plundering the planet*. Chelsea Green.
5. Berners-Lee, M., & Clark, D. (2013). *The burning question*. Profile Books.
6. McGlade, C., Speirs, J., & Sorrell, S. (2012). *A review of regional and global estimates of unconventional gas resources*. UK Energy Research Centre. http://www.ukerc.ac.uk/publications/a--review-of-regional-and-global-estimates-of-unconventional-gas-resources.html
7. MILES Project Consortium. (2015). *Beyond the numbers: Understanding the transformation induced by INDCs*. Paris: *Institut du développement durable et des relations internationales* (IDDRI). https://www.iddri.org/en/publications-and-events/study/beyond-numbers-understanding-transformation-induced-indcs#:~:text=The%20INDCs%20imply%20a%20significant,contributions%20and%20the%20global%20framework
8. EC. (2014). *Communication from the Commission to the European Parliament, the Council, the European Economic and Social Committee and the Committee of the Regions on the review of the list of critical raw materials for the EU and the implementation of the Raw Materials Initiative* COM/2014/0297. European Commission. http://eur-lex.europa.eu/legal-content/EN/TXT/?uri=CELEX:52014DC0297
9. Sahney, S., Benton, M., & Falcon-Lang, H. (2010). Rainforest collapse triggered carboniferous tetrapod diversification in Euramerica. *Geology, 38*(12), 1079–1082.
10. Crowther, T., et al. (2015). Mapping tree density at a global scale. *Nature, 525*, 201–205.
11. Vieira, S., Trumbore, S., Camargo, P., Selhorst, D., Chambers, J., Higuchi, N., & Martinelli, L. (2005). Slow growth rates of Amazonian trees: Consequences for carbon cycling. *Proceedings of the National Academy of Sciences, 102*(51), 18502–18507.
12. Margano, B., Potapov, P., Turubanova, S., Stolle, F., & Hansen, M. (2014). Primary forest cover loss in Indonesia over 2000–2012. *Nature Climate Change, 4*(8), 730–734.
13. Barichivich, J., Osborn, T., Harris, I., van der Schrier, G., & Jones, P. (2023). Monitoring global drought using the self-calibrating Palmer Drought Severity Index. *Bulletin of the American Meteorological Society, 104*(9), S66–S67.

Chapter 6
Ecological and Carbon Footprinting: One Planet to Live On

I have spread my dreams under your feet.
Tread softly because you tread on my dreams.
Extract from Aedh wishes for the cloths of heaven by W.B. Yeats, from: The Wind Among the Reeds (1898)
Walk lightly upon the earth and live in balance and harmony.

Anon—Native American Elder

6.1 Introduction

So far in this book we have discussed the numerous interconnected pressures on the planet and their actual and potential impacts. In order to visualise the cumulative impacts of most of these pressures, it is useful to identify one or more of the common denominators that allow a single number or rating to be determined to represent the totality of these impacts and enable comparisons and benchmarking for individual countries, activities, processes, buildings, equipment, people and many more.

The two main techniques in common use are ecological footprinting and carbon footprinting, which convert impacts into biologically productive area and equivalent carbon dioxide emissions respectively. Separate tools are also available for water footprinting and mapping biodiversity. Footprints are usually expressed on a per capita (person) basis for anything from an individual person through nations up to the entire planet. There are also assessment tools that attempt to rate environmental or sustainability performance for buildings, for example, such as the UK-based Building Research Establishment Environmental Assessment Method (BREEAM) and the US-based Leadership in Energy and Environmental Design (LEED) tools. These use internal benchmarks and weightings across a wide variety of factors in order to calculate a single number and rating that can be used for comparison between buildings at various stages of the design process through to the completed building. There are numerous versions of environmental assessment tool, some of which focus on specific building types, others designed for existing buildings and others which reflect the standards, legislation and practices of specific countries.

P. Appleby, *The Overburdened Planet*,
https://doi.org/10.1007/978-3-032-01510-5_6

In this chapter, we will focus on ecological and carbon footprinting, whist also briefly touching on other schemes such as water and biodiversity footprints and indices, looking at how they have evolved, how they are calculated and their application.

6.2 Ecological Footprinting

Ecological footprinting was developed by William Rees and Mathis Wackernagel at the University of British Columbia in Canada between 1990 and 1994 resulting in the 1996 book *Our Ecological Footprint: Reducing Human Impact on the Earth* [1]. In Chap. 1 of their book, the authors define ecological footprint analysis as 'an accounting tool that enables us to estimate the resource consumption and waste assimilation requirements of a defined human population or economy in terms of a corresponding productive land area'. Mathis Wackernagel went on to become the president of the Global Footprint Network (GFN) which produces regular Ecological Footprint reports for most countries.

The reports produced by GFN include for each country an estimate of 'biocapacity', which it defines as 'the ecosystems' capacity to produce biological materials used by people and to absorb waste material generated by humans, under current management schemes and extraction technologies'.

The Ecological Footprint as defined by GFN tracks six key ecosystem services demanded by the human economy as follows:

1. Plant-based food and fibre products—cropland;
2. Animal-based food and animal products—livestock/agricultural land;
3. Fish-based food products—fishing grounds;
4. Timber and forest products—forest and woodland;
5. Carbon dioxide absorption—carbon uptake land;
6. Shelter and infrastructure—built-up area.

Ecological Footprint and biocapacity are measured in global hectares where 'a global hectare is a biologically productive hectare with world average biological productivity for a given year'.[1] GFN estimated the per capita biocapacity of the Earth in 2011 by dividing 12 billion hectares of productive land by a population of 7 billion, giving 1.72 global hectares (gha). Its estimate for per capita Ecological Footprint for the same year was 2.7 which when divided by biocapacity gives the number of planets required to meet demand or 'Planet Equivalent' at 1.54, whilst the figure for the USA was 4.82. In other words, the planet was using approximately 18 months of resources in 1 year, whilst the USA used nearly 5 years' worth of resources over the same period.[2] This concept was adopted by the British sustainability specialists Bioregional in 2003 as the basis for 'One Planet Living' and a key

[1] http://www.footprintnetwork.org/en/index.php/GFN/page/glossary/#productivity

[2] https://data.footprintnetwork.org/?_ga=2.149179573.224510351.1723804463-1750309797.1723804463#/countryTrends?type=BCtot,EFCtot&cn=351

driver for the Beddington Zero Energy residential development in South London (BedZED). As well as incorporating energy efficient design and renewable energy generation this development incorporated many of the features that would be required to achieve very high scores under Ecobuild (the residential version of BREEAM and the predecessor to the Code for Sustainable Homes and the Home Quality Mark—see Chap. 7).

The Global Footprint Network uses accounting terminology to express the difference between biocapacity and Ecological Footprint such that if a nation's Footprint is larger than its Capacity then it is said to be in 'biocapacity deficit' or in credit if the reverse is true. These are also referred to as 'ecological deficit' and 'ecological reserve', respectively. According to the National Footprint Accounts published by GFN as part of the 2024 Public Data Package, the Ecological Footprint per capita for the USA in 2022 was 2.52 billion gha (global hectares) whilst its 'biocapacity' was 1.26 bn gha, leaving a deficit of 1.26 bn gha. The equivalent figures for the UK were 241 million, 69 million and 172 million, respectively; and for China were 5.28 bn, 1.17 bn and 4.11 bn (Fig. 6.1). Interestingly, whereas the deficit for the USA and UK had been mostly either steady or falling since the 1970s, China had increased from a small credit in the 1960s, with biocapacity only rising very slowly between 1961 and 2022. During this time Its Ecological Footprint increased by a factor of approaching tenfold (see Fig. 6.1), with the largest component due to the increase in its carbon footprint. Because of its massive population, at 5.28 billion gha in 2022 China had the largest Ecological Footprint of any country on the planet, more than double the US 2.52 billion gha, India was 1.47 bn gha, whilst the UK was 241 million.[3]

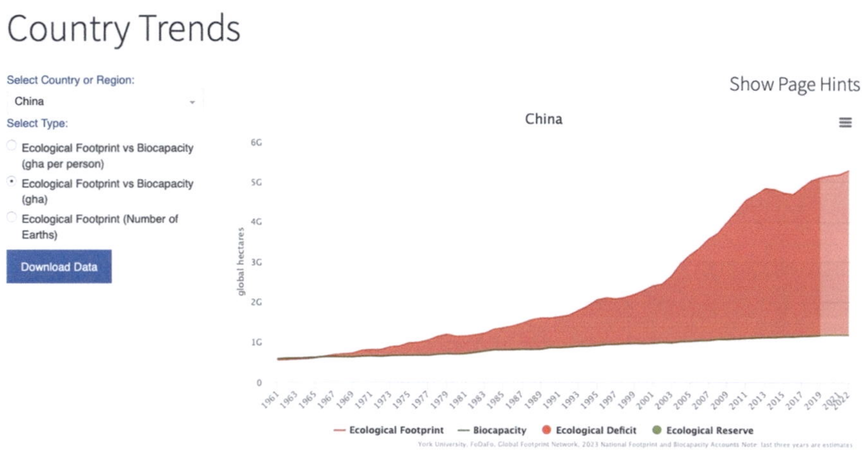

Fig. 6.1 Ecological Footprint and Biocapacity for China between 1961 and 2022 (https://data.footprintnetwork.org/?_ga=2.149179573.224510351.1723804463-1750309797.1723804463#/countryTrends?type=BCtot,EFCtot&cn=351)

[3] https://data.footprintnetwork.org/?_ga=2.149179573.224510351.1723804463-1750309797.1723804463#/countryTrends?type=BCtot,EFCtot&cn=351

A net ecological/biocapacity deficit for the whole planet is referred to as 'global overshoot'. This occurs when globally the demand for resources exceeds the supply available from the biosphere resulting in the depletion of non-renewable resources and natural capital and the accumulation of waste at harmful levels in the atmosphere, sea and land. Global overshoot is usually expressed in Planet Equivalents which, as we saw earlier, was estimated to be 1.54 in 2011. As can be seen in Fig. 6.2, GFN has estimated this to have increased to 1.6 by 2016, whilst predicting that it could continue to rise given a 'business as usual' scenario, reaching 2.0 by 2030. This graph also shows a more optimistic scenario, assuming a panoply of mitigation measures to reduce anthropogenic impacts, including a reduction in carbon dioxide emissions by 30%, and hence a lowering of Ecological Footprint and Planet Equivalent to 2010 levels.

In order to take account of all relevant environmental impacts, all of the parameters are converted to biological resources and uptake and hence land area expressed as global hectares using yield and equivalence factors that assume world average productivities. Using data from numerous national and international sources, yields are determined for the 'primary products' supplied by cropland, forest, grazing land and fisheries; whilst land that is built upon is assumed to be potentially productive land and thus has a negative impact on biocapacity. Equivalence factors for carbon dioxide emissions are determined taking account of the sequestration by large water bodies and forests. Primary products are the unprocessed biological materials as harvested, such as trees and crops. Equivalence factors are used to convert a specific

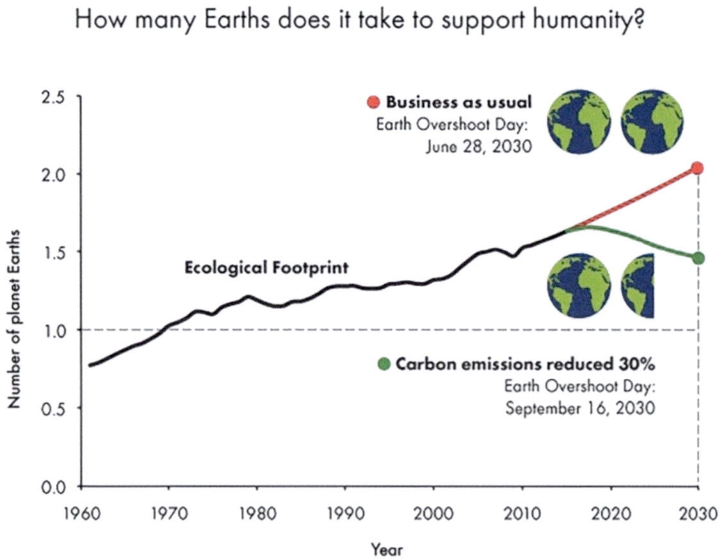

Fig. 6.2 Global overshoot in planet equivalents from 1960 to 2030 (From: http://www.footprintnetwork.org/en/index.php/GFN/page/world_footprint/)

activity or process into a 'universal unit of biologically productive area'[4] where the 'average productivity of all biologically productive land and water area on Earth' has an Equivalence factor of one.

Procedures and standards are also available to calculate Ecological Footprints for products and organisations for example [2].

It is important to note that, as with all metrics that attempt to integrate a large number of variables, the Ecological Footprint has its weaknesses. In a paper on the evolution of a 'Footprint Family' Alessandro Galli of GFN recognises the limitations of the Ecological Footprint. In expressing all impacts as areas of productive land, the Ecological Footprint rewards productivity over biodiversity: for example, when ancient woodland is replaced with a crop the higher the intensity of crop production the greater the reward. Similarly the procedure is blind to eco-toxicity, land degradation and soil erosion (such as might be associated with intensive farming), the life expectancy of resources, freshwater consumption and greenhouse gases other than carbon dioxide. Furthermore, Ecological Footprint 'relies on the assumption that all emissions are absorbed only by forests, neglecting carbon uptake by other biomes' [3].

Although these inadequacies may have little impact on the accuracy of global and national metrics, they increase in importance with decreasing scale: for example, when estimating Ecological Footprints for organisations, communities or individuals. Galli's Footprint Family comprises a combination of ecological, carbon and water footprints. Clearly there are overlaps between these footprints since Ecological Footprint expresses CO_2 emissions in terms of the area of land required to absorb them (but not other greenhouse gases) whilst Carbon Footprint will account for the carbon associated with water pumping and treatment, where applicable. It is also possible to determine the Ecological Footprint required for a utility to provide a given supply of water, whilst the availability of water influences the biocapacity of ecosystems.

In their paper from 2014 Galli and others stated that 'understanding the interactions between biodiversity, the drivers of biodiversity loss and humanity's (Ecological) Footprint is fundamental to slowing, halting and reversing the ongoing declines in natural ecosystems and populations of wild species' [4]. For example, Footprint analysis has been used to highlight that consumption of agricultural and forest products in some countries creates human-induced pressure on crop and forest ecosystems in other countries, and developing economies in particular.

Whilst the Ecological Footprint does give some indication of the pressures on ecosystem services and biodiversity it by no means gives the full picture. Indicators for the condition of these have been developed under the auspices of the UN and arising from the Convention on Biodiversity and the Intergovernmental Platform on Biodiversity and Ecosystem Services (IPBES). The so-called Aichi Biodiversity Targets, referred to in Chap. 2 in the section headed '*Agreements*', are being monitored by the Biodiversity Indicators Partnership (see *Biodiversity Footprint* section below).

---

[4] http://www.footprintnetwork.org/en/index.php/GFN/page/glossary/#productivity

Similarly the Ecological Footprint gives no indication of health and quality of life, whereas the human development index (HDI) incorporates a number of indicators that rate factors that contribute to these. Unlike the Ecological and other footprints, however, it is only applied to individual nations. Administered by the UN Development Programme (UNDP), the HDI comprises: life expectancy at birth; expected years of schooling; mean years of schooling; and gross national income per capita. The Human Development Report is published annually by the UNDP and includes a wealth of indices and statistics which supplement the HDI, including the Inequality-adjusted HDI, the Gender Inequality Index and Multidimensional Poverty Index. The Report also includes statistics on health outcomes, environmental sustainability and perceptions of well-being, including an 'Overall Life Satisfaction Index'.[5]

Box 6.1 Bioregional One Planet Living Principles

Health and happiness

Encouraging active, sociable, meaningful lives to promote good health and well being

Equity and local economy

Creating safe, equitable places to live and work which support local prosperity and international fair trade

Culture and community

Nurturing local identity and heritage, empowering communities and promoting a culture of sustainable living

Travel and transport

Reducing the need to travel, and encouraging walking, cycling, and low carbon transport

Land and nature

Protecting and restoring land for the benefit of people and wildlife

Local and sustainable food

Promoting sustainable humane farming and healthy diets high in local, seasonal, and organic food and vegetable protein

Sustainable water

Using water efficiently, protecting local water resources and reducing as flooding, drought (and water course pollution)

Materials and products

Using materials from sustainable sources and promoting products which help people reduce consumption

Zero waste

Reducing consumption, reusing and recycling to achieve zero waste and zero pollution

Zero carbon

Making buildings energy efficient and supplying all energy with renewables

[5] http://hdr.undp.org/en/content/human-development-index-hdi

BioRegional and the World Wildlife Fund developed a 'One Planet Living' calculator based on the ten principles set out in Box 6.1.[6] This allows individuals to complete a simple on-line questionnaire that enables an estimate of Ecological Footprint in global hectares and number of planets to be determined, along with carbon footprint in tonnes of CO_2 equivalent (see next section). The BioRegional 'Eco-footprint' translates all of the key issues covered in the GFN Ecological Footprint into lifestyle choices, along with a number of socio-economic factors that have no direct impact on ecological or carbon footprint, such as community engagement. Having been involved in the development of the One Planet Living programme WWF moved to a carbon based footprinting scheme similar to that integrated into the BioRegional calculator (see next section). The calculator also provides tips on how to reduce personal ecological and carbon footprints under each of the headings in Box 6.1.

6.3 Carbon Footprinting

As highlighted above carbon footprinting is used to indicate greenhouse gas emissions, from the individual to the global. Typically a carbon footprint comprises an estimate of the 'total greenhouse gas emissions caused directly and indirectly by a person, organisation, event or product'.[7] The footprint gives the total equivalent carbon dioxide emissions for all six of the Kyoto Protocol greenhouse gases: Carbon dioxide (CO_2), Methane (CH_4), Nitrous oxide (N_2O), Hydrofluorocarbons (HFCs), Perfluorocarbons (PFCs) and Sulphur hexafluoride (SF_6).

> A carbon footprint is measured in tonnes of carbon dioxide equivalent (tCO_{2eq}). The carbon dioxide equivalent (CO_{2eq}) allows the different greenhouse gases to be compared on a like-for-like basis relative to one unit of CO_2. CO_{2eq} is calculated by multiplying the emissions of each of the six greenhouse gases by its 100 year global warming potential (GWP).[8]

The Carbon Trust website provides access to a calculator tool for estimating the carbon footprint for an organisation, product, service or value chain. Its procedures are based on the international standards developed by the Greenhouse Gas Protocol, which is a collaborative initiative between the World Resources Institute (WRI) and the World Business Council for Sustainable Development (WBCSD) [5].

As illustrated in Fig. 6.3, the Greenhouse Gas (GHG) Protocol categorises emissions within three 'Scopes', where Scope 1 includes direct emissions associated with the activity for which a footprint is being estimated, such as company facilities and vehicles; Scope 2 covers indirect emissions from energy supplies and Scope 3 comprises the emissions from upstream and downstream activities from 'cradle to grave'.

---

[6] http://www.bioregional.com/oneplanetliving/

[7] https://www.carbontrust.com/resources/guides/carbon-footprinting-and-reporting/carbon-footprinting

[8] https://www.carbontrust.com/resources/guides/carbon-footprinting-and-reporting/carbon-footprinting

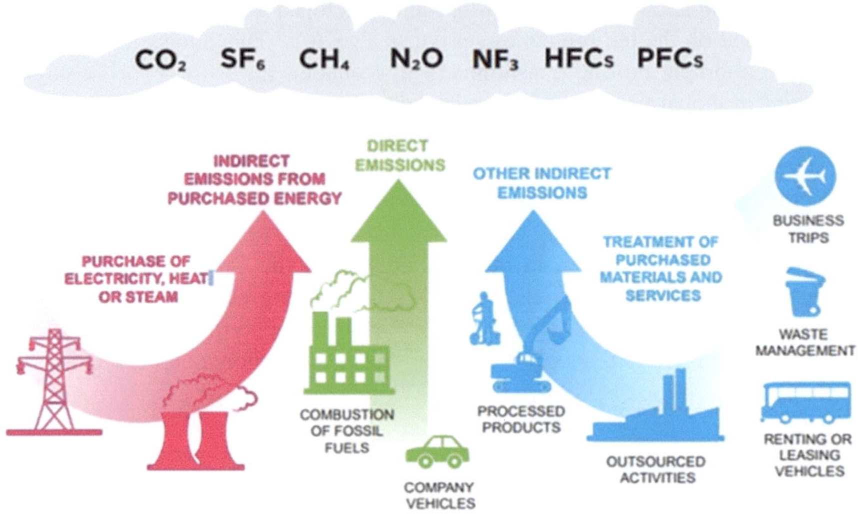

Fig. 6.3 Carbon footprint overview (From: https://commons.wikimedia.org/wiki/File:Carbon_Accounting_Scopes.png)

The GHG Protocol defines eight upstream emission sources and seven downstream, providing a comprehensive inventory of all of the GHG emissions associated with, for example, the manufacture, delivery and operation of a product or service.

The GHG Protocol calculation methodologies have been developed primarily to help companies report on their emissions as required in the UK, for example, by the *Companies Act 2006 (Strategic and Directors' Reports) Regulations 2013*. These regulations were introduced to satisfy EU Regulation 525/2013 which set out a 'mechanism for reporting greenhouse gas emissions', updating earlier requirements that were introduced following the publication of the Kyoto Protocol [6]. The USA, although being one of the few countries globally not to ratify the Kyoto Protocol, introduced its own mandatory Greenhouse Gas Reporting Programme, administered by the Environmental Protection Agency, under the 2008 Consolidated Appropriations Act.[9] This applied to facilities that emitted more than 25,000 tonnes of carbon dioxide equivalent per annum and included only direct and fugitive emissions of the six 'Kyoto' greenhouse gases associated with combustion and processes. Bearing in mind that a large coal-fired power station in the USA might emit 20 million tonnes of CO_2 in a year and a major steel works perhaps 10 million tonnes, the GHG Reporting Programme is likely to have captured most medium to large facilities in the USA.

In 2018 the ISO published its international standard on the development of a carbon footprint for products. ISO 16067: 2018 'doesn't tackle any management

[9] https://www.epa.gov/ghgreporting

systems, but it still plays a pivotal role in any company seeking to establish greater measures for sustainability—as it tracks the (greenhouse gas) emissions created throughout the entire lifecycle of a product, even after it is manufactured and sold'.[10]

At the other end of the spectrum, there are a number of carbon footprint calculators that allow individuals to estimate their personal footprint. One example of these is the 'WWF Footprint Calculator' which has been developed by the World Wildlife Fund in concert with the Stockholm Environmental Institute at the York and Leeds Universities.[11] This UK-orientated on-line tool brackets greenhouse gas emissions under the headings of food, home, travel and 'stuff' (consumables), including an allowance for the per capita emissions associated with 'government', public infrastructure and services.

6.4 Water Footprinting

As the title suggests the water footprint is the volume of water consumed for a specified activity, process, product, geographical area, company or individual. It can be given as a total volume of water consumed, or per capita, per tonne of production, per hectare of cropland or unit of currency; per annum or for a specified period. It may stand alone or be integrated with ecological and carbon footprints. As set out in Box 6.2, the Water Footprint Network has developed a rationale based on definitions of water resources categorised as green, blue and grey.

Box 6.2 Definitions of Water Resources (From: Water Network[12])

Green water footprint is water from precipitation that is stored in the root zone of the soil and evaporated, transpired or incorporated by plants. It is particularly relevant for agricultural, horticultural and forestry products.

Blue water footprint is water that has been sourced from surface or groundwater resources and is either evaporated, incorporated into a product or taken from one body of water and returned to another, or returned at a different time. Irrigated agriculture, industry and domestic water use can each have a blue water footprint.

Grey water footprint is the amount of fresh water required to assimilate pollutants to meet specific water quality standards. The grey water considers point-source pollution discharged to a freshwater resource directly through a pipe or indirectly through runoff or leaching from the soil, impervious surfaces, or other diffuse sources.

[10] https://greenly.earth/en-gb/blog/company-guide/iso-14067-meaning-standard-and-requirements

[11] https://footprint.wwf.org.uk

[12] http://waterfootprint.org/en/water-footprint/what-is-water-footprint/

Water that is delivered to consumers from reservoirs or underground sources is considered to be 'blue' water, whilst that required for plant growth will normally be 'green water' supplemented by 'blue' sources through irrigation when necessary. When building up an holistic inventory of water consumption, it is important to establish the impact of pollution on the water cycle. For example, most domestic, commercial, industrial and agricultural processes result in polluted water being discharged ultimately to natural sinks via drains or through run-off. These pollutants may be diluted to acceptable levels so that water can be re-used or the waste water may require filtration and chemical treatment before re-use. These processes may also impact carbon footprint, because of the energy required for pumping, chemical manufacture and the construction and operation of plant and buildings.

The Water Footprint Network (WFN) is an international non-profit foundation based in the Netherlands and established in 2008 by Professor Arjen Hoekstra who developed the concept of water footprinting whilst working at the UNESCO Institute of Water Education in 2002. One of the tools that the WFN has developed is a Global Water Footprint Assessment Standard that provides a methodology for conducting water footprint accounting and sustainability assessment [7].

According to Section 1.3 of the Water Footprint Assessment Manual: 'Broadly speaking, the goal of assessing water footprints is to analyse how human activities or specific products relate to issues of water scarcity and pollution, and to see how activities and products can become more sustainable from a water perspective' [7]. When either undertaking or interpreting a water footprint analysis, it is vital to establish the basis on which consumption is determined. For example, the WFN Manual recommends the following decision-making process:

1. Establish whether sources are green, blue or both and whether allowance has to be made for diluting pollutants.
2. Determine how far back along the supply chain water consumption data is required or available.
3. Establish the level of 'spatiotemporal explication' as defined by Hoekstra et al., i.e. whether scope is global (Level 1); national, regional, catchment-specific (Level 2); product-specific or site-specific (Level 3).
4. Determine period over which data is required and/or available.
5. Decide whether direct consumption only is sufficient, bearing in mind that indirect consumption can be significantly greater than direct.
6. For national footprints, decide whether exported consumption is to be accounted for; such as water consumed in other countries in the manufacture of goods supplied to the country for which a footprint is being estimated.

The WFN Manual provides separate procedures for estimating water footprints for growing crops and trees, products, consumers, geographically delineated areas, nations, catchments, river basins, administrative units and businesses. Some of these will incorporate the results from other estimates: for example, national footprints will include consumption associated with growing crops, manufacturing products and supplying consumers and businesses, both within the nation concerned and suppliers out with national boundaries.

Estimating the water footprint of agricultural, horticultural and forestry products is a complex process for which a number of competing models have been developed. For the estimation of evapotranspiration associated with growing crops and trees, the WFN Manual suggests using the 'irrigation schedule option' of the CROPWAT model developed by the UN Food and Agriculture Organisation (FAO). The calculation of total water footprint requires an estimate of evaporative losses associated with storing and transporting irrigation water, evapotranspiration during the growing cycle, volume of water incorporated into the crop and the grey water footprint. An example for growing sugar beet in a 1 ha irrigated field in the Valladolid region of Spain is given in Appendix II of the WFN Manual [7]. This estimates the green water footprint at 15 m^3/tonne of beet harvested, the blue at 82 and the grey at 22: a total WF of 120 m^3/tonne.

The grey water footprint for growing crops is determined from the chemical application rate for the most critical pollutant: which is the one that results in the highest water use (usually fertiliser). This is multiplied by the leaching run-off fraction: that is, that portion that is not taken up by the crop; and divided by the difference between the maximum allowable concentration in nature minus the background concentration found in the sink water body.

Appendix III of the manual then goes on to examine the water footprint of the ultimate refined sugar product obtained through processing locally-sourced sugar beet. The inputs and outputs associated with the manufacture of refined sugar at the factory in Valladolid are as follows:

- Water required for washing of beet—water filtered for re-use and soil recycled;
- Warm water used to separate sugar from pulp in diffusion towers—pulp sold for use as animal feed;
- Raw juice purified by addition of lime and CO_2, producing sugar pulp and lime fertiliser for agricultural use;
- Water evaporated from sugar pulp producing sugar and syrup, which is separated and sold on for use in alcohol industry;
- Refined sugar granules produced by seeding with fine sugar crystals.

For this sugar refining factory in Valladolid the water footprint for the manufacturing process was reported to be 107 m^3/tonne for green water, 570 for blue and 152 for grey: a total WF of 829 m^3/tonne. This includes the 120 m^3/tonne for the unprocessed beet referred to above. This means that for every kilogram of sugar produced from this factory in Spain more than 800 L of water is required. This compares with the global average of between 865 m^3/tonne and 1782 m^3/tonne for sugar derived from cane [8]. The WFN have estimated water footprints for a wide variety of crops, livestock and products, resulting in some extraordinary statistics such as the 15,415 L required to produce a single kilogram of beef and the 18,925 L for 1 kg of roasted coffee [9]. This is equivalent to around 3 and 4 months, respectively, of domestic water consumption per head for a typical British household to manufacture 1 kg of product.

Unlike the carbon footprint, however, the water footprint cannot directly translate into an environmental impact. A significant portion (the green footprint) may be derived from rainfall, so it is the availability of sufficient rain that dictates the impact; blue water supply on the other hand may have both long-term economic and energy (and hence carbon) implications; whilst pollution and chemicals in run-off may have both environmental and economic impacts. In order to provide a means of expressing these impacts in an integrated way, the WFN has developed a Water Footprint Sustainability Assessment (see Chap. 4 of its Manual) [7]:

> In order to get an idea of what the (water) footprint size means, one will need to compare the water footprint to the available freshwater resources …. in the same way as one needs to compare the ecological footprint to the available biologically productive space.

In other words sustainability of water supply is primarily a measure of water scarcity in terms of availability versus demand, with potential impacts on water quality and human health. The sustainability of water at the point of consumption, whether it be directly or through the life cycle of a product, is dependent on the sustainability at interlinked stages of the water cycle: from catchment through process, product, producer to consumer. As well as environmental impacts, social and economic impacts are also highlighted. Environmental sustainability is assessed in terms of water quality, river and groundwater flows and impact on ecosystems. Social sustainability is dependent upon the availability of 'fresh' (potable) water to meet basic human needs; whilst economic sustainability assessment determines on the one hand the balance between the cost benefit of the available supply versus the cost of extraction, storage, distribution and treatment; and on the other hand, the economic loss from water shortages to agriculture, industry, commerce and livelihoods, with knock-on impacts on health and welfare services, and so on. Clearly water stress and drought can have major impacts at a fundamental societal and geopolitical level, particularly when prolonged.

The WFN Manual discusses the application of a series of water footprint impact indices (WFII) which can be calculated separately for green, blue and grey footprints. These are aggregated and weighted measures of environmental impact calculated from the green or blue WF multiplied by the relevant Water Scarcity Factor (WS) and the grey WF times the Water Pollution Level (WPL). WS is the ratio of water footprint to water availability (WA) and WPL is estimated from the grey water footprint divided by the total run-off for the catchment under consideration. However the Manual's authors consider that 'water footprint impact indices are useful only as crude indicators of the environmental impact at catchment level; the aggregated indices (do not) contain spatial or temporal information. As a basis for formulating appropriate response measures, it is more useful to identify 'hotspots'…' indicating the location and/or time of year when water scarcity or pollution load needs remedial action.

There is also an International Standard for water footprinting. 'ISO 14046:2014 specifies principles, requirements and guidelines related to water footprint assessment of products, processes and organisations based on life cycle assessment (LCA)'.[13] This differs from the WFN approach in that it determines the

[13] https://www.bsigroup.com/en-AE/sustainable-development-goals/iso-14046/#:~:text=Water%20is%20an%20essential%20natural,environmental%20impacts%20related%20to%20water

environmental impact of water use through LCA but does not separate green, blue and grey footprints. It is part of the ISO's suite of environmental reporting standards and is designed primarily for organisations to assess and report their water footprint.

6.5 Biodiversity Indicators

We have already seen that, despite its name, Ecological Footprint does not include a discrete indication of biodiversity. There are, however, a number of metrics that have been developed for specific application. At a global and national level, the UN Convention on Biological Diversity (CBD) resulted in the development of Aichi Biodiversity Targets as part of a Strategic Plan for Biodiversity for the decade commencing in 2011. The Biodiversity Indicator Partnership (BIP) was established under the auspices of the UN Environmental Programme World Conservation Monitoring Centre (UNEP-WCMC) in 2007 and subsequently supported the development of indicators for the Aichi Biodiversity Targets rolled out through National Biodiversity Strategy Action Plans (NBSAPs).

The 20 Aichi Targets are grouped under six Strategic Goals (A to E)—see Box 6.3.

Box 6.3 Aichi Biodiversity Targets Strategic Goals[14]

Strategic Goal A: Address the underlying causes of biodiversity loss by mainstreaming biodiversity across government and society.

Strategic Goal B: Reduce the direct pressures on biodiversity and promote sustainable use.

Strategic Goal C: To improve the status of biodiversity by safeguarding ecosystems, species and genetic diversity.

Strategic Goal D: Enhance the benefits to all from biodiversity and ecosystem services.

Strategic Goal E: Enhance implementation through participatory planning, knowledge management and capacity building.

In 2005 the BIP published guidance that enabled individual countries to establish biodiversity indicators tailored to their particular challenges [10]. One such indicator is forest cover, which can be easily represented on a map such as shown in Fig. 6.4, which graphically indicates deforestation in Madagascar.

One important indicator has been developed by the World Wildlife Fund (WWF) and the Institute of Zoology. 'The Living Planet Index (LPI) is a measure of the state of the world's biological diversity based on population trends of vertebrate species from terrestrial, freshwater and marine habitats. The LPI was adopted by the

[14] https://www.cbd.int/sp/targets/

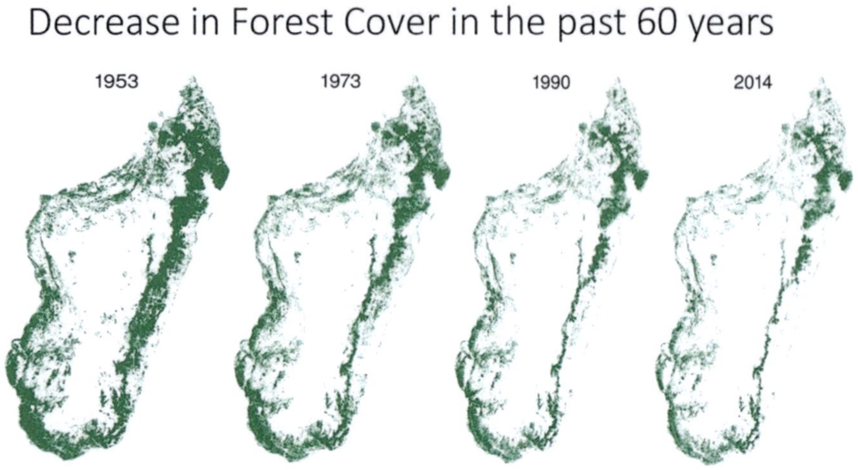

Fig. 6.4 Loss of forest cover in Madagascar (Image based on study by [11]; From: https://player.slideplayer.com/89/14260900/slides/slide_13.jpg)

Convention of Biological Diversity (CBD) as an indicator of progress towards its 2011–2020 targets and can play an important role in monitoring progress towards the post-2020 goals and targets'.[15]

By 2017 The Living Planet Database (LPD) held time-series data for over 18,000 populations of more than 3600 mammal, bird, fish, reptile and amphibian species from around the world, gathered from various sources such as journals, online databases and government reports.[16]

Figures 6.5 and 6.6 show the downward trend in LPI within 95% confidence limits, both globally and for all three habitats since the baseline of 1970.[17] The weighted average drop in overall global LPI between 1970 and 2020 was 73%, including an alarming 81% decline in migratory freshwater species and 79% reduction in forest specialist species between 1970 and 2018.

The UK-based Wildlife Trusts' Biodiversity Benchmark 'is a standard for assessing and certifying an organisation's systems for achieving continual biodiversity protection and enhancement on its landholdings and their implementation'.[18] It rewards responsible land management whilst recognising efforts to welcome and encourage wildlife through enhanced permeability, acting on advice from suitably qualified persons and engaging effectively with community and employees.

The UK Building Research Establishment Environmental Assessment Method (BREEAM) includes an assessment of 'ecological value' which is an indicator of

---

[15] https://www.livingplanetindex.org

[16] https://www.zsl.org/science/indicators-and-assessments-unit/living-planet-index

[17] http://www.livingplanetindex.org/projects?main_page_project=LivingPlanetReport&home_flag=1

[18] http://www.wildlifetrusts.org/biodiversitybenchmark

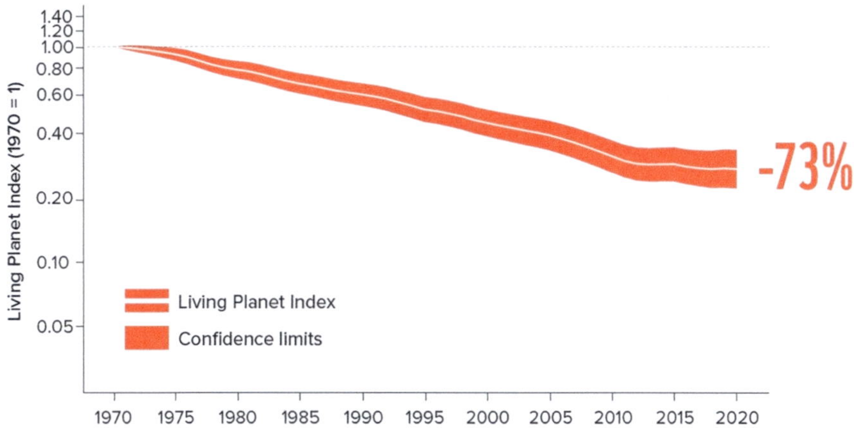

Fig. 6.5 Overall global Living Planet Index 1970 to 2020 [12] (Source: https://www.livingplanetindex.org/lpi)

biodiversity based on an area-weighted average of plant species or taxa for a site's broad habitat types.[19] Using data from the British Countryside Survey for 2007 BRE provides average ecological values for different habitat types expressed as total number of species, based on a plot size of 200 m^2. For example arable or horticultural land typically contains 10.25 species whilst land that has been abandoned for more than 20 years could harbour 19.9 species on average. It should be noted that these values are not area dependant and that any area of land can be expected to have the same number of species as the 200 m^2 sample plot. Of the habitat types listed by BRE, with around 43 species, calcareous grassland is by far the richest. However, according to a report on *Plant indicators for habitats in good condition* from the UK-based Centre for Ecology & Hydrology of the Natural Environment Research Council:

> Total plant species richness can be deceptive as a measure of biodiversity. Higher species numbers may be an indication of nutrient enrichment or disturbance. The map in this publication (see Fig. 6.6) is based on plant species that are positive indicators, or characteristic species, of a particular habitat. Their presence indicates that a habitat is in good condition.[20]

The map at Fig. 6.6 shows the percentage of potential indicators within the habitat types present in the mapped locations. These are indicators of good ecosystem health using a grid of 1 km^2, based on data from the 2007 Countryside Survey and habitat indicators compiled by the Botanical Society of the British Isles. No habitats include anywhere near all of the species that are listed in the Society's Common Standards, hence even the most species rich areas of England only fall within the 8–14% band indicated on the map.

---

[19] http://www.breeam.com/BREEAMInt2016SchemeDocument/#11_landuse/le02.htm%3FTocPath%3D12.0%2520Land%2520Use%2520and%2520Ecology%7C_____2

[20] https://eip.ceh.ac.uk/naturalengland-ncmaps/reportsData

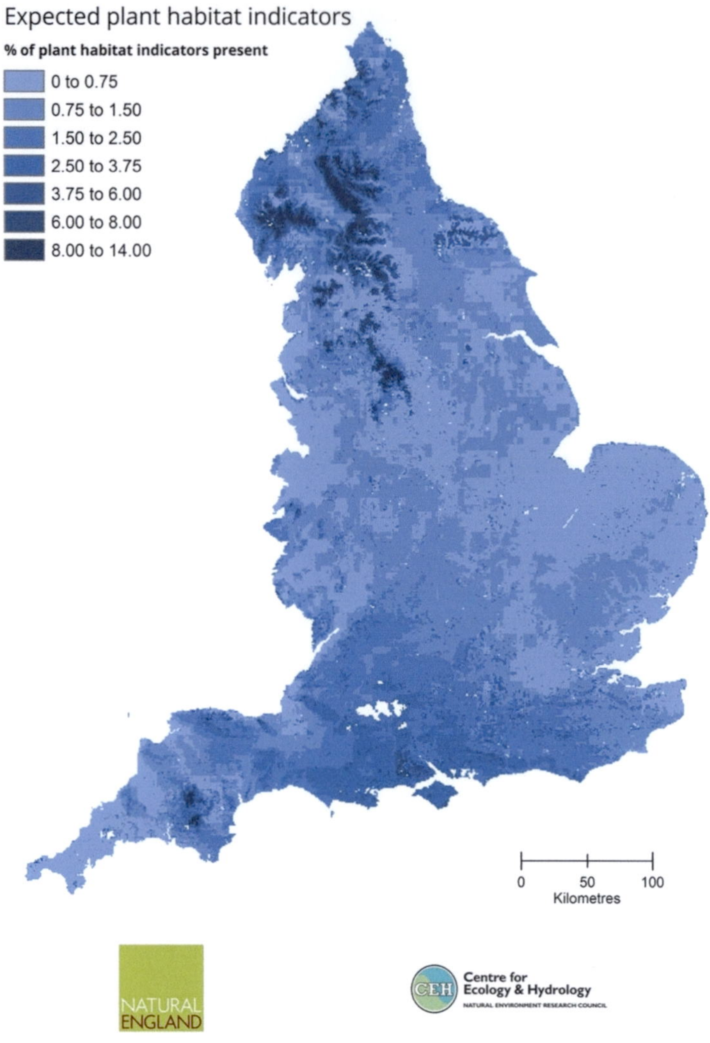

Fig. 6.6 Plant Habitat Indicators for England (Copyright Natural England; From: Maskell et al. [13], copyright Natural England)

A different definition of biodiversity footprint has been adopted by the Dutch Government in terms of a reduction in mean species abundance (MSA) or biodiversity loss.[21] This footprint is expressed in thousands of square kilometres of biodiversity loss due to food production, wood and paper, clothing and textiles, biofuels and other land use, although excluding the direct impacts of climate change. The Dutch

[21] http://www.clo.nl/en/indicators/en1464-dutch-biodiversity-footprint

Government has used this measure to track biodiversity loss since 1990, indicating a peak loss 11% higher than in 1990 by 2000, falling back to 5% by 2010.

The UK's Joint Nature Conservation Committee (JNCC) developed a biodiversity footprint to report on the impacts on biodiversity on foreign direct investment in 2009.[22] This incorporated a suite of markers including habitat change, over exploitation, invasive species, pollution and climate change impacts adopted from the Millennium Ecosystem Assessment (see also section headed *Agreements* in Chap. 2).

Although this Footprint has not been widely used the JNCC is also responsible for the UK's Biodiversity Indicators, based on the Aichi Biodiversity Targets referred to above. These do not provide a single footprint but separate performance indicators under 24 headings used to track both long and short-term changes by indicating whether there have been improvements, degradation or no change for each heading or sub-heading.[23]

The JNCC has reported on trends in these indicators annually since 2007. They are divided into five categories comprising: conservation and economic issues; management and pressures; habitat and species status; biodiversity and ecosystem services; record keeping and government expenditure. The impacts of climate change are tracked through identifying trends in the removal of greenhouse gases by UK forests and the 'Spring index' which indicates the timing of biological spring events by recording the number of days after 31 December that key events occur. These are the first flowering of hawthorn and horse chestnut, the first flight of the orange-tip butterfly and the first sighting of the swallow. Data is available for these sightings from 1891 to 1947 and from 1999 onwards. Although the average start dates fluctuate wildly from year to year the mean trend between 1999 and 2015 was reported to be some 7 days earlier than between 1891 and 1947.

References

1. Wackernagel, M., & Rees, W. (1996). *Our ecological footprint: Reducing human impact on the earth*. New Society Publishers.
2. Global Footprint Network. (2009). *Ecological footprint standards 2009*. Global Footprint Network.
3. Galli, A., Wiedmann, T., Erin, E., Knoblauch, D., Ewing, B., & Giljum, S. (2011). *Integrating ecological, carbon and water footprint: Defining the footprint family and its application in tracking human pressure on the planet*. 7th Framework Programme for Research and Technological Development. One Planet Economy Network. https://www.ecologic.eu/node/4218/printable/pdf
4. Galli, A., Wackernagel, M., Iha, K., & Lazarus, E. 2014 Ecological footprint: Implications for biodiversity. *Biological Conversation*. https://www.researchgate.net/publication/258261958_Ecological_Footprint_Implications_for_biodiversity

[22] https://hub.jncc.gov.uk/assets/3f86becf-69cb-4bc1-9358-723248272983

[23] https://jncc.gov.uk/our-work/uk-biodiversity-indicators/

5. Greenhouse Gas Protocol. (2011). *Corporate Value Chain (Scope3) accounting and reporting standard*. http://www.ghgprotocol.org/standards/scope-3-standard
6. European Union. (2013). *Regulation (EU) No 525/2013 of The European Parliament and of The Council of 21 May 2013 on a mechanism for monitoring and reporting greenhouse gas emissions and for reporting other information at national and Union level relevant to climate change and repealing Decision No 280/2004/EC*. EU. http://eur-lex.europa.eu/legal-content/EN/TXT/PDF/?uri=CELEX:32013R0525&from=EN
7. Hoekstra, A., Chapagain, A., Aldaya, M., & Mekonnen, M. (2011). *Global water footprint assessment manual*. Earthscan. http://waterfootprint.org/en/resources/publications/water-footprint-assessment-manual-global-standard/
8. Mekonnen, M., & Hoekstra, A. (2010a). *Value of Water Research Report Series No 47: The green, blue and grey water footprint of crops and derived crop products*. UNESCO-IHE. http://waterfootprint.org/media/downloads/Report47-WaterFootprintCrops-Vol1.pdf
9. Mekonnen, M., & Hoekstra, A. (2010b). *Value of Water Research Report Series No 48: The green, blue and Grey water footprint of farm animals and animals products*. UNESCO-IHE. https://www.waterfootprint.org/resources/Report-48-WaterFootprint-AnimalProducts-Vol1.pdf
10. Biodiversity Indicator Partnership. (2005). *Guidance for national biodiversity indicator development and use*. UNEP World Conservation Monitoring Centre. https://www.bipindicators.net/system/resources/files/000/002/191/original/Framework_Brochure_UK_0311_LOWRES_%281%29.pdf?1481634262
11. Vieilledent, G., et al. (2018). Combining global tree cover loss data with historical national forest cover maps to look at six decades of deforestation and forest fragmentation in Madagascar. *Biological Conversation, 222*, 189–197. https://www.sciencedirect.com/science/article/pii/S0006320718301125
12. WWF. (2024). *Living planet report 2024—A system in peril*. WWF.
13. Maskell, L., Henrys, P., Norton, L., & Smart, S. (2016). *Model estimates of expected diversity of positive plant habitat condition indicators*. NERC Environmental Information Data Centre. https://doi.org/10.5285/cc5ae9b1-43a0-475e-9157-a9b7fccb24e7

Chapter 7
Buildings and Construction: New Build Versus Retrofit

Buildings represent the largest energy-consuming sector in the economy, with over one-third of all final energy and half of global electricity consumed there. As a result, they are also responsible for approximately one-third of global carbon emissions. With an expected population increase of 2.5 billion people by 2050, and given improvements in economic development and living standards, energy use in the buildings sector is set to rise sharply, placing additional pressure on the energy system.

From the Forward to Transition to Sustainable Buildings: Strategies and Opportunities to 2050 [1]

7.1 Introduction

So far in this book we have been mainly discussing impacts, such as the changes that are occurring to this planet and its atmosphere, as well as associated geopolitical phenomena. The next few chapters will be examining the drivers for these impacts, broadly speaking: buildings, structures, infrastructure, industry, transport, agriculture, food production and land use. These represent the main influences that humans have had on earth, drawing on natural resources and changing the face of the planet.

In this chapter, we will consider the cradle to grave environmental impacts of buildings and structures: their operation and maintenance and the embodied impacts of the materials used in their construction. In this context we will examine the arguments around refurbishment and retrofitting existing buildings and structures against demolition and building anew.

The lifetime environmental impacts of a building or a structure comprise some combination of those involved in the extraction, manufacture, demolition and disposal of materials, as well as operation and ongoing maintenance. Within this lifetime there may be a number of changes in use, refurbishments, restacking, upgrades, extensions and the like. For a structure which is not primarily a shelter for people,

P. Appleby, *The Overburdened Planet*,
https://doi.org/10.1007/978-3-032-01510-5_7

such as a bridge or a dam, the balance between embodied and operational impacts will be more heavily weighted towards the embodied. Although there will usually be some operational requirements, such as power for lighting and equipment, the life cycle impacts of materials will dominate.

There have been numerous publications providing guidance on the sustainable design, construction, retrofit and operation of buildings, such as those by the current author [2, 3]. In this chapter, we will distil the essence of what makes buildings and structures sustainable and explore the decisions required when balancing operational versus embodied impacts. Although whole life carbon dioxide emissions are frequently the dominant issue in sustainable design, there are numerous other factors under the headings of environmental, social and economic sustainability. Many of these issues are shared with other human endeavours, such as travelling, food production and manufacture; however, it is important to define the framework for sustainable design, construction and operation of buildings and structures so we can understand what is involved in setting benchmarks and predicting and measuring performance in relation to the criteria listed in Box 7.1. These are all life cycle issues that relate to construction, operation and disposal.

Box 7.1 Life Cycle Sustainability Criteria for Buildings and Structures

Environmental

- CO_2 equivalent (greenhouse gas) emissions
- Water use
- Waste to landfill
- Pollutant emissions to atmosphere, water courses and land
- Light and noise pollution
- Ecological impacts
- Flood risk
- Impact on wind environment
- Impact on vehicle movements
- Health and well-being of occupants and neighbours
- Sourcing, manufacture and delivery of materials

Socio-economic

- Employment
- Amenities and facilities
- Security

Many of the criteria listed in Box 7.1 are inter-related: for example, greenhouse gas emissions are associated with the treatment and distribution of water, the collection, treatment, recycling and storage of waste, vehicle movements, construction activities and embodied carbon of materials. On the other hand, water may be consumed in the processing of waste, management of soft landscaping, ongoing building cleaning and maintenance, manufacture of materials and construction activities such as cement mixing and suppression of dust. Similarly, waste is generated from building operation, construction activities and materials manufacture. There are also links between the health and well-being of occupants and the energy, and associated carbon emissions, required to maintain comfort conditions. As well as CO_2 emissions from vehicle movements, boilers and at power stations, there are other emissions to atmosphere such as oxides of nitrogen and particulate matter from burning fossil fuels. There are also potential emissions of a cocktail of pollutants to atmosphere, water courses and land from waste disposal and treatment. Air, light and noise pollution both arising from and entering into a building and associated landscape and vehicle movements can impact on the health and well-being of neighbours and occupants, respectively. Furthermore, changes in wind environment, generated by a tall building, for example, can create unpleasant and sometimes dangerous conditions for pedestrians as well as discomfort for balcony users.

As we will see later in this chapter ecological impacts of a development can be either positive or negative depending on the ecological value of a site prior to construction and how many quality species are added by the new soft landscape. Landscape has the potential to ameliorate the effects of air, light and noise pollution both on the building and its occupants as well as neighbours. When considered cumulatively over a large area, soft landscaping absorbs CO_2 and other pollutants from the atmosphere and contributes to the mitigation of global temperature rise whilst locally reducing the so-called heat island effect. Landscaping can also reduce flood risk by incorporating features that store water and attenuate the impact of storms on downstream drains and water courses. The natural filtration provided by the roots of plants such as reeds and grasses can also be used to reduce pollution in run-off water. They can even be used for the off-grid filtration of sewage and industrial waste. Trees can also provide shade, provide a visual barrier and privacy, whilst also filtering out some airborne pollutants.

As well as operational carbon emissions, air quality, noise, light pollution and wind environment, health and well-being is also potentially impacted by access to green spaces and plants. Whilst choice of building and cleaning materials can impact air quality depending on the degree of outgassing and solvent emissions. Reliance on motor vehicles will be linked to some extent to the proximity of local amenities such as shops, schools, health centres and recreational spaces. This can impact on health through the potential for walking, cycling and sports activities. However, pedestrian and cycle routes tend to be avoided if they are not safe, hence 'secure by design' measures are a vital weapon in the designer's arsenal.

In most instances, the challenge is not simply to design a building or structure in isolation but to consider a building or group of buildings as part of an existing community or complex, which requires the masterplanning of a group of buildings within a landscape, all within the context of a larger area and its constraints. In which case it is important to consider the cumulative impacts of the buildings and associated infrastructure, landscape and facilities.

All of these interconnections pose a challenge to the creation of a structure for this chapter, hence the following sections will attempt to be holistic: for example, by incorporating all those aspects of building construction and operation that generate or absorb greenhouse gases under the heading of *Energy and greenhouse gases.*

This chapter will conclude with a discussion on retrofitting existing buildings versus demolition and new build, along with building on greenfield versus brownfield sites.

7.2 Energy and Greenhouse Gases

It is clear that buildings and structures make a major contribution to climate change through lifetime CO_2 and other greenhouse gas (GHG) emissions. As populations have grown and become more urbanised, and with standards of living on the increase, buildings have become more heavily serviced and more structures have been required as part of a burgeoning infrastructure. At the same time buildings are becoming more energy efficient and power generation steadily decarbonised. The balance between CO_2/GHG emissions associated with the operation and construction of buildings is changing as building fabric and services become more energy efficient and low carbon and renewable technologies become more prevalent. Operational emissions include those associated with environmental services such as heating, ventilation, air conditioning and lighting; and the so-called unregulated emissions—such as from office equipment, industrial processes and white goods. In the UK unregulated emissions are those that are not directly a function of building design and hence not included in the estimate of CO_2 emissions required for the purposes of assessing compliance with Building Regulations.

Good practice in building design is to adopt a combination of insulation, air tightness and passive design measures to reduce dependence on mechanical means to control the indoor environment. This is sometimes referred to as the 'fabric first' principle. However, as this approach reduces the carbon emissions associated with controlling indoor environmental conditions the significance of embodied carbon over the lifetime of the building increases.

A strategy for the design of buildings to minimise operational carbon should be based on the following hierarchy:

- Passive design features and energy efficient fabric;
- Energy efficient building services;
- Renewable energy generation—either building-integrated, on-site or facilitated off-site.

UK Building Regulations have been heading in this direction. With the publication of the Future Homes Standard in 2025 and subsequent enhancements to Part L of the Building Regulations, as well as promotion of the voluntary UK Net Zero Carbon Buildings Standard, all of which encouraged designers to incorporate the above features into the design of buildings, but without relying on off-site measures and assuming decarbonisation of the electricity supply. The Net Zero Carbon Buildings Standard also incorporates embodied carbon targets.[1]

[1] https://www.nzcbuildings.co.uk/_files/ugd/6ea7ba_1ef36b6835de46668f2ad8b589ff1b93.pdf

7.2.1 Passive Design

Passive design uses 'free' energy from natural energy sources, such as wind and sunlight, along with the thermodynamic properties of the building to reduce its carbon footprint. Examples include natural ventilation and daylight to mitigate or reduce the need for artificial cooling and lighting respectively.

The principles of designing a multi-storey building with an atrium to drive natural ventilation are shown in Fig. 7.1, extracted from the *CIBSE Application Manual on the design of natural ventilation for non-domestic buildings* [4]. For natural ventilation to be successful, the summertime temperatures in the occupied areas must not drift above acceptable levels for prolonged periods; openings to outdoors must not allow the ingress of pollution or noise at levels that create discomfort; and openings must be controllable. In winter months, when overheating is rarely a problem, sufficient air must be able to enter the building to prevent excessive pollution accumulating from internal sources, whilst minimising energy demand from infiltration.

Whether a building can be naturally ventilated or requires air conditioning for at least part of the year, passive design measures can be used to both improve performance and reduce energy consumption. The principle passive design measures that are available for controlling thermal environment and ventilation employ thermal mass and solar shading; whilst daytime illumination levels and quality are dependent upon the design of transparent elements in the fabric and the colour and

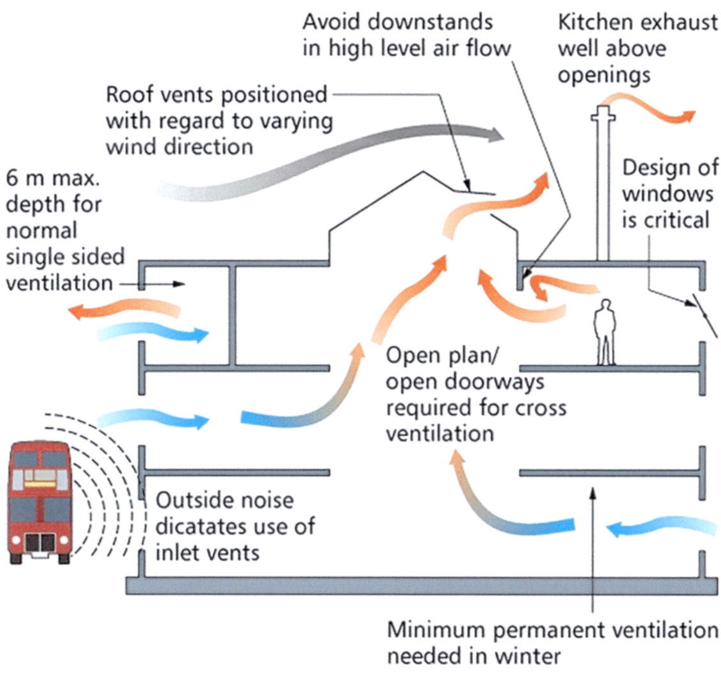

Fig. 7.1 Simplified section through an office building showing principles of natural ventilation (Source: [4, 5])

positioning of internal elements and shading. Good daylighting combined with lighting controls can reduce reliance on artificial lighting, thus lowering heat gains and electricity usage.

Exposed thermal mass can significantly reduce summertime temperatures. For example, a building that has exposed concrete soffits and night ventilation can expect to experience peak daytime temperatures between 2 and 4 °C lower during warm summer days compared with a building having lightweight suspended ceilings, whilst delaying the time at which the peak occurs by up to 6 h.

Figure 7.2 shows how exposed soffits can be used to create sculptured ceilings that provide a number of environmental benefits. This internal view of the Powergen headquarters in Coventry, designed by Bennetts Associates and completed in 1994, shows coffered slabs that not only provide thermal mass for night cooling but also create reflectors for uplighting with integral acoustic baffles that absorb sound, reduce reverberation time and mitigate against the risk of noise disturbance between neighbouring workstations.

The Passivhaus scheme[2] for the design of energy efficient dwellings was developed in Germany in the late 1980s. Although there had been numerous individual energy-efficient houses built in the USA and Europe following the energy crisis of the early 1970s, Passivhaus was the first attempt to develop a standardised approach to their design and construction. Essentially these are super-insulated houses with high levels of air tightness and mechanical ventilation with heat recovery to ensure good indoor air quality in winter (see below).

7.2.2 *Energy Efficient Building Services*

Depending on location and microclimate a building that is well insulated and employs many of the passive design measures referred to above may be able to operate for much of its life requiring little space heating, no mechanical cooling and ventilation and minimal artificial illumination during daylight hours. Good practice in the energy efficient design of building services is dealt with in detail in numerous publications, such as the CIBSE[3] Guides and and ASHRAE[4] Handbooks, along with other guidance from these professional bodies based in the UK and the USA, respectively. These publications are used extensively internationally, particularly in the English-speaking world.

Figure 7.3 provides an indication of the contribution each of the main energy consuming services makes to overall carbon dioxide emissions for a typical air conditioned office building built in the UK in the early 2000s. A similar building built in 2025 should have significantly lower regulated emissions, hence the proportion

[2] https://www.passivhaustrust.org.uk

[3] Chartered Institution of Building Services Engineers

[4] American Society of Heating, Refrigerating and Air-conditioning Engineers

Fig. 7.2 Exposed concrete coffered slabs at powergen headquarters, coventry (Source: Bennetts Associates architects; photograph copyright: Peter Cook/VIEW)

associated with office equipment, computer room, catering and 'other' would typically be higher: perhaps double the percentage indicated in Fig. 7.3.

Similarly for homes, the figure of 57% attributed to heating in Fig. 7.4 is significantly reduced as thermal performance of the fabric and heating system improves to meet tighter Building Regulations requirements. Non-regulated emissions

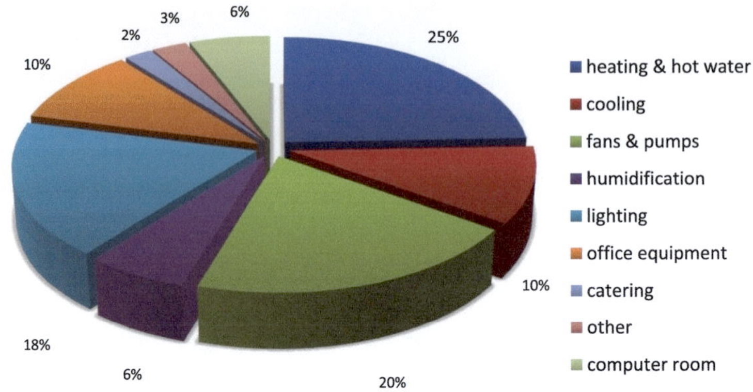

Fig. 7.3 CO_2 emissions associated with energy use in a typical UK air conditioned office building prior to 1999 (Image by Author using data from: Energy Consumption Guide 19 [6])

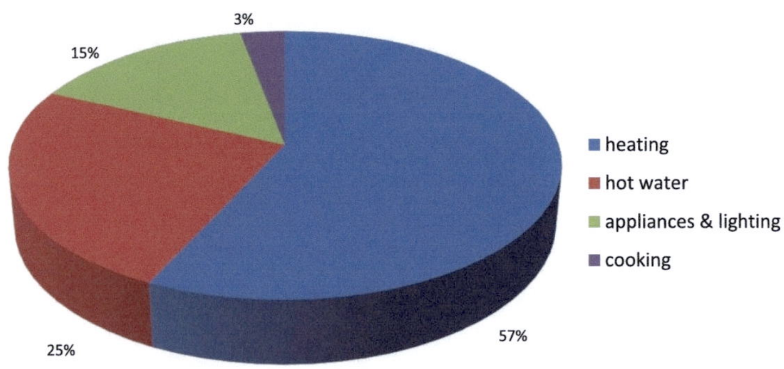

Fig. 7.4 Typical energy use in UK dwelling meeting 2006 Building Regulations (Image by Author using data from: Low Carbon Transition Plan DECC in 2009[5])

associated with appliances and cooking may exceed 50% of the total and, if the aim of zero regulated carbon is achieved, then unregulated emissions represent the only carbon dioxide emissions associated with a dwelling, other than embodied emissions (which are not regulated). However, zero operational carbon emissions might require some combination of on- and off-site renewable energy generation as well as a decarbonised electricity supply. Clearly if all regulated and unregulated demand is powered by electricity from a decarbonised grid, then the dwelling will have zero operational carbon dioxide emissions.

For a given building with associated thermal performance and comfort requirements, the building services design must be optimised to minimise energy bills and carbon dioxide emissions. Other factors come into play, such as controllability, maintainability, plant space, noise, capital and lifetime costs. Although a naturally

[5] https://www.gov.uk/government/publications/the-uk-low-carbon-transition-plan-national-strategy-for-climate-and-energy

ventilated building with perimeter heating may offer the simplest and cheapest solution, this may not always be an option in noisy and polluted city centre sites, particularly where there are sedentary activities and relatively high internal heat gains.

There is of course no single model for building services that can be applied across all building types and climate conditions. A terraced house built in England in 2017, for example, may have required only an array of radiators supplied with hot water from a gas-fired condensing boiler—that also provides instantaneous hot water for washing—and 240-V electrical distribution serving lighting, white goods and other equipment; whilst a large hospital in the Middle East required air conditioning throughout and a complex system of piped gases, plumbing and electrical services.

Hot and humid climates having relatively small variations in external thermal conditions between night and day present the fewest opportunities for passive design: for example, the thermal mass of a building offers limited opportunities for night cooling. Hence energy saving has to be achieved through reductions in electricity demand resulting from the operation of fans, pumps, refrigeration, lighting and other electrical equipment as well as the clever use of automatic control and building energy management systems (BEMS).

The choice and positioning of heat emitters has historically been based on a combination of optimising comfort and energy consumption. For traditionally constructed centrally-heated dwellings in the UK, for example, emitters would typically be located beneath windows to offset the downdraught from the cold surface of glass and outdoor air entering via trickle vents and infiltration. As buildings have become better insulated, and in particular with high performance triple-glazed windows, improved air tightness and ventilation being delivered by mechanical ventilation with heat recovery (MVHR), the surface area and heat output of heat emitters has shrunk and their location become less important.

7.2.2.1 Heating, Ventilation and Air Conditioning

The most carbon efficient method for generating heat for a particular building will depend on the seasonal and diurnal thermal loads, balance between heating and domestic hot water loads, availability of a gas infrastructure and carbon intensity of electricity supply. In countries where there has been major decarbonisation of the electricity supply, such as France and Norway, there is a strong argument for heating with electricity—either directly with electric heaters or indirectly using heat pumps. In areas where there is no gas infrastructure, then the choice might be between oil, solid fuels and electricity. Cost will inevitably play a significant role in this decision since kWh costs of electricity in the UK are typically three to four times higher than for gas. However, this ratio varies enormously across EU countries from 1.25 in Denmark to 4 in Italy, for example.[6] In some countries, particularly in the

[6] Based on average prices to industrial users in 2017 from: https://www.businesselectricityprices.org.uk/europe/

developing world, the infrastructure for gas is not as extensive as for electricity, whilst in some warmer parts of the world the only demand for heating is to provide hot water for personal washing and some industrial applications.

When deciding on the method of heating for a building, options include: direct electricity, night storage electric heating, heat pumps (air-to-air, ground-source, water-source), or low pressure hot water heating from gas, oil or solid fuel fired boilers. It is also possible to use the waste heat from electricity generation to provide space and water heating: referred to as combined heat and power (CHP) or cogeneration. Although packaged CHP units have been developed that are suitable for individual dwellings, the most common application is for commercial buildings and district energy schemes, up to and including waste heat from power stations. Because they work most efficiently when all of the waste heat produced from generating electricity is used they are best suited to applications where there is a steady requirement for heat coinciding with electrical demand: for domestic hot water, swimming pool or industrial processes, for example.

Boilers were originally developed to generate steam—hence the name—used to power trains, cars, boats and turbines to generate electricity. Their use in transportation dates back to as early as the late eighteenth century, whilst central heating systems using steam distributed through pipework started to be employed in the early nineteenth century in England, about the same time as the first hot water heating systems appeared. Although these first systems used pipework as their principal heating surface the first radiators were developed in Russia in the 1850s and became common towards the end of the nineteenth century.

The earliest boilers used coal or wood and, although oil and gas-firing came to dominate the central heating sector when air pollution became a major health hazard, manifested as smog, for example, in Britain in the early 1950s. Despite this, modern wood chip boilers continue to be offered as a 'low carbon' option in the twenty-first century.

Since being developed in the Netherlands in the late 1970s condensing boilers have become the norms for residential heating applications globally. Although oil-fired versions are available, the majority are fired using natural gas. In the UK, the Building Regulations were modified in 2005 such that all new boilers for dwellings had to achieve an efficiency that could only be achieved using gas-fired condensing boilers. Oil-fired condensing boilers were included in 2007.

Condensing boilers achieve a higher efficiency than conventional boilers by the incorporation of an additional heat exchanger in the flue gases through which return water from the heating system passes at a temperature that condenses the water from the flue gases thus increasing heat transfer and boiler efficiency by 20–30%. This requires return water temperatures lower than the 70 °C that has been typical for low temperature hot water heating. Since the heat output of heat emitters is dependent upon mean surface temperature and surface area, larger emitters are required for condensing boilers to operate at optimal conditions.

By definition air conditioning incorporates comfort cooling when heat gains drive internal temperatures above a prescribed level. It may also include a degree of humidity control in summer and winter. There are many different types of air conditioning system: ranging from simple packaged air conditioning units through to sophisticated multi-zonal systems based on centralised variable air volume (VAV) technologies or chilled beams, for example. Where a building has to be sealed to exclude noise and air pollution then a ducted air system can be installed, using filters to remove particulates, in-duct sound attenuators to reduce noise levels and fans to circulate air through the building. This system may simply be used to supply ventilation air to occupants, whilst limiting temperature rise without recourse to mechanical cooling. Where cooling is required, this would normally be zoned to provide thermostatic control as heat gains vary in different parts of the building. Winter heating, if required, can be provided with zonal reheat coils or mixing boxes in the ducted system or through a separate heating system, feeding perimeter radiators, for example.

Historically air conditioning has employed refrigeration to cool air supplied to rooms through forced convection. In its simplest form, this has something akin to a domestic refrigerator at its heart: using a vapour compression refrigeration cycle comprising an electrically driven compressor, a condenser, evaporator and expansion valve (Fig. 7.5). These are connected in a closed circuit loop containing a refrigerant fluid that evaporates at temperatures as low as 20 °C. Refrigerant vapour is compressed, which also raises its temperature to a level that can be cooled in a condenser. After which the refrigerant liquid is passed through a restrictor (expansion valve) which lowers its temperature to a level suitable for recycling back to the evaporator. When the roles of the evaporator and condenser are reversed, this becomes a heat pump, whereby the condenser can be used either to directly heat air

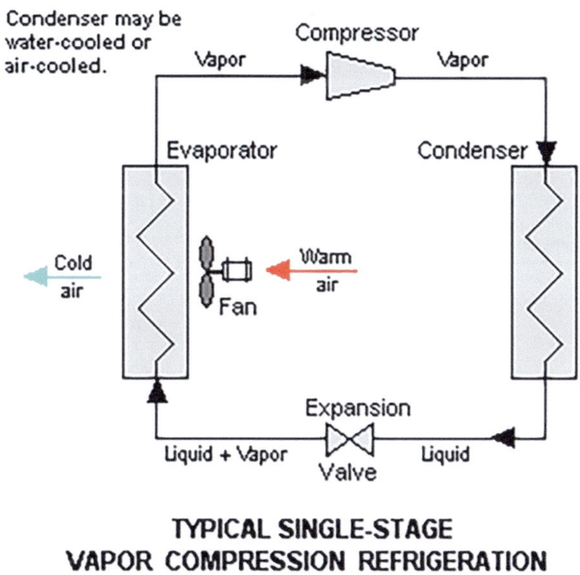

Fig. 7.5 Schematic of vapour compression refrigeration cycle (Source: Keenan Pepper; From: https://commons.wikimedia.org/wiki/File:Refrigeration.png)

supplied to a building or heat water for circulation to room heaters, for example. The evaporator can cool outdoor air directly or be connected to ground loops (ground-source heat pump) or a heat exchanger in a water body. These functions can be reversed between cooling and heating functions by the incorporation into the refrigeration of a four-port valve to change the direction of refrigerant flow so that the condenser becomes the evaporator and vice versa enabling winter heating as well as summer cooling from the same system.

The arrangement shown in Fig. 7.5 is usually referred to as direct expansion when the refrigerant coil directly cools room air and the condenser is used to reject heat directly to outdoors. Direct expansion air conditioning units may contain all these components in a single box that passes through an external wall, or compressor and condenser may be installed as a separate unit with refrigerant pipework connected to the evaporator unit on-site. This same circuit is at the heart of everything from a domestic refrigerator to "chillers" (refrigeration plant) providing chilled water to air handling units that supply cool air to large buildings. In the latter case the condenser may be air cooled or connected by a water circuit to a cooling tower, providing heat rejection through both convective and evaporative heat transfer.

Vapour compression refrigeration is inherently both costly to run and environmentally damaging as it requires electricity to drive the compressor and employs refrigerants which historically have been both damaging to the ozone layer (high ozone depletion potential—ODP) and have high global warming potential (GWP). Since the Montreal Protocol new refrigeration plant has contained zero ODP refrigerants, although GWPs remain high. Absorption chillers on the other hand use water as a refrigerant combined with chemicals such as ammonia or lithium bromide which are used as absorbents and have zero ODPs and GWPs.

Instead of a compressor, absorption refrigeration uses a 'generator' or 'desorber' which is a heat exchanger that can use heat from a suitable source, such as that rejected from electricity generation, geothermal heat, hot water from a boiler or direct

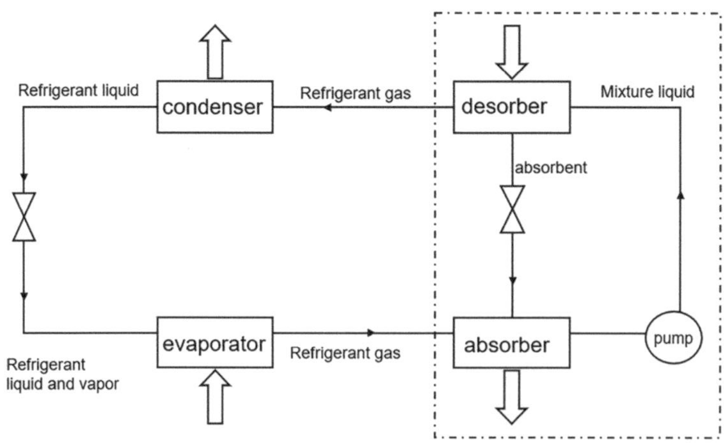

Fig. 7.6 Basic absorption refrigeration cycle (Source: https://commons.wikimedia.org/wiki/File:Absorption_heat_pump_configuration_%28refrigeration%29.jpg)

gas-firing. The refrigerant in this case is water which is mixed with the low boiling point chemical in the 'thermal compressor' shown in Fig. 7.6. The mixture of water and absorbent has a much lower boiling point than water and hence the application of relatively low grade heat at the generator forces vapour to circulate through the condenser where it rejects heat whilst becoming a liquid dropping in pressure through an expansion valve and becoming a cold liquid which evaporates once more when subjected to warming in the evaporator before returning to the absorber.

Although cooling towers are generally the most efficient method of rejecting heat from larger air conditioning systems (see Fig. 7.7), they present significant maintenance challenges because of their ability to scrub pollutants from the air, whilst scale and corrosion can affect heat exchange and create niches for bacterial growth. Badly maintained and contaminated cooling towers have been associated with a number of serious outbreaks of Legionnaire's disease, a type of pneumonia contracted through inhaling the bacterium *legionella pneumophila*, sometimes resulting in fatalities. Figure 7.8 shows the inside of a neglected cooling tower having a significant accumulation of contaminants in the pack fill material located above a cooling tower pond.

For a building that must be sealed to prevent the uncontrolled incursion of polluted outdoor air, it is necessary to provide sufficient filtered 'fresh' air to dilute internally generated pollutants in the air inhaled by occupants to an acceptable level. Minimising the emissions of indoor pollutants needs to be addressed during the design of the building and its services, materials specification, ongoing maintenance and cleaning. Historically the amount of outdoor air supplied to an occupied space has been based either on the number of occupants—in the case of office buildings, for example—or to make up the air extracted to bring exposure to airborne industrial pollutants to acceptable levels. When it is decided to mechanically ventilate a

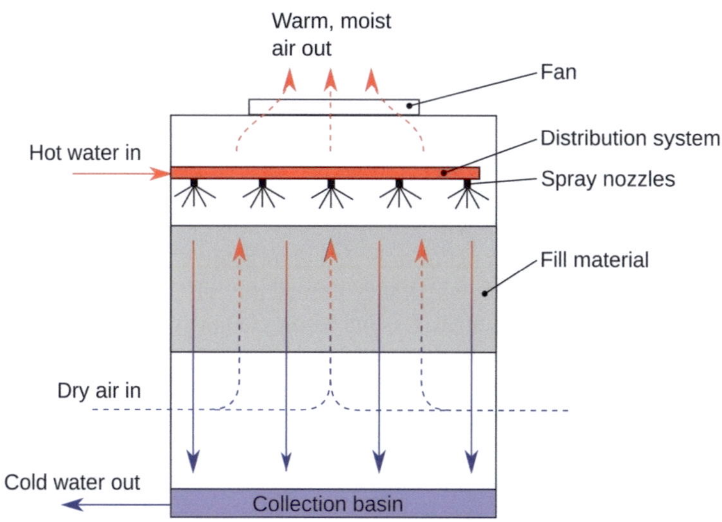

Fig. 7.7 Schematic section through typical induced draught cooling tower (Source: https://basicmechanicalengineering.com/what-is-cooling-tower-with-schematic-diagram/)

Fig. 7.8 Internal view of badly soiled pack/fill material above a cooling tower pond (https://www.nijhuisindustries.com/uk/industries/commercial)

dwelling, such as the mechanical ventilation heat recovery (MVHR) systems employed in a Passivhaus, the usual strategy is to supply sufficient air to make up the air extracted from bathrooms, toilets and kitchens.

Unfortunately in most buildings, there are numerous sources of odour apart from the occupants. For example in a typical office, there may be a combination of volatile organic compounds (VOCs) from solvents used in the manufacture of carpets, furniture, soft furnishings and decorating materials. VOCs are also released from cleaning chemicals and from mould and other micro-organisms that may be associated with poorly maintained or inaccessible parts of the building and its services. Odours may also be detected from the ozone emitted from photocopiers, laser printers and electric motors. Airborne dust and micro-organisms will also be present. Although not visible unless in very high concentrations, dust particles may act as an irritant on the eyes, nose and throat, particularly at low humidities. Some particles, such as those found in tobacco smoke and asbestos may damage human health once inhaled. Micro-organisms may be present most commonly as airborne fungal spores, bacteria and viruses, some of which may have an allergenic or more rarely pathogenic effect on those who inhale them.

Studies at the Technical University of Denmark (TUD) in the late 1980s found that the total strength of odour reported by subjects was between 6 and 7 times that which would have been expected from body odour alone [7]. Fanger's team developed a unit of perceived odour which they termed the 'olf', equivalent to the average perceived odour from a human. It is interesting to note that at that time tobacco smoking was allowed in Danish offices and assembly halls, but that this accounted for only 2 olfs, whereas the odour emanating from the ventilation system increased odour by 3 olfs, and materials within the occupied space contributed to an increase in odour of between 1 and 2 times that from body odour (1–2 olfs).

There are obvious energy penalties in having to deal with this sort of pollution burden, even in the absence of tobacco smoke. It is essential that the ventilation plant itself does not become a source of pollution. This requires attention to detail both at the design stage and in carrying out ongoing maintenance. Workers at the

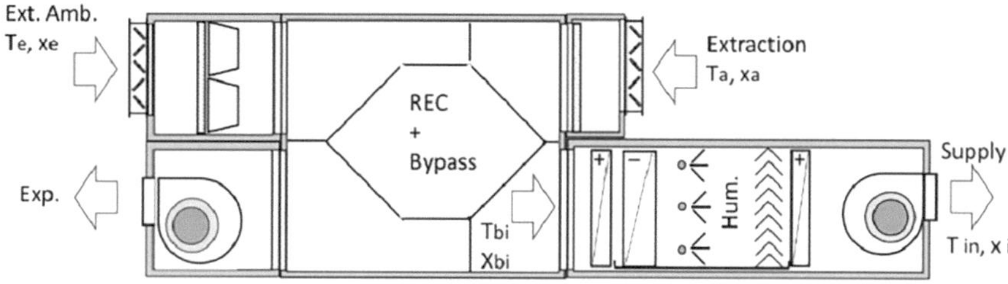

Fig. 7.9 Typical air handling plant schematic (From: https://www.researchgate.net/figure/Layout-of-an-Air-Handling-Unit-equipped-with-a-sensible-heat-recovery_fig1_336747174)

TUD discovered that poorly maintained air handling systems emitted odours and allergens from dirt and mould accumulated in filters, drip trays under cooling coils and other internal surfaces, particularly where both dirt and moisture were able to accumulate [8].

Designers should ensure that the equipment required for supplying air to a building is protected from the accumulation of dust on internal surfaces by installing filters upstream and as close to the intake point as possible, whilst taking steps to avoid filters becoming clogged with ice in winter. Where return air from the building is mixed with outdoor air as shown in Fig. 7.9 another layer of filters should be provided upstream of heat exchangers.

The entire internal surface area of the air handling system should be accessible for cleaning. Cooling coils generally are used to take moisture out of incoming air as well as reduce its temperature and hence incorporate a tray to catch moisture and carry it away to a drain. These trays can be a site for the accumulation of sludge and microbial growth and hence require regular inspection and cleaning. Adding moisture to the supply air in order to control room humidity in winter adds a further potential contamination risk, depending on the type of humidifier employed. Like cooling towers, spray humidifiers also act as air washers, resulting potentially in an accumulation of sludge in the collection tank that holds the water sprayed over a cooling coil or matrix. Steam humidifiers are inherently more hygienic and are normally installed in a straight run of supply air ductwork downstream of the air handing plant. However, if the length of straight ductwork is inadequate, then condensation can occur resulting in potential mould growth.

Following the work on odour carried out by the TUD and studies reported by Seppänen et al. in 1999[7] and Wargocki et al in 2006[8], many authorities have revised their recommendations for outdoor air supply rates for sedentary occupation. Some have concluded that 'ventilation rates below 10 L/s per person result in lower air

[7] https://pubmed.ncbi.nlm.nih.gov/10649857/

[8] https://www.researchgate.net/profile/Pawel-Wargocki/publication/242306573_The_effects_of_classroom_air_temperature_and_outdoor_air_supply_rate_on_performance_of_school_work_by_children/links/0a85e538e4d21ee6be000000/Theeffects-of-classroom-air-temperature-and-outdoor-air-supply-rate-on-performance-of-school-work-by-children.pdf

quality and worsening health problems, (whilst) risk of sick building syndrome is reduced, and perceived air quality is improved, when the ventilation rates increase from 10 L/s to about 20 L/s per person' [9]. Despite this the default ventilation rate stipulated by Approved Document F Volume 2 of the Building Regulations for offices in England having no 'significant levels of pollutants other than body effluents/odour' is 10 L/s per person or 1 L/s/m^2 [10]. A European Standard published in 2007 [11] provides a flexible basis for determining minimum outdoor air supply rate, depending on the susceptibility of the occupancy, whether they are adapted to ambient odours and the contribution to odour from the building itself. However, in a comparison of ventilation rates adopted by European countries carried out as part of a European Commission project aiming to harmonise ventilation standards across Europe, it was evident that there was a wide variation in standards, especially for the office environment: ranging from 5.6 L/s/person in Poland to 25 L/s in Germany and Hungary. The latter two countries had adopted the highest standard from EN15251, taking into account typical emissions from internal sources found in office spaces [12]. There are of course energy penalties to be paid for increasing outdoor air supply rates. Doubling the rate doubles the energy required for winter heating/humidification and summer cooling/dehumidification (if required). This can be partially mitigated by high efficiency heat recovery between the exhaust and intake air. Efficiencies for sensible heat transfer of 80–85% are possible with thermal wheels and plate heat exchangers, for example, whilst some devices also allow the transfer of latent heat via condensation or a membrane [13].

The method by which air is distributed through a space has a significant impact on both energy consumption and the quality of air in the occupied zone. Air supplied from high level in a room requires momentum to encourage heat exchange with the air below, whilst also mixing with stale air rising from the occupied zone. When tackling maximum heat gains this air has to be cooled to between 10 and 14 °C. Because, unless deflected downwards, the air from ceiling diffusers adheres to the ceiling surface (known as the Coanda effect), there is a risk that supply air will be drawn into extract points before full heat exchange has occurred (see Fig. 7.10). In contrast, delivering air with low momentum close to the floor requires significantly lower fan energy, whilst a supply air temperature of 19–20 °C is required to avoid draughts. This is usually referred to as displacement ventilation or buoyancy-driven air supply since the room air is displaced by the supply air, rises through the room by natural convection created by the heat sources therein and is extracted at high level. This encourages stratification in the room, with a maximum gradient of 3K between head and feet to avoid discomfort, and with a corresponding gradient in pollutant concentrations (see Fig. 7.11). The higher supply air temperature allows free cooling for longer periods in temperate climates.

Displacement ventilation using low velocity outlets was originally developed in Scandinavia to provide outdoor air supply for industrial applications, whilst a system employing outlets recessed into a suspended floor was developed in Germany about the same time for use primarily in offices. Most of these systems supply 100% outdoor air throughout the year with heat recuperation equipment installed between the exhaust and outdoor air intake, thus enabling outdoor air supply rates that exceed minimum requirements year round.

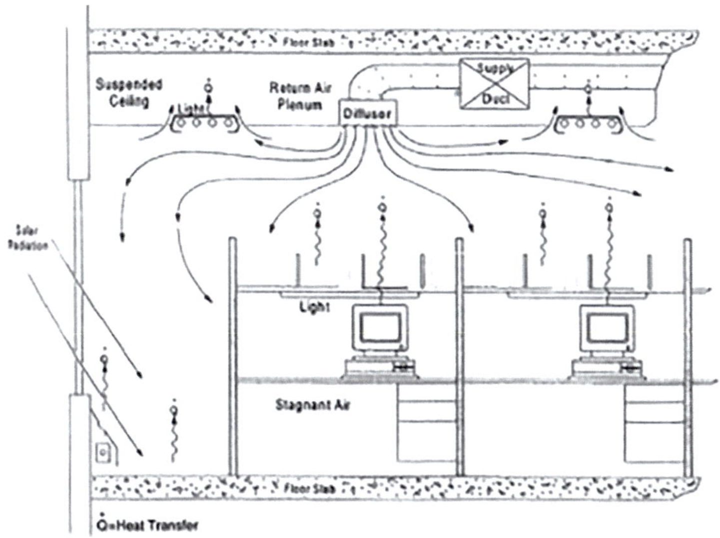

Fig. 7.10 Air supply from ceiling diffusers. (Source: http://what-when-how.com/energy-engineering/underfloor-air-distribution-ufad-energy-engineering/)

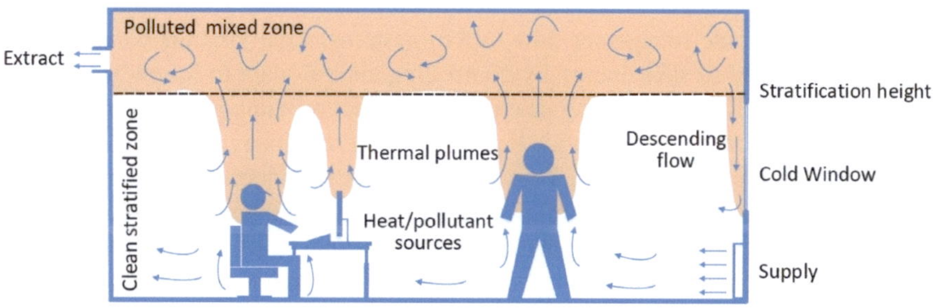

Fig. 7.11 Displacement ventilation—typical flow patterns (Source: [14]; From: https://www.mdpi.com/1996-1073/14/4/952#)

There is no space here to describe the many different types of air conditioning system that have evolved over the years. There are numerous sources of design guidance, the most authoritative of which are published by CIBSE and ASHRAE [13, 15]. An air conditioning system is an assembly of components the design of which requires an understanding of operation under the full range of likely external climate, internal heat loads and occupant requirements in order to create a comfortable environment year round at lowest energy consumption and optimum whole life cost. These will typically vary from one part of a building to another and hence an analysis of zonal performance and control requirements is required. Most developments in multi-zone air conditioning have focussed on how to tackle varying loads simultaneously, whilst ensuring that there is sufficient outdoor air delivered to occupants throughout a building.

One of the greatest challenges is to be able to provide heating in some areas when others require cooling, using the least possible amount of energy. Combining perimeter heating with central air conditioning runs the risk of some areas being heated and cooled simultaneously, for example, when south-facing perimeter zones experience solar gain at the same time as north facing zones experience a heat loss. This becomes less of an issue for buildings that have a façade that is well insulated against heat loss, sealed against infiltration and protected against solar radiation. On the other hand, internal heat gains from lighting and equipment have fallen with the development of LED lamps for commercial use and more energy efficient computing and other office equipment.

There are two main categories of air conditioning used in commercial buildings globally: all-air systems that supply treated air from a central air handling plant, and local air conditioning with air handling plant that supplies 100% outdoor air. Historically the most popular type of all-air system has been variable air volume (VAV), which adjusts to zonal heat gains by varying the volume of air supplied to a zone rather than the supply air temperature, thus avoiding wasteful reheat of air that has been cooled at the central plant. Unfortunately the dampers required to vary the airflow into the space require high inlet pressures to function properly and hence high pressures to be developed at the supply fans, offsetting some of the saving achieved by avoiding reheat. To maximise the energy saved, fans have to have motors that are controlled in a way that reduces electrical demand as air flow rate falls, through a device such as an inverter that is designed to reduce frequency of alternating current to the fan motor.

Recent developments in local air conditioning for multiple zones have mostly focussed on variable refrigerant flow (VRF) systems. Like VAV air conditioning, these are designed to save energy by controlling central plant to follow simultaneous demand, whilst also enabling heat rejected whilst providing local cooling to be recovered at central plant to provide heating elsewhere. This is achieved by extending the principal of the reversible heat pump to multiple indoor coils, each one connected to a reversing valve that allows either high temperature refrigerant vapour distributed from a central compressor to be supplied to the indoor coil, which becomes a condenser that heats the zone served, or low temperature liquid refrigerant to be supplied to the indoor coil, which becomes an evaporator, thus cooling the zone served. Central compressors are controlled to allow for reductions in simultaneous demand by inverter control of motor speed.

7.2.2.2 Lighting and the Visual Environment

The visual environment is created by a combination of façade engineering, interior design and artificial lighting. A good design combines aesthetics, functionality, visual acuity and energy efficiency, whilst avoiding glare from the sun and other light sources. Ideally daylight should be available at a level that mitigates against the use of artificial lighting across the surfaces that require illumination during daylight hours. However, where there is manual control of solar shading and lighting, with no separate control over perimeter zones, it is not unusual to find lights are left on throughout the day.

The quality of the indoor environment, whether it is for work or residential use, can be much enhanced by good lighting design and the complimentary availability of daylight. Indeed, since 1989, European Union countries have been directed to require that 'Workplaces must as far as possible receive sufficient natural light and be equipped with artificial lighting adequate for workers' safety and health.' [16]. In the UK, these requirements are implemented through the 1992 Workplace Regulations.

The integration and control of daylight and electric lighting, the selection of efficient lamps and the minimisation of heat gain from lighting are key components of low energy design. The contribution of lighting to overall carbon emissions for a building varies enormously, depending on a combination of application, optimum lamp type, luminaire design and location, the availability of useful daylight and the application of control technologies.

The contribution of lighting to carbon dioxide emissions has steadily decreased since the invention of electric incandescent lamps towards the end of the nineteenth century. Most office buildings in the UK built since the 1950s have been illuminated with fluorescent lamps, which have a luminous efficacy 3 or 4 times higher than incandescent lamps. For new buildings lighting related energy consumption has steadily decreased through a combination of lower lighting levels, improved lamp efficiency and intelligent controls. As electrical demand has reduced so too has the amount of energy required to offset the heat gain from lamps for air conditioned buildings, whilst fortuitous heating in winter has also reduced. The contribution of lighting to the CO_2 emissions for a typical air conditioned office building from the 1990's is around 18% [6]. It is likely that this percentage has remained around this mark through the first two-decades of the twenty-first century for new office buildings, primarily because other energy demands, such as IT equipment and heating and cooling demand, have fallen proportionately.

According to a 2017 report from the UN Environment Programme 'Electricity for lighting accounts for approximately 15 per cent of global power consumption and 5 per cent of greenhouse gas (GHG) emissions. High-efficiency lighting technologies offer up to an 85 per cent improvement in energy efficiency compared with conventional lighting technologies, while providing an equal or better quality of light' [17].

The rapid improvement in luminous efficacy and reduction in costs of light emitting diode (LED) lamps has led to a revolution in the sector, resulting in a phasing out of incandescent (tungsten) lamps for all applications, whilst halogen lamps and fluorescent tubes could no longer be sold in European Union countries from September 2023.[9] For example, products launched by manufacturers such as Cree, Samsung and LG in 2016/17 were claimed to achieve between 200 and 300 lumens/W luminous efficacy, compared with around 100 lm/W for a modern fluorescent lamp with high frequency ballast. According to analysis carried out by IHS Markit on the contribution LED lighting retrofits made to global CO_2 emissions a saving of 570 million tonnes had been achieved in 2017 or 1.54% of total global emissions.[10]

---

[9] https://ec.europa.eu/info/energy-climate-change-environment/standards-tools-and-labels/products-labelling-rules-and-requirements/energy-label-and-ecodesign/energy-efficient-products/lighting_en

[10] https://news.mongabay.com/2022/11/led-lights-could-contribute-to-massive-carbon-reductions/#:~:text=In%202017%2C%20analysts%20at%20IHS,162%20coal%2Dfired%20power%20plants

Fig. 7.12 Roof-mounted solar photovoltaic panels (Source: https://commons.wikimedia.org/wiki/File:Solar_panels_on_a_roof.jpg)

7.2.2.3 On-Site Energy Generation and Renewables

Traditionally buildings have used mains electricity to meet electrical energy demands, whilst heating has been drawn from a range of energy sources from solid fuel to electricity via gas and oil. In order to reduce the carbon dioxide emissions associated with energy generation, a number of low carbon and renewable energy technologies have been developed for use on site. Some of these, such as combined heat and power plant, are used in place of or to supplement conventional boiler plant, whilst also generating electricity; others, including photovoltaic and solar hot water panels, can be integrated into or mounted on the building fabric (see Fig. 7.12).

Some renewable energy technologies generate electricity by harnessing the potential energy in nature, such as sunlight and wind (photovoltaics and wind turbines respectively).[11] Other technologies use solar energy to heat water (solar hot water panels), whilst others use either the heat found in rock underground in some areas to heat water or create steam to drive a turbine to generate electricity (geothermal energy). The relatively constant temperature underground can also be exploited to provide cooling in summer, or upgraded via a heat pump to provide heating in winter (ground-source heat pump—GSHP)—although this does require electricity to drive a compressor for the refrigeration cycle. Depending on the carbon intensity of the electricity supply GSHPs would usually be associated with lower carbon footprint than a gas-fired boiler with the same thermal load, whilst costing no more to run.

Biomass boilers have also tended to be categorised as renewable technologies. This is because the CO_2 emitted whilst burning biomass is theoretically equal to that absorbed during the growing process. However, this ignores any greenhouse gas

[11] https://www.researchgate.net/publication/268487729_Potential_and_Challenges_for_Building_Integrated_Photovoltaics_in_the_Agder_Region/figures?lo=1

emissions associated with processing and transporting the material. Manufacture of wood pellets requires some combination of drying, chipping, grinding, palletisation, cooling and sieving, depending on the state of the raw material. Much of the world's supply of biomass comes from North America, whilst the greatest demand is in the UK and Europe. Burning biomass emits fine particles, oxides of nitrogen (NO_x) and other pollutants, depending on source. Some estimates indicate that the global warming potential of biomass on average is only 60% less than the average for fossil fuels. Biomass in the form of logs has been used as a fuel for millennia, although modern biomass boilers typically use pellets (see Fig. 7.13): which can be derived from the waste products of timber production; milling and manufacture of wood products; timber items that cannot be re-used (post-consumer wood waste); crop wastes and harvested or coppiced trees.

A report from the UK Committee on Climate Change (CCC) from December 2018 recommended that Government policies should 'Limit support for bioenergy use in buildings to biomethane produced from anaerobic digestion and other niche uses (as part of hybrid heat pumps systems in hard to treat off-gas homes, local combined heat and power systems and small-scale district heat networks)—whilst minimising air quality impacts'. It does not foresee a role for biomass boilers apart from for 'hard-to-insulate rural properties where heat pumps are not viable' [18].

Integrating wind turbines into building designs can be problematical for a number of reasons. Wind resources inland, and particularly in built-up areas, tend to be significantly lower than off-shore, whilst integrating them into building structures is difficult because of the noise and vibration generated. Although the three 9 m diameter horizontal axis wind turbines built into the Strata residential tower in the Elephant & Castle, London (Fig. 7.14) are located to catch the higher wind speeds in this exposed location, the rotational speed has had to be limited to reduce noise and vibration to apartments below.

Photovoltaic (PV) materials convert light to direct current (DC) electricity through photon-to-electron energy transfer within the dielectric materials from which they are manufactured. Because they react to a broad spectrum of visual radiation they generate electricity even when the sky is overcast, although peak output occurs when the sky is clear and the angle of the sun's rays is closest to normal to the panel surface. DC electricity generated by the cells can be converted to alternating current (AC) by passing it through an inverter in order to contribute to the power supply for a building. Surplus electricity can be fed back to the grid as AC or stored on-site in rechargeable batteries as DC electricity.

PV materials fall into two main categories: poly- or monocrystalline cells and thin film panels. Within these categories there is a wide range of materials and configurations that result in significant variations in efficiency, cost and appearance. Efficiency is determined by manufacturers according to an international standard published by the International Electrotechnical Commission (IEC) as part of the IEC 60904 suite of standards for Photovoltaic Devices [19]. Efficiency is measured under standard test conditions, with a light source generating 1000 W/m^2 simulating the spectral composition of sunlight, and efficiency calculated from the measured electrical output per square metre (W/m^2) divided by 1000. Efficiencies range from

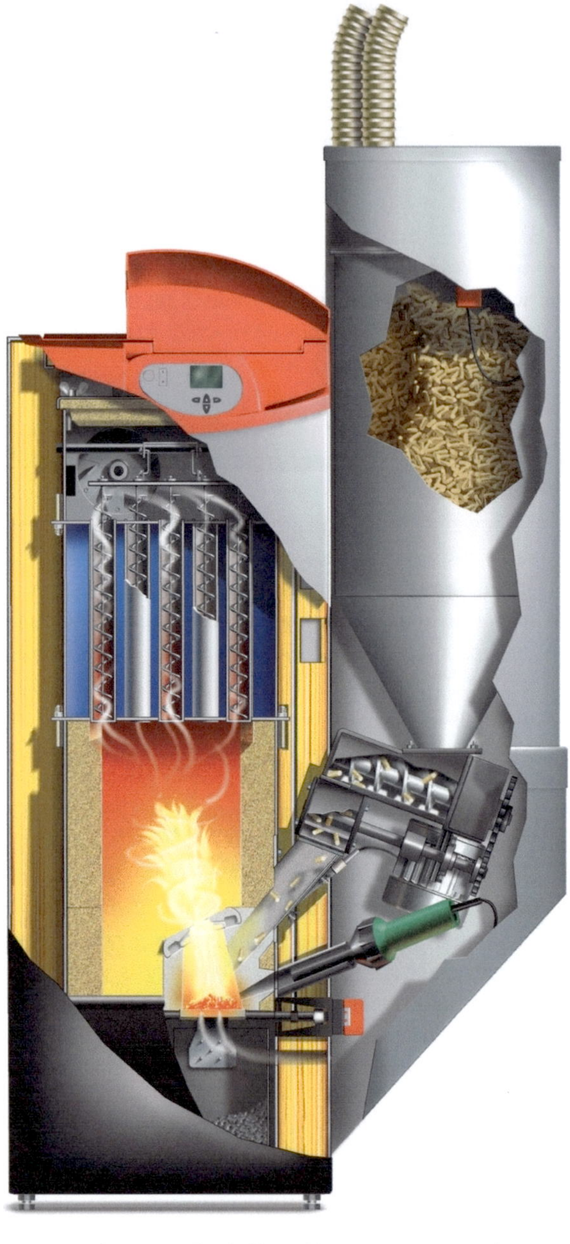

Fig. 7.13 Internal view of wood pellet boiler with integral hopper (Source: GUNTAMATIC Heiztechnik GmbH; From: https://commons.wikimedia.org/wiki/File:Pelletheizung_Biostar_Schnittbild.jpg)

Fig. 7.14 Wind turbines integrated into the roof of the Strata tower, London (Source: Author)

7% to 12% for thin film panels to 20% for monocrystalline modules with the potential for 30% or more for the latest generation of perovskite-silicon tandem cells.[12]

Crystalline PV cells are manufactured by 'growing' silicon ingots using the same process as employed in the manufacture of semiconductors, then slicing them into thin wafers, typically consisting of 100—150 mm square or rectangular plates, subdivided by metal contact grids, which collect the current. Monocrystalline silicon cells are made from a single crystal and are more efficient than polycrystalline cells, which are made from silicon cast in a mould, allowing larger cells to be formed. Thin film PV—also referred to as second generation photovoltaics—is manufactured by depositing a thin layer of the dielectric material in vapour form within a vacuum onto a substrate such as glass, plastic or metal. There are a number of substances used that have suitable dielectric characteristic including amorphous silicon (a-Si), cadmium telluride (CdTe) and copper indium gallium sellenide (CIGS). Thin film manufacturing techniques allow for faster and cheaper production and hence

[12] https://www.oxfordpv.com/perovskite-silicon-tandem

can be more cost effective per unit of electrical output despite lower efficiencies than crystalline products.

Organic photovoltaic solar cells (OPVs or OSCs), which fall within the third generation PV category—use extremely thin conductive organic polymers printed onto a substrate that can be flexible and transparent, allowing mass production and unobtrusive incorporation into facades and glazing systems. International research effort has focussed on improving efficiency and life expectancy in order to develop commercial products across a range of applications. Whereas CdTe and CIGS PV use rare and toxic materials, OSCs use very small quantities of carbon-based materials. In the conclusions set out in an undated paper from Chicago University, the author states that:

> Solar energy is a booming field, and with good reason; it is a completely renewable technology with great potential waiting to be harvested. However, the current technologies do have room for improvement. Traditional silicon solar cells are well established but are bulky and could become expensive if the price of silicon peaks. The usual competitors to silicon cells, CdTe- and CIGS-based devices, use rare and toxic elements. OPVs have the potential to solve all these issues. The problems left to be addressed in the OPV world are well known and solving them is receiving much attention in the field. It is our hope that, once these concerns are addressed, OPVs will take their place as an important renewable energy technology for a greener world [20].

Historically the most common method of installing photovoltaic panels, either from new or retrofitted, has been roof mounting as shown in Fig. 7.12. A sloping roof with a southerly orientation provides simpler and less obtrusive mounting, with the potential for flush installation, such as with PV slates that look very similar to the traditional variety (see Fig. 7.15).

PV slates are typically manufactured from monocrystalline cells having an overall efficiency lower than the equivalent monocrystalline panels because of the reduced effective area required to allow terminals to be unobtrusively mounted on the back of the slate. Cost per kWh output tends to be higher, hence application is typically limited to buildings for which solar panels would be unacceptable aesthetically, for example, where there are planning restrictions.

Thin film PV media can be incorporated into roofing systems, such as the standing seam roof shown in Fig. 7.16. This is less obtrusive and more robust than bracket-mounted crystalline panels. The film can be applied to an existing standing seam roof or supplied as a package for a new roof. Typically these flexible film 'peel and stick' systems use amorphous silicon thin film media.

Crystalline PV media have been integrated into window systems by incorporating transparent elements into the PV panels, such as at the Solar Office at Doxford International in Sunderland). This early example of a solar façade (commissioned in 1998) uses polycrystalline PV on a glass substrate with transparent borders giving a tartan effect, as well as fully glazed sections in the façade to improve daylighting.

A logical development from the Doxford Solar Office has been the application of thin-film technologies to glass in order to create a transparent material for use in glazing systems that can generate electricity, reduce solar gain and provide thermal insulation. One such product is shown in Fig. 7.17. This particular product is

Fig. 7.15 Photovoltaic slates integrated into traditional slate roof (Image Courtesy of PV Slate by GB-Sol)

Fig. 7.16 Thin film PV in a metal standing seam roof (Source: https://en.m.wikipedia.org/wiki/File:Thin_film_solar_on_standing_seam_metal_roof.jpg)

available in a range of transparencies ranging from opaque to 'high' transparency. The latter having a Visible Light Transmittance (VLT) of 28.4% for double glazing, compared with around 45% for a typical bronze tinted double glazed system.

Fig. 7.17 Solar photovoltaic glass system (Source: https://www.onyxsolar.com/documents/onyx-solar-asi-solutions-brochure-sqm.pdf)

The system shown in the figure uses amorphous silicon as the PV medium and has an efficiency of a very low 2.8% for the glazing having the greatest VLT. There are also cadmium telluride-based products which, according to manufacturer's literature, have slightly better performance. It is worth noting, however, that because a portion of solar energy is converted to electricity, these systems have very good solar heat gain coefficients (SHGC), whilst costs are claimed to be comparable with conventional glazing systems having similar thermal performance [21].

Solar hot water systems similarly use solar radiation, but to heat water instead of generating electricity. Solar hot water panels work with both direct and diffuse solar radiation, although because the intensity of diffuse radiation is a fraction of direct and solar altitude angles are lower in winter than in summer, in the UK for example the contribution that solar panels make to offsetting hot water demand in winter is typically 25–30% of that in summer.

The most common use for heat from solar hot water panels is for domestic hot water, although they can be used for heating swimming pool water and inter-seasonal underground thermal storage. The most common types of heat exchanger are the evacuated tube solar collector and the flat plate collector. The former comprises an array of annealed glass tubes within which a vacuum is created and either a heat pipe or pipe loop contained. The heat pipe contains a low boiling point fluid such as alcohol which evaporates when heated and migrates to the header end of the array (see Fig. 7.18) where it heats a circulating fluid containing antifreeze, if necessary.

Flat plate collectors comprise an array of tubes embedded in a dark coloured absorber plate, insulated between and to the rear of the tubes and covered in a glass or plastic sheet. Typically a water/glycol mix is circulated through the tubes and to a heat exchanger in a domestic hot water cylinder.

Fig. 7.18 Evacuated tube solar collector (Source: Norbert Nagel; From: https://commons.wikimedia.org/wiki/File:Solar_heating_system_-_Thermosolaranlage_-_Mörfelden-Walldorf_-_Germany.jpg)

Flat plate collectors are significantly cheaper than the evacuated tube type, whilst the latter is up to twice as efficient per unit surface area. This is partly due to its larger surface area exposed to the sun as well as its greater heat exchange efficiency. Despite the larger footprint of the flat plate collector required to achieve the same performance it tends to have a shorter payback period than the evacuated tube [22].

Heat pumps are frequently categorised as renewable technologies, although since they generally use electricity as their prime source of energy, this is debatable—being dependent upon decarbonisation of the electricity supply. When comparing them with alternative technologies, it is important to estimate potential lifetime carbon emissions as well as factoring in advantages that are project specific. For example, heat pumps do not require a gas supply or alternative fossil fuel delivery and storage such as oil or biomass. Reversible heat pumps enable summer cooling, although with commensurate energy/carbon penalties. Energy consumption is strongly dependent upon the seasonal variations in condensing temperatures. Air-source heat pumps are the least efficient because the winter and summer temperature differences are greatest; whereas ground-source heat pumps (GSHPs) benefit from a relatively constant ground temperature; whilst if condenser coils are laid in the bottom of a water body the seasonal variations are somewhere between the two, depending on the water depth and flow rate. GSHPs tend to be more costly than air-source because of the need for site condition investigations, ground works such as the digging of trenches or drilling of bore holes and the cost of lengthy ground coils.

Fig. 7.19 Slinky coil laid in trenches serving a ground-source heat pump (Source: https://commons.wikimedia.org/wiki/File:3-ton_Slinky_Loop.jpg)

The choice between bore hole and trench coil systems may depend on the space available and cost implications. Boreholes only require a small footprint but cost significantly more than a system laid in trenches. The latter can be a simple pipe loop or an array of 'slinky' coils as shown in Fig. 7.19, which require less than half the length of trench required by a simple pipe loop.

When compared with a gas-fired central heating system, the seasonal efficiency (CoP)[13] of a ground-coupled system needs to be greater than the ratio between the annual average carbon intensities of electricity and gas, which for the UK is around 2, although this is steadily falling as electricity generation is being decarbonised. CoP depends primarily on flow temperature, and can be as high as 4.0 if flow water can be kept below 40 °C. Hence heat pumps tend to be most compatible with underfloor heating, which demands flow temperatures of around this level to avoid overheated feet.

Although geothermal energy has historically been used to provide steam for electricity generation and district heating (e.g., Reykjavik, Iceland), the stable thermal characteristics of the ground can be used directly to cool buildings, either by drawing water from an aquifer and pumping it through heat exchangers within the building or by installing pipe loops within bore holes or structural piles. The earth can also be used to store energy inter-seasonally to pre-heat water for winter heating using heat rejected during the summer, and vice versa. Aquifer thermal energy storage (ATES) involves the drilling of a number of boreholes or 'wells': with some dedicated to storing excess heat in the summer (hot wells) and others for storing 'coolth' in the winter (cold wells); reversing the flows in winter from the hot wells

[13] CoP = Coefficient of Performance = the ratio of heat out to electrical power in

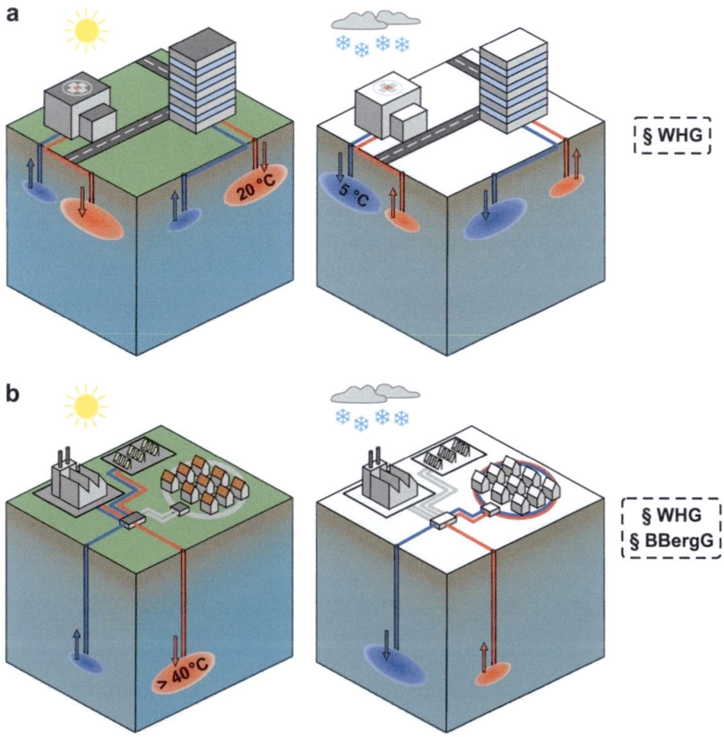

Fig. 7.20 Principles of aquifer thermal energy storage (ATES) with heat pump (Source: [23])

and in summer from the cold wells. Figure 7.20 shows the principle of this system providing both winter heating and summer cooling.

The lower images (b) illustrate the scenario when ATES is combined with solar hot water panels to augment the heating provided for residential development. It is interesting to note that around 85% of the 3500 ATES schemes installed globally have been in the Netherlands [23].

Electricity can be produced on site by means of a generator powered by an internal combustion (IC) engine, Stirling engine or turbine. These have historically been adopted for buildings with no access to an electricity supply or for emergency standby. Generating electricity in this manner results in surplus heat that can be used for other purposes, such as heating hot water. This is referred to as combined heat and power (CHP) or cogeneration. When surplus heat is also used via an absorption refrigeration plant to provide chilled water (see above) the plant is referred to as combined cooling, heating and power (CCHP) or trigeneration.

CHP plants vary in electrical outputs from less than 1 kWe to hundreds of MWe, that is, from small-scale domestic to power stations, whilst CCHP plants tend to be towards the middle of this range, serving individual air conditioned buildings or district heating and cooling systems. Most of the technologies employed have been around for decades and have been adapted from power stations, jet engines, road vehicles and fuel cells. Internal combustion engine CHP generators run on gas, diesel or fuel oil and are available up to around 4.5 MWe. Domestic scale plant typically uses a wood pellet or gas-fired Stirling engine, which works in a similar manner to an IC engine but comprises a cylinder casing that contains two pistons, one of which (the 'displacer' piston) is heated externally and is driven by expanding gas within the sealed cylinder which, as it cools at the further end of the cylinder drives the piston back the way it came. The second or 'power' piston is attached to the displacer at one end and the power train at the other, generating electricity in the process.

It is axiomatic that fossil fuel fired CHP and CCHP are not renewable technologies and their carbon efficiency compared with a conventional gas-fired boiler and mains electricity supply depends primarily on the carbon intensity of mains electricity. Hence in countries such as Norway and France, where low or zero carbon sources make a large contribution to electricity generation, CHP/CCHP becomes less attractive. As carbon intensities fall in the UK with an increased contribution from renewable energy sources this is also becoming the case.

7.2.3 Embodied Carbon and Other Environmental Impacts

As referred to earlier in this chapter, when buildings become more energy efficient the proportion of lifetime carbon emissions that is contributed by embodied energy becomes greater. The carbon emissions associated with the construction, maintenance and demolition of a building constitute its embodied carbon. These include the carbon emissions associated with the extraction, manufacture, delivery, maintenance, replacement (if required) and demolition of materials and plant. Over the lifetime of a building there may also be extensive refurbishment and retrofitting of fabric, equipment and plant, all of which contribute to the whole life carbon emissions.

Buildings and structures have a wide spectrum of environmental impacts beyond carbon emissions. Some of these are caused by other emissions, mostly during material extraction and manufacture, such as: those to land and water resulting in damage to ecosystems (ecotoxicity); eutrophication of water resources creating oxygen depletion; emissions to atmosphere of oxides of nitrogen (NOx) and volatile organic compounds (VOCs) creating photochemical oxidation and ozone; landfill or incineration of waste; and acidification of the water cycle and oceans (see Sect. 3.4 in Chap. 3). Other impacts include depletion of natural resources such as fossil fuels and other virgin irreplaceable materials, water use and toxicity to humans. To this list could also be added emissions of ozone depleting substances such as chlorinated and brominated gases, although these have been mostly phased out as a growing number of nations have ratified *the Montreal Protocol on Substances that Deplete the Ozone Layer* since it entered into force in 1989.

Unlike operational energy consumption and carbon emissions, no country has a mandatory or legal requirement to determine or meet specified targets for cradle to grave embodied carbon, although since 2012 the Dutch Building Decree has required a statement of life cycle environmental performance to be submitted for homes and offices larger than 100 m^2. As well as carbon emission—expressed as global warming potential—this assessment evaluates the environmental impact of each of the materials used in the construction of the building across a wide range of headings. These are derived from the series of standards that have been developed for harmonising the assessment of environmental impacts associated with construction products and buildings internationally. 'The resulting CEN/TC 350 series of European Standards has been developed and published with cross-industry involvement and consultation, giving a common methodology for providing Environmental Product Declarations (EPD) across Europe in EN 15804 [24] and linking to a common methodology for the assessment of building Life Cycle Assessment (LCA) in EN 15978 [25], (which in turn uses the principles of LCA set out in ISO 14044 [26])'. There are numerous EPD programmes in Europe and around the world that use the EN 15804 standard' and offer compliant EPD.[14]

The UK-based Building Research Establishment (BRE) has developed an environmental rating system based on BS EN 15804 for use as part of its BREEAM suite of environmental assessment protocols.[15] This uses a weighting system to convert performance against the impacts referred to above into so-called 'Green Guide' ratings, divided into bands ranging from E to A+, where E represents those materials or products with the greatest environmental impact and A+ the least. The Guide also rates performance under each of the separate impacts using the same bands.

The information in the Green Guide has been used by BRE and others to develop a tool that can be used in conjunction with Building Information Modelling (BIM) to model whole life environmental impacts and costs. The Integrated Material Profile and Costing Tool (IMPACT) also allows benchmarking against typical building types and integration with BREEAM:

> IMPACT takes quantity information from the BIM and multiplies this by environmental impact and/or cost 'rates' to produce an overall impact and cost for the whole (or a selected part) of the design. The results generated by IMPACT allow the user to analyse the design to optimise cost and environmental impacts, and compare whole-building results to a suitable benchmark to assess performance, which can be linked to building assessment schemes.[16]

A similar web-based tool has been developed in the USA by the National Institute of Standards & Technology (NIST) using the acronym BEES (Building for Environmental and Economic Sustainability).[17]

---

[14] https://www.rics.org/news-insights/wbef/environmental-product-declarations-industry-need-epds

[15] BREEAM is the Building Research Establishment Environmental Assessment Method. For more information see: https://www.breeam.com

[16] https://bregroup.com/products/impact

[17] https://ws680.nist.gov/Bees2

Prediction of the embodied impacts associated with building materials and services, including carbon emissions, requires not only information from the manufacturers on the life cycle of their products, but also an understanding of the emissions associated with delivery to site. The same product may need to be delivered to a site in close proximity to the factory or conveyed half way round the world. Much work has been done in estimating embodied carbon from 'cradle to (factory) gate' for a wide range of building materials and a number of databases have been developed to be used by construction professionals to assist in the estimation of total building carbon footprints. For example the Inventory of Carbon & Energy (ICE) database developed by researchers at the University of Bath provides 'cradle to gate' data for a broad range of construction materials, based primarily on materials procured in the UK.[18] This data includes the carbon equivalent of greenhouse gas emissions but excludes carbon emissions associated with delivery to site, and is for individual materials only, apart from limited data on windows, photovoltaic modules and roads, which allow for carbon associated with assembly, as well as for each component material. No information is provided for plant or equipment, building services, on-site works, maintenance, demolition or disposal.

In 2021 CIBSE published a calculation methodology for *Embodied Energy and Building Services*, from a UK perspective.[19] CIBSE has suggested that building services typically contributes between 4% and 25% of the total embodied carbon for buildings, with naturally ventilated lightly serviced buildings being towards the bottom end of this range and heavily serviced buildings, such as hospitals and office buildings comprising mainly of trading floors, contributing the most. The UK Net Zero Carbon Buildings Standard, published in 2025, incorporated embodied carbon and whole life carbon assessment as part of the protocol.[20]

The determination of carbon footprints for buildings, construction products, materials and building services requires the same fundamental methodology as products and other entities. Guidelines for this were published by the ISO in 2018 using in turn the principles for LCA set out in ISO 14044 (ISO in 2018)[21]—see Sect. 6.3 in Chap. 6.

In terms of environmental impact, cement and steel are the two most important materials in construction, for commercial buildings and civil engineering projects in particular. For the UK more than 90 million tonnes of concrete was used in building and civil engineering construction in 2022, of which about 10% was cement, whilst steel used in construction that year was around 900,000 tonnes.

As we saw in the Sect. 4.8 in Chap. 4, the global carbon footprint of cement is significant and growing. Because of the enormous volumes of concrete used for infrastructure projects globally its environmental impacts are particularly

[18] http://www.circularecology.com/embodied-energy-and-carbon-footprint-database.html#.XGVkrq10fuQ

[19] https://www.cibse.org/knowledge-research/knowledge-portal/embodied-carbon-in-building-services-a-calculation-methodology-tm65

[20] https://www.nzcbuildings.co.uk/_files/ugd/6ea7ba_1ef36b6835de46668f2ad8b589ff1b93.pdf

[21] https://www.iso.org/news/ref2205.html#:~:text=ISO%2016745%2C%20Sustainability%20in%20buildings,GHG%20emissions%20arising%20from%20the

important. As well as being the principle binding ingredient in concrete, cement is used in mortar, plaster and stucco render of buildings. The CO_2 emissions associated with concrete manufacture are typically between 0.1 and 0.180 $kgCO_2/kg$ with 14% Portland cement, whilst cement contributes around 0.9–0.95 $kgCO_2/kg$ or up to 88% of the total emissions from concrete.[22]

According to a paper published in 2018 by the Norwegian Center for International Climate Research (CICERO) global cement production increased 30-fold between 1950 and 2016, with China contributing to 73% of the growth since 1990 [27]. According to the analysis carried out by CICERO the total CO_2 emissions from cement production in 2016 contributed around 8% of total anthropogenic carbon emissions. The process of manufacturing Portland cement is highly energy intensive: around 60% of emissions occur through the chemical reaction that occurs in converting the quarried limestone ($CaCO_3$) mixed with clay into $CaO + CO_2$ (calcium oxide and carbon dioxide) through heating to around 600 °C. The remaining carbon emissions are associated with further heating the mix with suitable additives to 1500 °C to form clinker—through a process known as calcination—the initial crushing of the limestone and the grinding of clinker, which ultimately forms the very fine powder characteristic of cement.

In 2018 Portland cement in its various forms represented around 95% of global production, whilst 'supplementary cementitious materials' (SPMs) have been developed as a replacement or supplement for cement in order primarily to reduce carbon emissions. The most common of these are ground granulated blast furnace slag (GGBS) and fly ash—waste products from iron foundries and coal-fired power stations respectively, also referred to as industrial pozzolans. Replacing 30% of Portland cement with fly ash reduces the carbon footprint by around 30% (to around 127 $kgCO_2$/tonne). According to a 2017 report from the Department for Business, Energy and Industrial Strategy (BEIS), although there were surplus quantities of these SPMs being produced in China and India at the time, a steady shrinking of coal-firing and iron manufacture in the UK was resulting in a major reduction in their domestic availability and hence increased reliance on imports [28]. Naturally occurring pozzolans, such as volcanic ash and pumice (*Pozzolana*), were used in cement and mortar by the Romans and more recently, where volcanic material is abundant, such as in parts of the USA, China and some European countries, as an SPM to Portland cement.

As we saw above at between 9 and 10 million tonnes per annum cement production in the UK in recent years has been around ten times the weight of steel used for construction. This includes reinforcing bars, steel frames and decking, as well as that used for building services. The average embodied carbon of steel, allowing for around 50% recycled content, has been around 1.46 $kgCO_{2eq}/kg$ of product, which is roughly 1.5 times that for cement per kilogramme.[23]

[22] http://www.circularecology.com/embodied-energy-and-carbon-footprint-database.html#.XGVkrq10fuQ

[23] http://www.circularecology.com/embodied-energy-and-carbon-footprint-database.html#.XGVkrq10fuQ

An ISO standard for determination of water footprint was published in 2014 [29]. As we saw in Chap. 6 the volumes of water required for manufacture, processing and, where applicable, cultivation of products can be out of all proportion to the size of the finished article. The profile of water use for construction follows a very similar pattern to energy consumption and associated carbon emissions. For most materials water is consumed at every stage from cradle to grave, whilst it is important to recognise that water is also used in electricity generation and the production of fuels for heating and transportation [30].

All materials used in the construction of buildings and other structures have a water footprint to some extent but, because they dominate in terms of the total product used in construction, cement and steel tend to make the greatest contribution to the total water footprint of buildings and structures.

The water footprint of cement is highly dependent on whether there are significant concentrations of mercury in the raw materials used in processing, and particularly where gypsum is a significant ingredient, which is a by-product of flue-gas desulphurization. According to an analysis published in 2018 this typically results in a grey water footprint that is around 100 times that for the blue water footprint [31]: that is, 210 L/kg compared with 2.0–2.6 L/kg.

From the analysis reported by Gerbens-Leenes, the amount of water required for steel manufacture can also be highly dependent upon the grey water footprint which, at between 1500 and 2270 L/kg, compares with blue water footprints of 77 and 12 L for chromium-nickel unalloyed steel and unalloyed steel, respectively. These grey water footprints are based on diluting cadmium in water discharged to waste to an acceptable level during manufacture. However, modern methods of steel production have evolved to reduce cadmium discharges, whilst other pollutants, such as phenol, cyanide, ammonia and PAHs are significantly reduced by water treatment processes such as biochemical oxidation and activated carbon filtration.

Energy and water are also used on-site during the construction process which, when combined with impacts on air quality, aural environment and soil and water quality, can be many times greater than the impact of the final building or structure (see Sect. 7.8 Sustainable *Construction* below).

7.3 Noise and Air Quality

We have already discussed the importance of external noise and air quality on the quality of the indoor environment, the feasibility of natural ventilation, the positions of openings in the fabric and the design of ventilation and air conditioning systems. Noise is defined as unwanted sound and the level at which it causes discomfort will depend on the activity within a building and the duration and characteristics of the sources. Exposure to the ingress of air pollutants on the other hand can have long term health impacts without obvious discomfort: the main pollutants of concern emitted by road vehicles for example are oxides of nitrogen (NO_x) and fine particles

($PM_{2.5}$), both of which are odourless, although they may be accompanied by malodorous VOCs and ozone.

There are numerous potential sources of noise and pollution that can impact buildings, both outdoors and within. Aural environments and pollutant concentrations outdoors are generally the result of emissions from multiple sources, with major variations in air quality depending also on weather conditions such as wind speed and direction, temperature gradients and, to a lesser extent, humidity. Air quality and the aural environment both vary with industrial activity, transport infrastructure and movements. These will be examined in more detail in Chap. 8 and 9.

The noise at the outside surface of a building may therefore be the result of these various sources which, depending on the relative decibel levels, may contribute to a higher level based on logarithmic addition. For example, two equal sound pressure levels at the outside of a building result in a decibel level 3 dB higher than each. The greater the difference between the levels the lower the contribution from the lower levels such that when the lowest level is 20 dB below the highest it will have no impact on the total. The extent to which these sounds enter a building and impact on the comfort and health of its occupants depends to a large extent on the fabric of the building and the size and location of openings. Similarly the size and location of openings will play a major role in the extent to which external pollutants enter a building.

The aural environment within a building is the result of internal and external sources. In the UK, there are a number of legislative and good practice levers that are available to ensure that this environment is acceptable to most occupants for most of the time. These levers treat external and internal noise as well as noise transmission through fabric internally and from neighbouring properties separately. In many ways, the residential environment is the most sensitive, particularly during the night. For this reason WHO Guidelines on Environmental Noise have a particular focus on dwellings, particularly with regard to noise from transportation (see Chap. 8) whilst guidance from the UK Ministry of Housing, Communities and Local Government on 'how planning can manage potential noise impacts in new development'[24] in relation to the 2019 revisions to the National Planning Policy Framework (NPPF) references BS8233:2014. This standard recommends maximum noise levels in dwellings (see Table 7.1, which is Table 4 from the British Standard) [32].

Table 7.1 Indoor ambient noise levels for dwellings

Activity	Location	07:00 to 23:00	23:00 to 07:00
Resting	Living room	35 dB $L_{Aeq,16\,h}$	–
Dining	Dining room/area	40 dB $L_{Aeq,16\,h}$	–
Sleeping (daytime resting)	Bedroom	35 dB $L_{Aeq,16\,h}$	30 dB $L_{Aeq,8\,h}$

From BS8233:2014 Guidance on sound insulation and noise reduction for buildings [32]

[24] https://www.gov.uk/guidance/noise%2D%2D2

There is no equivalent to Table 7.1 in US standards, although guidance to the US Housing and Urban Development (HUD) standards for external noise from air, rail or road traffic indicates a goal of 45 dB as a maximum indoor level, although no reference is made to the avoidance of sleep disturbance in setting this level.

Acceptable levels of noise transmission through party walls, floors and ceilings into rooms used for habitation are set out in the Approved Document E (ADE) to the 2010 English Building Regulations which sets out minimum standards for sound insulation between adjacent 'dwelling houses, flats or rooms for residential purposes' [33] (similar criteria apply to other nations within the UK).

The sound insulation criteria set out in ADE are divided into those setting out attenuation through walls, floors and staircases for new houses, flats and rooms for residential purposes, such as student accommodation and prison cells; and maximum impact noise on floors and staircases. Slightly different criteria apply to existing buildings where there is a proposed material change of use.[25] The minimum attenuation requirement ranges from 43 to 45 dB, whilst the maximum permissible impact noise ranges from 62 to 64 dB. The criteria also apply within dwellings and specifically between bedrooms and rooms containing a WC and other rooms as well as internal floors. The default method of guaranteeing the performance of walls and floors as set out in ADE is for a post-construction test to be carried out; however, a proposed specification can be deemed to satisfy if it follows exactly an appropriate construction detail as set out in the most recent edition of *Part E Robust Details*.[26]

The nearest equivalent protocol adopted across the US is the *International Building Code*: Section 1206 of which sets out criteria for Sound Transmission.[27] These refer to American Standards for both laboratory and field testing of airborne sound transmission between rooms ASTM E90 and E336 respectively.[28] These rate the acoustic performance of internal partitions and floors against a Sound Transmission Class (STC) based on plots of decibel reductions for 16 frequencies between 125 and 4000 Hz. For example, an STC of 50 represents a condition at which loud sounds emanating from one side of a partition or floor are faintly heard by the recipient on the other. This is the minimum STC standard for fabric encasing a 'sleeping unit' that has been tested using the ASTM E90 laboratory test. Sleeping units here include hotel rooms as well as student accommodation and bedrooms in apartments and houses. The *International Building Code* (IBC) also sets out standards for impact noise or 'structure-borne sound' using laboratory or field tests according to ASTM E492, using similar definitions for audibility as airborne sound transmission, with a recommended maximum impact insulation class rating of 50

[25] Material change of use is defined in Part 2 of the Building Regulations 2010. see: https://www.legislation.gov.uk/uksi/2010/2214/part/2

[26] www.robustdetails.com

[27] https://codes.iccsafe.org/content/IBC2018P6/chapter-12-interior-environment

[28] ASTM is the American Society for Testing and Materials which publishes a suite of standards for testing the sound transmission through partitions and floors in buildings.

based on laboratory tests. For both airborne and structure-borne noise assessed by field tests, a lower rating of 45 is acceptable, bearing in mind there are likely to be other sources of noise within the sleeping unit being assessed. However it is interesting to note that the US-based International Code Council in its guidance published in 2010 recommended much higher ratings for both airborne and impact field tests on completed buildings with an acceptable performance of 52 and preferred performance of 57.[29]

The aural environment in non-residential buildings is subject to a wide range of criteria which depend primarily on the activities for which the building is being designed. The measures required to achieve these criteria will depend on external and internal noise sources, services strategy and the acoustic properties of internal surfaces. The noise perceived by occupants in any one building will vary depending on their proximity to sources, including activities within the space, sound from building services and external noise entering through openings or transmitted through solid fabric.

Internationally there have been numerous criteria published to define aural comfort and the acoustic environment across a wide range of applications, based on sound emanating from external sources and building services, or a combination of the two. There is not space here to detail criteria for all of these applications, but the primary references for the UK are the CIBSE Guide A volume, Chap. 1: *Environmental criteria for design* [34] and BS 8233:2014—*Guidance on sound insulation and noise reduction for buildings* [32]. For the US refer to the ASHRAE Handbook, Chap. 48: *Noise and vibration control*. Confusingly these all use different scales to represent acceptable levels of noise, however these all model the response of the human ear to sound and helpfully the CIBSE guidance provides a distillation of all the others, leaving it to the designer to choose which is applicable to their needs.

The NC and NR rating systems are applied to background noise which is pretty constant over the exposure time and are generally used to rate the noise levels resulting from building services. Figure 7.21 shows the octave band plots for the Noise Criterion (NC) system which are very similar to the NR curves, whilst being between 5 and 6 dB lower than the equivalent dBA plot. The solid black line in Fig. 7.21 shows a typical plot of sound levels in a room which meets NC45. As we will see in Chap. 8 time averaged A weighted ratings ($L_{Aeq,T}$) are generally used to represent and measure the noise from traffic for a specified period of time, T. Unlike the sound generated by building services, traffic noise tends to vary with time.

[29] https://www.iccsafe.org/wp-content/uploads/asc_a117/Supporting_doc_8-15-5_ICC_G2-2010_Guideline_for_Acoustics-A1171_Committee.pdf

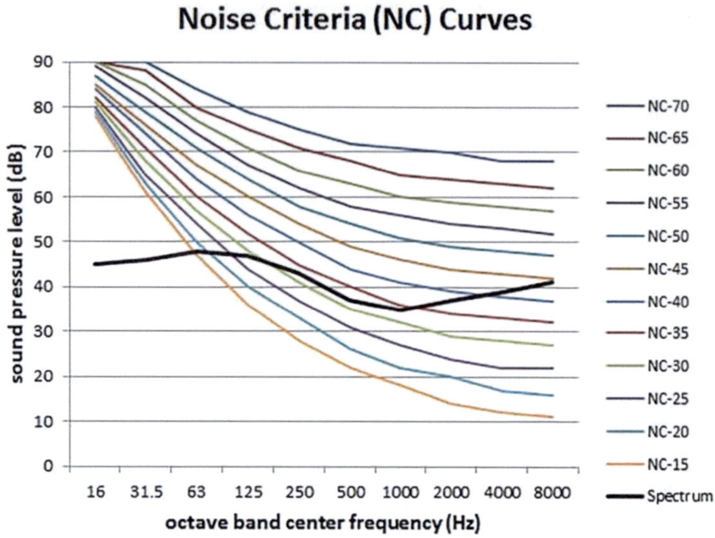

Fig. 7.21 Noise criterion curves (From: https://commons.wikimedia.org/wiki/File:Noise_Curves_Graph_Spectrum_with_NC.jpg)

7.4 Water Use and Drainage

We saw in Chap. 5 how access to fresh water is dependent upon both socio-economic conditions and the incidence of drought. Demand from the occupants of buildings also varies considerably across the planet. For example average per capita household water consumption for the USA in 2006 was 575 L/day which compares with around 20 L/day for Ethiopia, whilst a UK household was on average using about 150 L/day per person FAO in 2020[30]. The excessive consumption figures for the USA indicate endemic waste behaviours, which can be found across the developed world, despite the drought conditions experienced by many countries, which are steadily worsening as average temperatures rise and extreme weather incidents increase.

Water that is going to be used for human consumption has been the subject of World Health Organization (WHO) normative guidelines since 1957. By 2018 104 countries and regions had instigated standards, regulations or directives incorporating reference to the parameters in the WHO guidance. This sets out a preventive risk management approach covering catchment to consumer through the instigation of Water Safety Plans that include aesthetic, chemical, microbiological and radiological parameters as well as health-based targets.[31] The European Union published its *Drinking Water Directive* in 1998, resulting in regulations being introduced across the Community, such as the *Water Supply (Water Quality) Regulations 2000*

[30] https://data.apps.fao.org/aquastat/

[31] https://www.who.int/water_sanitation_health/water-quality/guidelines/en/

(England & Wales). Similarly the USA has National Primary Drinking Water Regulations, which came into force in 2009, covering similar territory.

In order to meet these water quality standards, there will normally be a requirement to treat the water, depending on the quality of the source. It is technically feasible to treat waste water from sewers or water from the sea to a suitable quality for drinking; however, the cost and energy demand of the plant required can be substantial. One such waste water recycling plant has been operating in Orange County California since 2008.[32] The 'Groundwater replenishment system' (GWRS) takes waste water that has already been treated in a conventional sewage treatment plant using a combination of bar screens, grit chambers, trickling filters, activated sludge, clarifiers and disinfection to achieve a quality suitable for discharge to water bodies such as rivers or the sea, and adds another three stages of treatment. Bacteria, some viruses and ultra-fine particles larger than 0.2 microns are removed in micro-filters that comprise hollow fibre polypropylene straws, whilst finer particles are removed through reverse osmosis (see below). Finally, further purification is carried out through a combination of ultra-violet light and the addition of hydrogen peroxide which between them are capable of removing trace quantities of organic compounds such as 1, 4-Dioxane (a potentially carcinogenic solvent) as well as all remaining viruses and bacteria.

Of course the vast majority of water on this planet is to be found in the seas and oceans. Although seawater is not fit for human consumption, it is possible to remove the salt through a process called desalination. There are a number of techniques available for this, ranging from distillation through to reverse osmosis. Put simply the distillation process uses heat to evaporate the seawater which is passed over a cool surface on which the now fresh water condenses, leaving a concentrated brine to be discharged back to the sea. This phenomenon was first noted by Aristotle in his treatise *Meteorologica*, written in c340 BCE and was used by seafarers from the early 1600's by piping the steam from a kettle into a cool bottle. This technique increased in scale over the years into centralised plant supplying drinking water for ships and even remote islands. Most of the larger plants used steam to indirectly boil the seawater, with some combined with steam turbines used for generating electricity.[33] These generally use multiple stages, the two most common being multiple effect distillation and multi-stage flash distillation. Small-scale desalination stills are available using solar energy to evaporate seawater which condenses in turn on a glazed or transparent plastic lid. These simple devices use minimum or zero non-renewable energy and are used to provide drinking water for small sailing boats or as an emergency supply in remote locations.

Reverse osmosis involves the forcing of contaminated water or brine from a region of high contaminant or salt concentration through a semipermeable membrane to a zone of low concentration through application of a pressure in excess of the osmotic pressure. This pressure must be developed by a pump to overcome the resistance to flow of the membrane requiring significant electrical power and hence energy, although typically between two and five times lower than the thermal energy

---

[32] https://www.ocwd.com/gwrs/the-process/

[33] https://www.oas.org/dsd/publications/unit/oea59e/ch21.htm

required for distillation. Carbon emissions may be similar, however, depending on the carbon intensity of the electricity supply and whether waste heat can be recovered for the distillation option. The carbon footprint of mains water obtained from reservoirs and underground sources not requiring specialist filtration, along with associated wastewater varies depending on the power required for pumping and the embodied carbon of water treatment chemicals and processes. Methane and Nitrous Oxide also contribute to total emissions as they are released during wastewater treatment and the downstream use of by-products such as fertiliser. Data from Anglian Water for 2006/7 published in a 2013 report by CIWEM indicate that around 55% of the carbon footprint for its region in the UK was from electricity consumption, whilst 35% was due to methane and nitrous oxide emissions from sludge deposited on land.

It is clear from the above that water can be a costly and energy intensive resource, potentially subject to restricted access and interruption, even in the developed world. The greatest demand in residential buildings such as homes, hospitals and hotels tends to be for domestic use including bathing, toilet flushing, laundry, dishwashing, catering, cleaning, garden maintenance and vehicle washing. Other high demand applications include restaurants, sports centres, car washes, evaporative cooling towers and industrial processes.

According to a report by the Energy Saving Trust from 2013: in the UK:

> …showers are now the largest user of water in the home. Across Britain, we use in the order of 840 billion Litres each year, and spend around £2.3 billion on heating water for showers. We can report that, every year, more than 740 billion Litres—enough to fill 300,000 Olympic-sized swimming pools—is flushed down the loo. And our dishwashers and washing machines use 360 billion Litres each year, costing households £1.6 billion in electricity bills (Energy Saving Trust, 2013).

If we look at water use for a typical British home in 2010, shower use (25%) and toilet flushing (22%) dominated water consumption Energy Saving Trust in 2013[34]. In this same year Regulation 17 K of the Building Regulations limited water consumption for new dwellings in England and Wales to 125 L/person per day based on demand prediction using a standard water efficiency calculator incorporated into the amended Approved Document G of the 2010 Building Regulations [35]. The analysis carried out for a 2018 report by Ofwat on water use in the UK also shows that average water consumption per head increased steadily from 85 L/day in 1961 to 155 L/day in 1996, although falling from 155 L/day in 2004 to 140 L/day in 2016.[35] Estimates published by the Energy Saving Trust in 2024 indicate that there has been no decrease in household water consumption since 2016, with 'each household in the UK (using) around 345 L of water a day, on average. That's 145 L per person, every day. Also, around 12% of a typical gas heated household's energy bill is from heating the water for showers, baths and hot water from the tap'.[36]

[34] https://database.waterwise.org.uk/wp-content/uploads/2019/10/Energy-Saving-Trust_At-Home-With-Water.pdf

[35] https://www.ofwat.gov.uk/wp-content/uploads/2018/05/The-long-term-potential-for-deep-reductions-in-household-water-demand-report-by-Artesia-Consulting.pdf

[36] https://energysavingtrust.org.uk/advice/saving-water-at-home/

Ofwat report also indicated that between 1976 and 2015, the percentage contribution of personal washing increased and toilet flushing decreased, whilst other uses stayed roughly the same. It has been postulated that this has been due to a trend away from a weekly bath to daily showering. A survey in 2013 referred to in the Ofwat report indicated that 70% of the population have a full body wash at least daily, mostly by showering, and that more than 50% of the population never have a bath [36]. This being particularly the case for the younger demographic, who also spend an average of 20 min in the shower. The popularity of power showers has also contributed to this trend.

On the other hand, washing machines, dishwashers and WC cisterns have become more water efficient, although the replacement interval for WCs is generally much longer than for white goods—typically around 9 years for both washing machines and dishwashers. Toilet cisterns can last almost indefinitely, although some of the internal components may require replacing after as little as 5 years. Most WCs installed prior to 2001 in the UK had cisterns of 9 L or larger after which a *Specification for WC Suites* was introduced under the 1999 *Water Supply (Water Fittings) Regulations* which required a maximum single flush of 6 L and dual flush no more than two thirds of maximum [37].

The potential for reducing water volume used for personal washing is largely in the hands of the individual. However, there are a number of design measures that can reduce overall water consumption. Water efficient shower heads are available that either supply water in pulses or mix water with air (aeration). A standard bath typically contains about 160 L of water, which is more than the average daily per capita water consumption in the UK. However, low water volume baths, which are narrower at the tap end, are available that typically require only 70 L for a comfortable water level.

The 1999 Water Fittings Regulations also allowed the installation of WC cisterns with button-operated drop and flap valves in the UK, having been widely used in Europe and the USA for many years. This has resulted in a system more vulnerable to failure than the lever operated siphon flush. Push button operation has allowed a simpler dual flush operation, but there have been widely reported problems with valves sticking and allowing continual seepage of water from the cistern into the bowl. This can be overcome with the fitting of a delayed action inlet valve. Exciting research into 'non-stick' surface treatments by scientists at Penn State University in the USA could result in significantly less water being required for flushing of solids from a WC pan. They have developed a spray that can 'reduce cleaning water consumption by c90% compared with untreated surfaces due to its extreme repellency towards liquids, bacteria and viscoelastic solids' [38]. Although the aim of this research has been to develop a product for field application, particularly in the developing world, clearly there are major implications for wider application and significant reductions in water use globally.

It is evident from research carried out by the Energy Saving Trust that water use behaviour of dwelling occupants is influenced by metering and hence an understanding of the cost impact of water use. Based on responses from 100,000 households between 2010 and 2012 it found that '43% of respondents have a water metre in their homes. These properties used around 3% less water than unmetered properties' (Energy Saving Trust, 2013). Where this feedback loop is missing—such as for hotels and holiday lets—there is little motivation for guests to reduce water consumption.

Water use in non-residential applications depends to a great extent on the activities and processes undertaken. For example, offices may have toilet facilities for occupants with associated flushing of WCs and urinals and hand washing; or there may be showers for cyclists and runners; or anything from kitchenettes through to café/restaurant facilities with associated food preparation and dishwashing. Some may also have air conditioning with evaporative cooling towers and humidifiers. Many factories will have a similar water use profile to offices unless there are processes that consume water, such as food processing or washing of dirty root crops for example. Schools and other educational buildings may have similar water use characteristics to offices with catering, but also have shower use associated with sporting activities.

The water consumption profile in hospitals is partly domestic and partly process-driven, with a combination of hotel-type usage, including laundry, catering, patient bathing and WC use, a variety of surgical and treatment uses and office-type use for staff. Cleaning and maintenance will usually consume greater volumes than other applications. In the UK hospital water use is typically twice that for a 3-star hotel with a swimming pool and three times that for an office building. In both hospitals and hotels whether laundry is carried out in-house or contracted out has a significant impact on total water use, although the consumption is simply transferred to the contractor in the latter case.

For a more detailed review of high use applications such as restaurants, laundries, swimming pools, whirlpool spas and car washes refer to Chapter 3.9 of [2] and Chapter 3.5 of [3]. For design guidance, refer to *CIBSE Guide G: Public Health and Plumbing Engineering* for the UK [39] and various guides from the American Society of Plumbing Engineers. Many other countries have their own plumbing guidance whilst practices and legislation vary from country to country.

7.5 Waste Water Recycling and Rainwater Harvesting

There is nothing new about capturing rainwater to supply water to building occupants or for irrigation purposes. Rainwater capture in ancient civilisations such as that found in India, China, Mesopotamia and across the Roman empire, was largely part of a communal infrastructure, whilst the modern application of rainwater harvesting refers to the capture of rainwater from individual building roofs and the use within the same building or for landscape irrigation.[37]

Recycling the so-called grey water on a building-wide scale is a relatively recent development. Water that has been used for personal and clothes washing can be

[37] https://www.stormsaver.com/rainwater-harvesting-/

collected and re-used for a purpose compatible with the quality of the resultant waste water. In theory it is possible to filter and purify any building waste water on site to a quality suitable for drinking, however, the cost and energy required to achieve this is usually prohibitive. Hence typically waste water is collected from baths, showers, hand washbasins and washing machines for use in flushing toilets and/or irrigation. It is not generally feasible to recycle waste water from kitchen sinks and dishwashers because of the grease load. Indeed in some parts of the world, such as California in the USA, it is illegal to do so in residential applications.[38]

With appropriate filtration and water treatment grey water can be used for supplying washing machines. According to the Environment Agency's 2011 guide to grey water use in residential properties untreated grey water should not be stored for more than a few hours and hence: 'All systems that store grey water have to incorporate some level of treatment, as untreated greywater deteriorates rapidly in storage' [40].[39]

Grey water is usually warm and nutrient-rich and hence provides ideal conditions for the multiplication of bacteria if stored for any length of time. It is, however, possible to use untreated grey water for irrigation, although it is advisable not to use it on crops intended for human consumption. When used for toilet flushing, and especially when supplied to a washing machine, it is necessary to remove particulates and prevent microbial growth. The contaminants in grey water will typically include soaps and detergents, skin scales, hair, dirt, plastic and natural fibres. The concentration of these need to be reduced to a level at which biocidal control is effective, which can be through dosing with a disinfectant such as chlorine or bromine, or exposure to ultra-violet (UV) light. The effectiveness of UV disinfection depends greatly on filter efficiency since larger particles tend to protect some microorganisms from UV light. However, UV is preferable to chemical treatments, which have detrimental environmental impacts on waste water downstream when discharged to natural water courses, whilst also requiring greater user interventions.

Removal of what can be quite a high contaminant load poses a potential maintenance challenge, particularly for individual households where skilled maintenance staff are not at hand. Mechanical systems typically incorporate multiple tanks to allow settlement and removal of large particles and skimming off of light surface debris and soap, along with chemical or UV disinfection. Packaged equipment such as the Hydraloop system shown in Fig. 7.22 can be incorporated into bathroom/toilet designs for individual homes or modular systems.

Key: T1 - Greywater inlet tank T2 - Bioreactor tank with moving bed bioreactor (MBBR) T3 - Storage tank which holds recycled water T4 - Tank for backup water connection, complete with safe air gap.

Biological systems take advantage of the phenomenon that involves bacteria and fungi in the breaking down of potential harmful contaminants into harmless components via complex chemical reactions. Aerobic bacteria require oxygen to thrive, and this is provided by either mechanical means or naturally through the use

[38] https://centralcoastgreywater.org/getting-started/

[39] http://www.feeltheheatuk.com/downloads/ecoplay_full_brochure.pdf

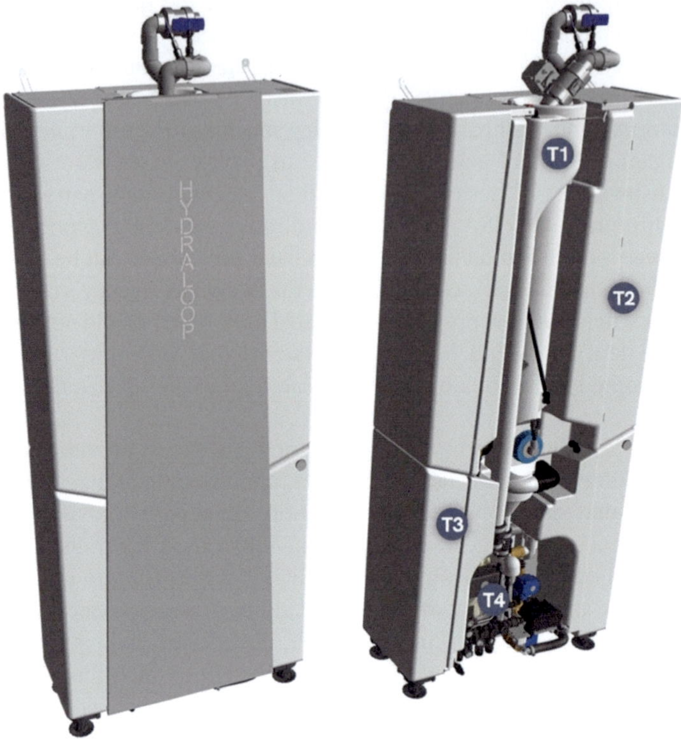

Fig. 7.22 Packaged grey water system for single dwelling or modular system. Key: T1—Greywater inlet tank, T2—Bioreactor tank with moving bed bioreactor (MBBR), T3—Storage tank which holds recycled water, T4—Tank for backup water connection, complete with safe air gap (Image: Hydraloop; From: https://info.hydraloop.com/docs/recycle-ready-guide-hydraloop-h300-collection25/)

of reeds which channel atmospheric oxygen to their roots to survive. Systems that are integrated into surface landscaping typically incorporate reed beds. These systems can process large contaminant loads, treating for example sewage and industrial chemical wastes very effectively. Smaller versions of these systems have been used to treat grey water or in combination with septic tanks for off-grid applications.

Reed beds generally require a sloping site into which a rectangular pit is excavated and lined, filled with humus-rich compost and planted with the reeds (see Fig. 7.23). The size and flow rate along the bed must be appropriate for the required wastewater discharge rate, and the reeds can take a couple of years to become fully established. The land take may make reed beds impractical for individual houses, but they may suit groups of dwellings, particularly where access to conventional sewers is not possible. Most importantly they require minimal energy to operate or construct and provide long term enhancement to local biodiversity.

Packaged systems have been developed that use the principles of the reed bed within a series of tanks using microbial cultures grown on a suitable medium or by passing the water through membrane bioreactors (MBR) with multiplication

Fig. 7.23 Newly planted domestic gravel reed bed system for septic tank effluent treatment prior to percolation (Source: FH Wetland Systems; https://www.lowimpact.org/lowimpact-topic/reed-beds/)

encouraged by bubbling oxygen through the water. Coarse particles are typically removed using a backwash filter, and sludge is allowed to collect for automatic removal from the base of the tanks and final microbial control achieved through UV disinfection, where necessary.

Rainwater harvesting installations range from a simple water butt storing rainwater from the roof of a dwelling to large storage systems that incorporate filtration and water treatment that allows harvested rainwater to be used for toilet flushing as well as irrigation. Although the contaminant load in rainwater should be lower than for grey water it may still be necessary, once leaves and other detritus are removed, to provide biocidal treatment if it is to be used for toilet flushing, car washing or similar.

One arrangement for rainwater harvesting is shown in Fig. 7.24. The storage can be underground as shown or in a basement and may comprise a single compartment, oversized drainage pipes or geocellular tanks as used in storm water attenuation and infiltration. The latter has the advantage of requiring shallower excavations, although with a larger footprint than for the type shown in Fig. 7.24.

Storage tanks for rainwater harvesting systems are normally sized based on the volume of rainwater flows, which depends on the roof surface area, its slope and surface characteristics. For example, only around half of the water that falls on a flat roof will be available for harvesting, that is, a drainage factor of 0.5. For green or landscaped roofs this figure is even lower, whilst for a smooth, sloping roof the drainage factor can be as high as 0.9. Typically storage volumes are estimated by multiplying the predicted average annual rainfall volume by 0.05 and by the drainage factor.

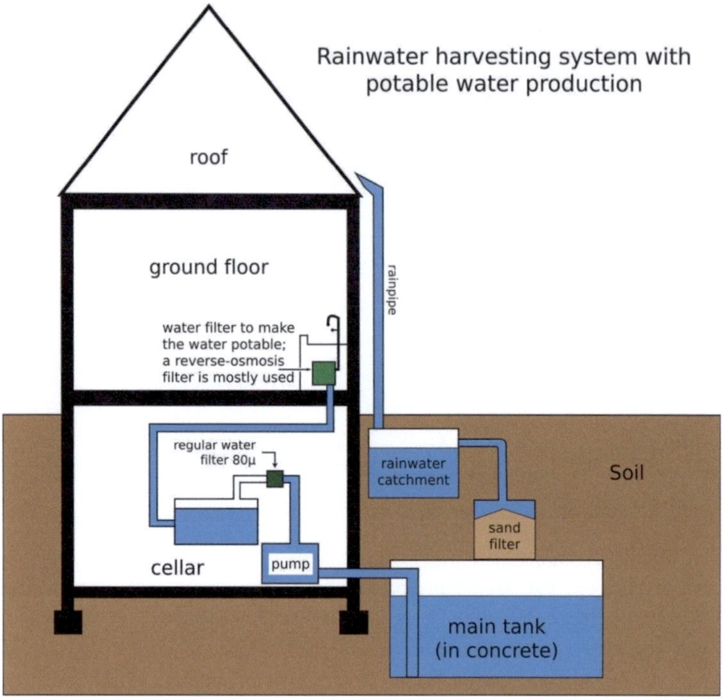

Fig. 7.24 Schematic of rainwater harvesting system (Source: https://commons.wikimedia.org/wiki/File:Rainwater_harvesting_system.JPG)

Further guidance is available for the design of both grey water recycling and rainwater harvesting systems in the CIBSE Knowledge Series guide on Reclaimed Water [5]. Across the US many states and city authorities have published their own guides and requirements, whilst across mainland Europe, where uptake of these systems has been very popular, most countries have their own guidance and legislative requirements.

7.6 Landscaping, Biodiversity and Sustainable Drainage

Every building has a setting: viability and planning constraints in towns and cities may dictate that facades have to be located close to boundaries, with little space for an external landscape, whilst in less constrained settings there may be the opportunity to create a green landscape that links with its surroundings. However, for constricted sites, roofs, terraces and even walls can be designed to provide some combination of amenity, ecological niches and rainwater attenuation.

Depending on the application, external spaces have numerous competing and potentially conflicting demands. These may include the provision of space for access and circulation routes, car parking, refuse bin storage and user-specific requirements such as play space, eating areas and the such. The challenge is to achieve a balance between amenity and ecological value; hard and porous surfaces, whilst mitigating flood risk. A living landscape can have a significant impact on both indoor environmental conditions and external microclimate, whilst also absorbing atmospheric carbon dioxide and other air pollutants.

Although the percentage of the earth's land surface area occupied by buildings and their associated landscape is relatively small, buildings do tend to be concentrated into distinct groupings and frequently alongside rivers. Hence they have historically created islands of low biodiversity, high flood risk and higher summer temperatures. Tall buildings can interfere with the migratory routes of birds, whilst the pollutants in run-off and sewage overflow can create lifeless water bodies. Therefore the potential for introducing living landscapes into built-up areas introduces benefits out of proportion to the areas involved.

The concept of 'biodiversity net gain' was developed by the International Union for Conservation of Nature (IUCN) in the 2010s to support the so-called 2020 Aichi Targets as well as the Sustainable Development Goals (see Chap. 2, Sect. 2.2). In the EU 2020 Biodiversity Strategy, the European Union set a target to help stop the loss of global biodiversity through incentivisation, including a commitment that 'Incentives will be provided to attract private sector investment in green infrastructure and the potential of biodiversity offsets will be looked into as a way of achieving a 'no net loss' approach' [41]. These set broad objectives taking in all aspects of the natural world under human influence. The successor to this sets similar targets for 2030, recognising that those set in 2011 were largely not achieved.

In 2020 the British Government focussed specifically on the built environment by publishing an Environment Bill for England and Wales which included at Sect. 7.6 a requirement for new developments to achieve 'biodiversity net gain' of 10% compared with the existing site, through either on-site measures or off-site means in the locale [42]. The requirement is assessed using a tool published by Natural England which publishes a suite of Biodiversity Metrics which assess pre- and post-construction biodiversity under the headings of distinctiveness, condition, strategic significance and habitat connectivity.[40]

According to the on-line *biological dictionary*, and in its broadest sense 'Biodiversity is a term which describes every living organism within a single ecosystem or habitat, including numbers and diversity of species and all environmental aspects such as temperature, oxygen and carbon dioxide levels and climate. Biodiversity can be measured globally or in smaller settings, such as ponds'.[41] A biodiverse ecosystem is a healthy ecosystem, creating the synergistic interconnectedness that supports life on earth. Hence an individual landscape associated with a new development should be part of a matrix that links and supports healthy and biodiverse ecosystems across a large area. Hence it is not just the number of

[40] https://publications.naturalengland.org.uk/publication/6049804846366720

[41] https://biologydictionary.net/biodiversity/

different species of flora and fauna that are found on a specific site (ecological value) after development that is important but also how they link with neighbouring habitats.

This is recognised when assessing the land use and ecology credits as part of a BREEAM Assessment for example, through establishing the 'Capacity and feasibility for enhancement of the ecological value of the site and, where relevant, areas within the zone of influence'.[42]

No matter what area of land is available for soft landscaping on and around a building a strategy should be developed that includes species that are suitable for the location. This means incorporating species that are native to the area. The National Biodiversity Network has developed a database for the UK that gives a list of native plant species that have been identified within a specified radius of a given postcode.[43]

Although it may seem logical that the greatest improvements in biodiversity should be achievable when developing a brownfield site, in fact sites that have been neglected for many years may have been colonised by a wide variety of species. For example, industrial derelict land that has been unused for 10–20 years and colonised typically by tall herb vegetation can have an ecological value significantly higher than farmland that has been used for single crop arable farming for 5 years or more. On the other hand a site which requires an existing building to be demolished within a hard landscape setting would be ripe for ecological enhancement. At the other end of the scale are sites that are resident to mature trees and hedges, watercourses, wetland areas and wild flower meadows, contributing to a high ecological value and local biodiversity. Indeed it is likely that these features will be the subject of conservation orders or planning restrictions and hence limit development opportunities.

Another reason why some sites may be unsuitable for development is that they are prone to flooding. As discussed in the Sect. 2.4 in Chap. 2 climate change is increasing the risk of flooding in many areas. This is exacerbated further when porous surfaces are replaced with impermeable structures such as buildings, roads and hard landscaping from which precipitation runs off before it can be absorbed into the ground.

Globally there are many locations that experience floods periodically. Some of these floods occur when there has been above average rainfall, others following storm surges or earthquakes creating large waves or tsunamis that penetrate inland. Apart from for earthquakes, the risk of these events occurring within a specified time frame can be predicted and in the UK and other European nations flood risk mapping has been a requirement since the EU Floods Directive was published in 2007 [43]. For example the UK has been divided into Flood Zones where Zone 1 represents the lowest risk and Zone 3 the highest (see Box 7.2).

---

[42] https://bregroup.com/news/breeam-news-version-6-1-launched

[43] https://records.nbnatlas.org/explore/your-area

Box 7.2 Flood Zones as Set Out in National Planning Policy Framework (NPPF) Guidance for England[44]

Flood zone	Definition
Zone 1 low probability	Land having a less than 1 in 1000 annual probability of river or sea flooding (Shown as light blue on the flood map in Fig. 7.2)
Zone 2 medium probability	Land having between a 1 in 100 and 1 in 1000 annual probability of river flooding; or land having between a 1 in 200 and 1 in 1000 annual probability of sea flooding (Land shown in mid blue on the flood map)
Zone 3a high probability	Land having a 1 in 100 or greater annual probability of river flooding; or land having a 1 in 200 or greater annual probability of sea flooding (Land shown in dark blue on the flood map)
Zone 3b the functional floodplain	This zone comprises land where water has to flow or be stored in times of flood. Local planning authorities should identify in their strategic flood risk assessments areas of functional floodplain and its boundaries accordingly, in agreement with the environment agency (Not separately distinguished from zone 3a on the flood map)

Figure 7.25 shows the flood risk map for Hebden Bridge in West Yorkshire, which was severely inundated following Storm Ciara in early 2020. Hebden Bridge has been flooded more than 50 times in the last 200 years, and remains a high risk zone to this day.

New development in England proposed in a flood risk zone within any of the categories requires a flood risk assessment within the context of any strategic assessment that may have been incorporated into the relevant local plan by the local planning authority. Consultation may be required with the Environment Agency and the Internal Drainage Board for development in Flood Zones 2 or 3. Only sites in Zone 1 larger than 1 hectare and having drainage issues or being otherwise vulnerable require a flood risk assessment.

As part of the their deliberations for planning applications a 'Sequential Test' is applied to the whole local authority planning area and in some cases neighbouring authority areas to identify sites for development away from high flood risk. An 'Exception Test' applies to the specific development and according to para 164 of the 2022 update to the NPPF:

> For the exception test to be passed it should be demonstrated that:
>
> (a) The development would provide wider sustainability benefits to the community that outweigh the flood risk; and

[44] https://www.gov.uk/guidance/flood-risk-and-coastal-change#flood-zone-and-flood-risk-tables

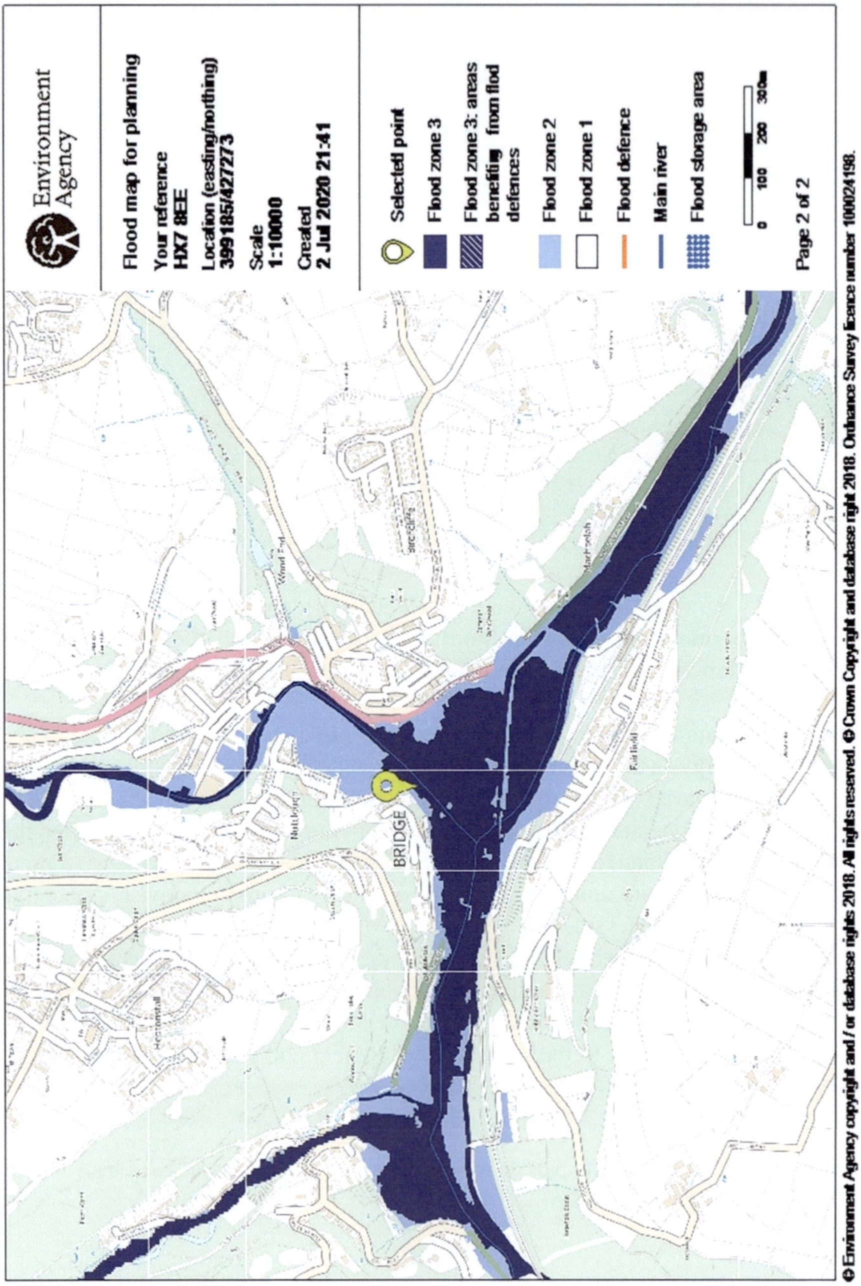

Fig. 7.25 Flood risk map for Hebden Bridge in Yorkshire, England[45] (https://flood-map-for-planning.service.gov.uk/map?cz=39916842712315)

[45] https://flood-map-for-planning.service.gov.uk

(b) The development will be safe for its lifetime taking account of the vulnerability of its users, without increasing flood risk elsewhere, and, where possible, will reduce flood risk overall.[46]

Although framed differently in other nations these principles apply to development across the whole of the European Union and flood risk maps are available for many countries beyond.

All new development has the potential to contribute to flood risk to a greater or lesser extent and, with this in mind, landscape should be designed to minimise impermeable surfaces and allow surface water to be retained on site and where possible absorbed into the water table beneath the site (infiltration) or attenuated and released slowly into downstream water courses (detention). The measures required will depend on the site underlying soil composition and permeability (ground composition) as well as the local topography and capacity of any downstream water courses. These can be achieved by incorporating natural features into the landscape strategy with only limited recourse to structural or manufactured components such as concrete retention and attenuation tanks or underground stormwater pipework whilst reducing the surface area of impermeable surfaces such as tarmac roads and paths. This approach creates what is widely referred to as a Sustainable Drainage System (SuDS) and has been a requirement for all 'major' developments in England for example since 2015.[47]

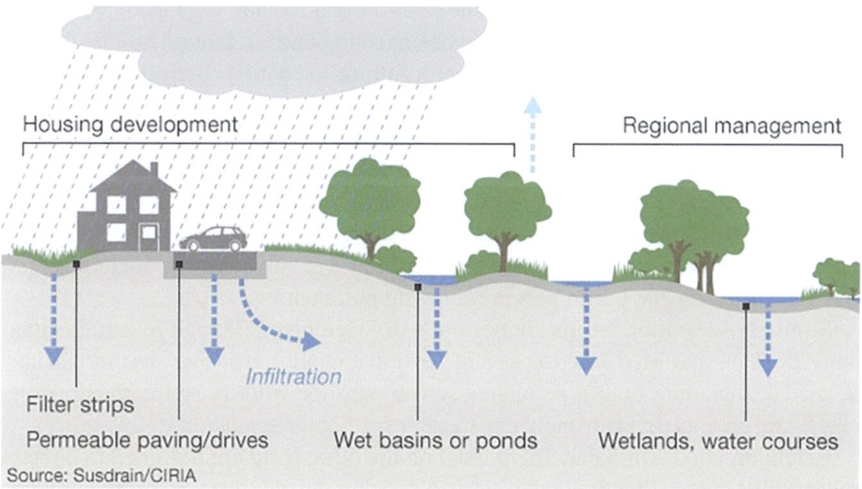

Fig. 7.26 Sustainable drainage management train (From: https://wiki.sustainabletechnologies.ca/wiki/File:72208733_sustainable_drainage_624.jpg)

[46] https://www.gov.uk/guidance/flood-risk-and-coastal-change#full-publication-update-history

[47] https://www.parliament.uk/documents/commons-vote-office/December%202014/18%20December/6.%20DCLG-sustainable-drainage-systems.pdf

Fig. 7.27 Typical swale (From: https://commons.wikimedia.org/wiki/File:Grass_lined_channel_NRCS.jpg)

The SuDS 'management train' shown in Fig. 7.26 replicates the processes that occur in a natural catchment between source and the receiving watercourse.[48] Every site will have a different level of complexity: ranging from simple source control whereby the soft landscape provides sufficient attenuation prior to infiltration into the water table through to a series of site controls that include vegetated channels or ditches (swales) depressions (detention basins) and ponds (retention ponds) that are designed to maximise the area available for infiltration whilst controlling run-off to a rate that doesn't create downstream flooding during storm events.

Swales are ditches that can either be dug on the level to form a detention basin or be used to channel rainwater into a downstream water course, detention basin or retention pond, whilst allowing infiltration along their full length (see Fig. 7.27). They can also perform both functions by incorporating dams at suitable locations. They can also form an excellent environment for vegetative growth, including species such as reeds that play a part in removing pollutants.

Similarly 'detention basins are normally dry (see Fig. 7.28) and in certain situations, the land may also function as a recreational facility. However, basins can also be mixed, including both a permanently wet area for wildlife or treatment of the runoff and an area that is usually dry to cater for flood attenuation'.[49]

Retention ponds (or balancing ponds) on the other hand are designed to contain water whilst controlling flow rates by storing water during storm events then releasing it gradually once the flood risk has passed (see Fig. 7.29). The water level will fluctuate, and ponds should cater for varying weather conditions. Reed beds are frequently used to control water quality.

[48] https://www.susdrain.org/delivering-suds/using-suds/suds-principles/management-train.html

[49] https://www.susdrain.org/delivering-suds/using-suds/suds-components/retention_and_detention/Detention_basins.html

Fig. 7.28 Typical detention basin (Image: Author)

Fig. 7.29 Typical retention pond

It is possible to extend local control beyond that provided by the soft landscaping by creating a vegetated green or brown roof. As well as being populated with species that are found locally and hence contributing to ecological value, they attenuate run-off volumes compared with a roof having a hard surface. Green roofs fall into two main categories: extensive and intensive. Extensive roofs are generally intended

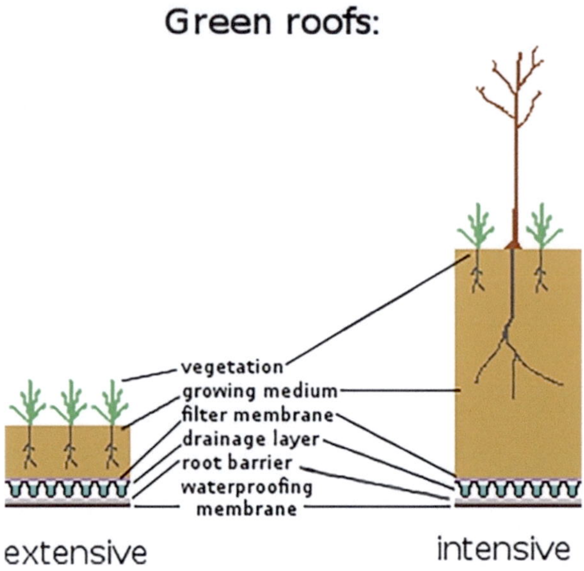

Fig. 7.30 Typical build-up of intensive and extensive green roofs

to support a vegetative layer such as sedum and not for human access and amenity, whilst intensive green roofs are essentially roof gardens, including paved or boarded areas for human access and landscaped areas that may include trees, cultivated borders or even vegetable plots.

Typical compositions for intensive and extensive green roofs are shown in Fig. 7.30.[50] In order to maximise ecological value, a species-rich vegetative layer should be planted as shown in Fig. 7.31. The growing medium can be soil or a suitable lightweight compost-based mixture and can be designed for lightweight structures. Intensive roofs such as the one shown in Fig. 7.32 on the other hand tend to be much heavier, requiring deeper growing medium and a structure suitable for the resultant loads.[51] As can be seen in Fig. 7.31, it is possible to incorporate a wide variety of species into the planted area.

For more detailed guidance for England, both Defra and the Local Authority SuDS Officer Organization (LASOO) have produced on-line documents with the same title, that is, *Non-Statutory Technical Standards for Sustainable Drainage: Practice Guidance.* The former was published in March 2015[52] the latter is undated[53] but provides more detailed guidance on both planning and technical considerations.

---

[50] https://greenrooftechnology.com/green-roof-finder/extensive-green-roof/

[51] https://www.optigruen.com/planning-aids/specialist-topics/intensive-roof-greening

[52] https://assets.publishing.service.gov.uk/government/uploads/system/uploads/attachment_data/file/415773/sustainable-drainage-technical-standards.pdf

[53] https://www.susdrain.org/files/resources/other-guidance/lasoo_non_statutory_suds_technical_standards_guidance_2016_.pdf

Fig. 7.31 Extensive green roof (Photo: YvesSch; From: https://commons.wikimedia.org/wiki/File:Green_Roof_%2815456078087%29.jpg)

Fig. 7.32 Intensive green roof on Chicago city hall (Photo: TonyTheTiger. https://commons.wikimedia.org/wiki/File:20080708_Chicago_City_Hall_Green_Roof.JPG)

7.7 Socio-Economic Impacts: Sustainable Communities

A community comprises a group of people who are somehow connected. The term is frequently used in connection with faith, ethnicity or political persuasion. However, in this chapter, we are concerned about communities that are linked through geography, whether it is a group of people who live in a housing development, a tower block or village. Although individual buildings can play an important role in a sustainable community, generally it is in the masterplanning of a development and its context that designers and planners have most to contribute.

There are numerous definitions for sustainable communities, but one from the 2003 Sustainable Communities Plan produced by the UK Department for Communities and Local Government (CLG) sums up their essence very well:

> Sustainable communities are places where people want to live and work, now and in the future. They meet the diverse needs of existing and future residents, are sensitive to their environment, and contribute to a high quality of life. They are safe and inclusive, well planned, built and run, and offer equality of opportunity and good services for all.[54]

The key phrase in this quotation is 'high quality of life'. The developer and local planning authority in choosing or allocating the site, and the masterplanning and design teams in producing a sustainable design, must balance a wide range of parameters within the physical and financial constraints imposed by location and the pressures of commercial viability. There are also numerous definitions of what constitutes a high quality of life. There have been a number of attempts to develop indices or ranking systems for comparing quality of life in countries or cities for example: including the Organization for Economic Co-operation and Development (OECD) Better Life Index[55] and the Economist City Liveability Score.[56] Although designed for international comparisons many of the factors considered in calculating these indices apply to individual developments and can be influenced or mitigated through the planning and design processes. These include environmental and social issues such as air quality and noise, healthy living, safe and secure setting, social connections, access to amenities and links with nature, for example.

We have already touched on a number of these issues in this chapter so in this section we are going to focus on the social and economic benefits that a development with good urban planning can bring to a community. A residential development cannot comprise homes in isolation. Safe, easily accessed and pleasant connections to all necessary amenities are required. Ideally it should be possible to walk safely, day and night, all year round to essential amenities, including, for example, a convenience store, pharmacy, post-office, bank/cash point, health centre,

---

[54] https://webarchive.nationalarchives.gov.uk/+/http://www.communities.gov.uk/archived/general-content/communities/whatis/

[55] https://read.oecd-ilibrary.org/economics/how-s-life/volume-/issue-_9870c393-en#page5

[56] https://www.economist.com/graphic-detail/2019/09/04/vienna-remains-the-worlds-most-liveable-city

primary school, children's play area, community centre, places of worship, café/ takeaway and pub/bar. It should also be possible to cycle or walk to from and between these facilities without coming into conflict with each other or with vehicles. The concept of 'walkable or low traffic neighbourhoods' and 'living streets' is gaining traction across much of the developed world.

'A walkable neighbourhood is typically seen to comprise high density, compact, mixed use, amenity rich, transit-orientated development, good pedestrian infrastructure and pedestrian-orientated streets' [44] and this has certainly been confirmed by the US-based Walk Score organization, which has calculated 'walk scores' for cities and neighbourhoods across the US, Canada and Australia[57] with the densest cities having the most amenities, such as New York city, achieving the highest scores for walkability. Although these cities also do well for transit facilities, they are frequently not the best at safely catering for cyclists.

New developments may be built anywhere from a city centre through an urban extension to a green field site surrounded by open countryside. Each of these scenarios poses different socio-economic challenges and needs to be assessed in terms of its context: constraints, infrastructure, connectivity, and amenities. Although city centres may offer easy access to transport nodes, amenities and places of work, the street environment may be noisy and polluted, whilst access to green spaces and child-friendly play areas may be limited. Apart from the environmental consequences of building on green field sites, it may be necessary to provide most or all of the amenities referred to above along with extending or improving public transport provision.

Safe travel to schools in particular—for all ages, all year round—is of primary importance. Children should be able to walk or cycle to and from school via well-lit and overlooked routes. Where these have to cross roads ideally, there should be a 20 mph (30 kph) speed limit with suitable pedestrian crossings, where speed limits are higher crossing should be, in order of preference, via a bridge, subway or traffic lights. Footpaths and cycleways should be of a generous width and have a suitable landscape setting to avoid creating narrow sections that might be perceived as threatening, provide cover for criminal activity or where cyclists and pedestrians could come into conflict.

Location is also intrinsically linked to safety and security. Incidence and type of criminal activity tends to be different between cities, suburbs and rural communities. They also tend to vary from nation to nation. Hence the measures required to achieve a secure design will differ accordingly. Back in 2003 the UK's now defunct Office of the Deputy Prime Minister (ODPM) stated that 'safety and security are essential to successful, sustainable communities. Not only are such places well-designed, attractive environments to live and work in, but they are also places where freedom from crime, and from the fear of crime, improves the quality of life.'

[57] https://www.walkscore.com/cities-and-neighborhoods/

For England and Wales the Police Crime Prevention Initiative (PCPI) publishes a suite of design guides under the heading of Secured by Design (SBD).[58] These include specific guidance for homes, schools, commercial buildings, hospitals, sheltered accommodation and car parks (Park Mark™) as well as technical guides for alarms and doorsets. Published in 2015, Approved Document Q of the Building Regulations for England 'contains guidance on security in new dwellings, including measures taken to avoid any unauthorised entrance to dwellings and flats within a building'.[59]

Apart from making individual buildings burglar-proof SBD addresses a range of masterplanning issues such as the design of roads, footpaths, car parking and associated landscape and lighting to minimise the opportunities for criminals to approach properties unobserved, find an easy route for escape or practise other illegal or anti-social behaviour. Although developed for use in the UK, SBD is equally relevant globally and, as well as informing the design of new developments, can help improve existing buildings, housing estates, mixed use quarters, and many more.

7.8 Sustainable Construction

The best laid plans of the design team can count for nought if they are lost in the procurement and construction processes. There are many reasons why a disconnect can occur: the contractor may have limited experience in delivering the principles and technologies required; the sustainable design features may be value engineered out of the finished product; there may be a gap between predicted and actual performance; there are inadequate quality checks during construction; or the design team is not novated to the contractor in a design and build procurement process and hence supervision and quality control may be weakened.

Most of these challenges can be tackled by incorporating key performance targets into the tender and contracting process. Although BREEAM and LEED assessment protocols allow for trade-offs to achieve a target rating they both require assessment of the delivered building to confirm the predicted rating. Trade-offs that dilute critical criteria such as whole life carbon emissions and water consumption can be avoided by assigning post-construction performance targets for them.

Predictions of operational energy consumption and associated carbon emissions during the design stage tend to significantly under-estimate actual performance. For the UK, this performance gap has been estimated as between a factor of 2 and 5.[60] This is partly because design stage estimates are typically based on the so-called regulated emissions, determined for Building Regulations compliance, which ignore those associated with energy usage outside the influence of the design team

[58] https://www.securedbydesign.com/guidance/design-guides

[59] https://www.gov.uk/government/publications/security-in-dwellings-approved-document-q

[60] https://www.cibse.org/media/a1skdgsi/cb11.pdf

(referred to as 'unregulated emissions') such as IT equipment, kitchen appliances, audio-visual equipment and the such. The calculations used for Building Regulations approval are simplified and not based on dynamic building simulation models: tools such as the Simplified Building Energy Model (SBEM) are not intended to be used as a design tool and really only allow a comparison with in-built benchmarks from a database of notional buildings.

Other reasons for the performance gap include: differences between predicted and actual occupant behaviour such as window opening and thermostat adjustment; construction defects, affecting air tightness, insulation effectiveness and controls set-up, for example; whilst unpredicted weather patterns can have a major impact on operational energy consumption.

The construction process itself can have significant environmental impacts, including noise and pollution from construction processes and traffic for example. A building site can be a hostile place and impacts on its neighbours and the wider environment can be much greater than that of the completed building, albeit over a shorter period of time. Impacts can start some distance from the site: we discussed those from the manufacture of construction products earlier in this chapter, including those from the delivery to site, which can also contribute to congestion of local roads with associated noise and pollution. Impacts can be considerably greater and over longer periods if demolition of existing structures and subsequent site preparation is required.

For detailed guidance on good practice for controlling and minimising these impacts refer to the latest edition of the CIRIA *Environmental good practice on site guide* [45]; whilst specific guidance on noise and vibration control has been published by the London Authorities Noise Action Forum.[61] These guides are equally valid for construction and demolition sites anywhere in the world, although legislation will vary from country to country of course.

Minimising environmental impact, particularly on neighbours is at the heart of the various considerate contracting schemes which evolved out of a scheme developed for the City of London in the late 1980's. The Considerate Constructors Scheme (CCS) and similar local schemes have since become widely adopted across the UK: for example, the City of London has introduced mandatory requirements in its *Code of Practice for Deconstruction and Construction Sites*[62] including for compliance with its own Considerate Contractor Scheme.

The CCS is based around a Code for Considerate Practice under the headings that require the contractor to:

- Care about appearance
- Secure everyone's safety
- Respect the community

[61] https://www.cieh.org/media/1251/london-good-practice-guide-noise-vibration-control-for-demolition-and-construction.pdf

[62] https://www.cityoflondon.gov.uk/assets/Services-Environment/code-of-practice-for-deconstruction-and-construction-sites-9th-edition-pollution.pdf

- Value their workforce
- Protect the environment.

One important component of minimising environmental impacts is the effective management of construction and demolition waste. A study by Transparency Market Research in 2018 predicted that a total of 2.2 billion tons of solid construction and demolition waste will be produced in 2025 compared with 1.2 billion tons in 2012.[63] These figures include waste associated with infrastructure and civil engineering projects such as roads, railways, tunnels and bridges. A significant proportion of this is re-used or recycled, however: in the UK, for example, of 66.2 million tonnes of construction and demolition waste generated in 2016 some 91% was recovered.[64] The EU 2008 Waste Framework Directive requires that 'Member States shall take the necessary measures designed to achieve that by 2020 a minimum of 70% (by weight) of non-hazardous construction and demolition waste excluding (uncontaminated soil and stones) shall be prepared for re-use, recycled or undergo other material recovery (including backfilling operations using waste to substitute other materials)' [46]. According to statistics published by the US Environmental Protection Agency 600 million tons of construction and demolition (C & D) 'debris' (waste) was generated in 2018—twice that from municipal solid waste—of which 90% was from demolition.[65]

In 2008 site waste management in England was regulated for projects having a value of more than £300,000, with specific requirements based around the production and use of a site waste management plan (SWMP), although the Site Waste Management Plans Regulations were repealed in 2013 as part of the Government's plan to reduce the so-called red tape. However, the production of a SWMP is recognised as good practice and is rewarded within the BREEAM assessment process and the Considerate Constructors Scheme. The BRE has developed the SmartWaste system for environmental site monitoring and reporting which includes a module for site waste management as well as managing the use of water, energy, transport and materials for construction, as well as biodiversity impacts.[66]

In the USA, requirements for construction waste vary from State to State: for the construction or demolition of federal buildings there is a minimum requirement to divert (recover) at least 50% of waste by weight, whilst credits are awarded under LEED for a minimum diversion rate of 70%.[67]

The principles adopted for construction waste management follow good practice for general waste reduction strategies: that is, reduce, re-use, recycle with the primary objectives of improving resource efficiency, reducing waste going to landfill and eliminating fly tipping.

---

[63] https://www.transparencymarketresearch.com/construction-waste-market.html

[64] https://www.gov.uk/government/statistics/uk-waste-data

[65] https://www.epa.gov/smm/sustainable-management-construction-and-demolition-materials

[66] https://www.bresmartsite.com/products/smartwaste/

[67] LEED is the Leadership in Energy and Environmental Design assessment protocol—the US version of BREEAM.

7.9 Retrofit Versus New Build

It has been estimated that in the year 2022 buildings worldwide had a total floor area of approximately 252.7 billion m^2, which is predicted to rise to 290.9 billion m^2 by 2030, an increase of around 15%.[68] In a 2010 report by the Low Carbon Construction Innovation and Growth Team commissioned by the UK Government it was predicted that, for the United Kingdom at least, between 72% and 75% of buildings that existed at that time will still be in use in 2050.[69] As energy efficiency standards have improved, coal-fired heating and electricity generation have been phased out in many countries and properties have been built with greater energy efficiency, average carbon dioxide emissions for both dwellings and commercial buildings have fallen. For example, according to the 2013 *UK Housing Energy Fact File,* the average household emissions fell from 9.2 tonnes CO_2 per annum in 1970 to 5.6 tonnes CO_2 in 2010.[70] However, an average home built to 2010 *Part L Building Regulations* standards creates 2.4 tonnes. When combined with the drive towards a zero carbon economy to bring climate change under control these statistics provide an ineluctable case for a major refurbishment programme to reduce the carbon footprint of the existing building stock in the UK.

Unfortunately not only are there millions of buildings that perform inefficiently thermally but many fall below acceptable decency standards: unsafe, uncomfortable and unhealthy for those living or working therein. The UN Statistics Division has reported estimates that in 2015 around 76.5% of the global population lived in cities, or around 5.6 billion people, about double that in 1975.[71] This rapid urbanisation, combined with a steady growth in population, has resulted in inadequate provision of decent and affordable housing. In 2018 23.5% of this urban population lived in slums: equivalent to a staggering one billion people.[72] Note that the UN defines slums as dwellings that meet one out of a list of five criteria, that is,

- Not durable or permanent and providing protection against extreme weather;
- Three people or more sharing the same room;
- Not having easy access to sufficient safe water at an affordable price;
- Inadequate sanitation;
- No protection against eviction.[73]

---

[68] https://www.statista.com/statistics/1414826/global-buildings-floor-area/#:~:text=In%20 2022%2C%20the%20floor%20area,meters%20during%20the%20same%20period

[69] https://www.gov.uk/government/publications/low-carbon-construction-innovation-growth-team-final-report

[70] https://assets.publishing.service.gov.uk/government/uploads/system/uploads/attachment_data/file/201167/uk_housing_fact_file_2012.pdf

[71] https://ec.europa.eu/knowledge4policy/foresight/topic/continuing-urbanisation/developments-and-forecasts-on-continuing-urbanisation_en

[72] https://unstats.un.org/sdgs/report/2019/goal-11/

[73] https://mirror.unhabitat.org/documents/media_centre/sowcr2006/SOWCR%205.pdf

Although the picture is at its bleakest in Eastern, South-Eastern, Southern and Central Asia, as well as sub-Saharan Africa, there remains a proliferation of substandard dwellings across the whole of the so-called developed world. In England, for example, the number of dwellings that fall below established decency standards has been tracked since 2001 using initially the Decent Homes Standard and subsequently the Housing Health & Safety Rating System (HHSRS). HHSRS was enshrined in law under the 2004 *Housing Act* by the *HHSRS Regulations 2005*. The HHSRS comprises a checklist of 19 issues which could impact the health, safety and quality of life of residents, ranging from damp and mould growth, through poor thermal conditions to risk of structural collapse [47].

According to a report on the British housing stock published by the Building Research Establishment in 2019 the percentage of English homes not meeting this decency standard in 2017 was 17%, or nearly four million homes: a definite improvement on the figure for 2007, which was 35%. A more detailed analysis, however, indicates that a quarter of privately rented accommodation in 2017 fell below decency standards compared with 12% of social rented [48]. This is primarily due to a programme of Government funded improvements to social rented homes that has been implemented by Local Authorities and registered social landlords (RSLs) in England since 1997. The BRE estimated that the total cost to the economies of England, Wales and Northern Ireland of poor housing could be as high as £15 billion per annum taking account of direct costs to the NHS, 'ongoing treatment beyond the first year, care costs, sometimes for the rest of the victim's lifetime, loss of economic potential (poorer educational achievement, loss of working days and career prospects) and the mental health cost of suffering and trauma'.

Of course it is possible to demolish existing poorly performing buildings and replace them with modern ones built to the latest energy efficiency standards. However there are many reasons why this may be the wrong solution, such as:

- Construction impacts associated with refurbishment tend to be lower, for example, avoiding demolition and associated site clearance and preparation, along with increased waste management, noise and pollution.
- Uncertainties surrounding site condition or potential costs and delays resulting from site investigations.
- Refurbishment programme times should be shorter and disruption can be reduced through phasing.
- Refurbishment should require less materials with lower associated costs and embodied carbon.
- In some countries, such as the UK, there can be significant tax benefits from refurbishing an existing building rather than replacing it with new.
- In England refurbishments that do not significantly increase accommodation are not subject to contributions to the cost of local infrastructure and affordable homes etc. through the Community Infrastructure Levy (CIL), Section 106 (planning) and Section 278 (highways) agreements.

On the other hand refurbishments can pose particular problems, including:

- Flexibility may be limited due to fixed floor-to-ceiling heights, constrained service routes and plant rooms, whilst opportunities for enhanced insulation, daylighting and natural ventilation may be restricted.
- Older buildings may harbour asbestos, which can add considerably to cost and disruption during the refurbishment process.
- Unforeseen problems, such as mould growth and structural defects, may be discovered during the refurbishment process that may cause delay and extra costs.
- Not all contractors can handle refurbishments and a premium may have to be paid for specialist contractors.

One overriding factor that may govern this decision is the quality and condition of the buildings concerned. Many of the slums that are still to be found globally are not fit for refurbishment and should be demolished and replaced with good quality homes and, in many instances, a completely new infrastructure to service them.

There is no single model that can be applied to all refurbishments, which may vary from no more than an internal refit to replacement of the entire structure and most of the fabric, apart from street-facing facades for example. Indeed for complex sites such as shopping centres and university campuses, redevelopment may include some combination of refurbishment, new build and relocation.

There are a number of reasons why the decision may be made to refurbish a property: it may be to improve the aesthetic value and facilities provided by an existing tired building, it may be to cater for a change of use, it may be to increase the value of a property for speculative development, or it may be to improve the thermal performance and comfort offered—perhaps taking advantage of incentives provided through Government sponsored schemes—or a combination of all these reasons.

A number of countries have introduced some level of support for refurbishment to improve energy performance of buildings or to encourage the installation of renewable energy technologies. These range from adventitious loan schemes, such as the State-funded US PACE financing programme[74] to partially subsidised schemes, such as the German CO_2 Building Modernisation Programme, which has loan and grant assisted variants.[75]

The UK has introduced numerous energy saving schemes over the years aimed primarily at householders. The Supplier Obligation (SO) was introduced by the Government in 1994 which imposed a responsibility on energy suppliers following liberalisation of the energy market in the early 1990s. Since its introduction the SO has taken various forms including since 2012 the Energy Company Obligation, with

---

[74] PACE is the Property Assessed Clean Energy program see: https://www.pacenation.org/what-is-pace/

[75] https://www.bmwi.de/Redaktion/EN/Downloads/factsheet-02-co2-gebaeudesanierung.pdf?__blob=publicationFile&v=1

energy suppliers offering their customers free loft insulation and fluorescent lamps, for example. Between 2012 and 2014 Green Deal finance provided low interest loans attached to the property, repaid through reduction in energy bills. In 2014 the Green Deal Home Improvement Fund (GDHIF) provided grants up to a maximum of £7600 to cover the cost of an energy assessment and installation of energy saving measures. This was replaced in 2020 by the Green Homes Grant covering two thirds of eligible improvements up to a contribution of £5000 or 100% of costs up to £10,000 for applicants on specified low income benefits.

In parallel with these schemes, the UK Government introduced a number of incentives to encourage the uptake of renewable technologies. The feed-in tariff scheme was launched in 2010, following a model that had been used in Germany since 1990, which enabled the sale of electricity by consumers from renewable sources to the suppliers at a specified tariff. This system has been adopted in various forms by numerous countries worldwide, but as the cost of the technologies such as photovoltaics has fallen the tariff has typically been reduced.

A Renewable Heat Incentive (RHI) scheme was introduced in the UK for non-domestic applications in 2011 and for domestic in 2014. Qualifying technologies included ground and air-source heat pumps, solar hot water panels and biomass fired heating, although the tariffs paid per kWh varied. The scheme worked in a similar manner to the feed-in tariff for renewable electricity except the amount paid was determined for predicted performance where metering was not possible. Tariffs were adjusted a number of times during the lifetime of the scheme, which initially favoured biomass to the detriment of heat pumps and solar thermal. In the UK government incentives have moved towards subsidising heat pump installations, a policy which was refreshed by a new government elected in 2024.

7.10 Urban Regeneration Versus Green Field Development

This section addresses the issues that fundamentally underly the title of this book: in an overburdened planet with a growing population, there is an increasing demand for space to accommodate the human population, the buildings to shelter them and the infrastructure to support them. As referred to above, according to an international standard for defining urban areas, adopted by the United Nations in 2020 and referred to as the 'Degree of Urbanisation', 76.5% of the world population lived in urban areas in 2015. Although this has led to a proliferation of slums in the developing world, there has been a trend for the better off to move to the suburbs and rural communities whilst commuting to work in towns and cities. Hence the pressure on urban fringes, rural areas and associated infrastructure continues to increase worldwide.[76]

Although cities generally offer a more sustainable lifestyle than rural communities their inhabitants tend to suffer from a wide range of disadvantages, such as high

[76] https://news.gallup.com/opinion/methodology/287189/approves-new-define-cities-urban-rural-areas.aspx

noise levels and poor air quality, greater incidence of violent crime, higher risk of road traffic accidents—particularly involving cyclists or pedestrians, higher land and property prices and greater stress levels in general. Country dwellers on the other hand tend to have to travel further to access amenities and to commute to work, hence have a commensurately higher carbon footprint than town dwellers. Older people living by themselves in particular may experience loneliness and particular difficulty in accessing amenities where there is limited access to public transport.

Of course development in the countryside will typically result in the destruction of natural habitats and can have an irreversible impact on biodiversity and in some instances threaten the survival of species. Most countries have legislation designed to protect particularly sensitive sites: such as the 1981 *Wildlife and Countryside Act* in England and Wales which sets out protections for Sites of Special Scientific Interest (SSSIs) which, like the National Nature Reserves, were established in 1949. There are some 5000 SSSIs and 305 National Nature Reserves in England and Wales. There are also European Directives that define Special Areas of Conservation, Special Protection Areas (for migratory birds) and Ramsar sites (wetlands).

Many cities throughout the world have expanded through a process of development along the main arterial routes into the city or bordering ring roads. This has led to 'urban sprawl' and in some cases erosion of the green belt. The designation of land to be retained as a 'green belt' around a city was first postulated in England in the nineteenth century and was at the heart of the garden city concept developed by Ebenezer Howard in the 1900s. It was enshrined in English Law in the Town and Country Planning Act of 1947 with the intention of combatting urban sprawl and retaining land for forestry, agriculture and outdoor leisure close to urban centres. The concept has been adapted by some other countries as their cities have grown. The amount of green belt in England in 2018 was 1.63 million hectares, some 10,000 ha less than in 2011. In 2016/17 alone 2890 ha of green belt was built on, around 60% of which was greenfield land.[77] Most of this erosion has been due to house building as part of an ongoing commitment by successive Governments to cater for increasing demand and population growth. Although the National Planning Policy Framework (NPPF) allows building on green belt in England only in 'very special circumstances', paragraphs 145 to 147 of the 2019 edition provide 14 circumstances under which development might be allowed.[78]

Building on previously developed or the so-called brownfield sites is supported by many local authorities but the economic and environmental pressures are not always easy to balance. Brownfield sites may have existing structures to demolish, services to disconnect or reroute, or require complex ground investigations. As discussed in the previous section demolishing existing buildings can pose a range of challenges, whilst the environmental conditions found in many urban locations may fundamentally impact the design and servicing of the development. This may result

[77] https://commonslibrary.parliament.uk/research-briefings/sn00934/

[78] https://assets.publishing.service.gov.uk/government/uploads/system/uploads/attachment_data/file/810197/NPPF_Feb_2019_revised.pdf

in buildings having to be air conditioned, with sealed envelopes to prevent the intrusion of external noise and air pollutants. This presents particular problems for dwellings, which in temperate climates have historically not required air conditioning. One solution to this, which has been used in hot climates for centuries, is the courtyard—with all of the main living accommodation facing onto a courtyard protected from external noise, pollution and direct solar gain, whilst incorporating soft landscaping and water features to provide some passive cooling.

Inevitably growing populations require more homes, workplaces, facilities and infrastructure and a major proportion of this growth will be catered for by either expanding existing towns and cities or building new ones. Hence there is a vital need for sustainable urban planning, utilising the good practice in construction and energy reduction referred to above as well as integrated public transit and the provision of safe routes for pedestrians and cyclists as outlined above and in the next chapter.

References

1. International Energy Agency. (2013). *Transition to sustainable buildings: Strategies and opportunities*. IEA. https://www.iea.org/reports/transition-to-sustainable-buildings
2. Appleby, P. (2011). *Integrated sustainable design of buildings*. Earthscan.
3. Appleby, P. (2013). *Sustainable retrofit and facilities management*. Routledge.
4. CIBSE. (2005a). *Application manual AM10: Natural ventilation in non-domestic buildings*. Chartered Institution of Building Services Engineers.
5. CIBSE. (2005b). *KS01: Reclaimed water*. Chartered Institution of Building Services Engineers.
6. Action Energy. (2003). *Energy consumption guide 19: Energy use in offices*. Action Energy. https://www.osti.gov/etdeweb/biblio/21007991
7. Fanger, P., Lauridsen, J., Bluyssen, P., & Clausen, G. (1988). Air pollution sources in offices and assembly halls, quantified by the OLF unit. *Energy and Buildings, 12*, 7–19.
8. Pejtersen, J., Bluyssen, P., Kondo, H., Clausen, G., & Fanger, P. (1989). Air pollution sources in ventilation systems. In: E. Kulic et al. (Eds.), *Proceedings of CLIMA 2000, Sarajevo. The 2nd World Congress on Heating, Ventilating, Refrigerating and Air Conditioning* (Vol. 3, pp. 139–144). Sarajevo.
9. Clements-Croome, D. (2008). Work performance, productivity and indoor air. *The Scandinavian Journal of Work Environment & Health (Supplement), 4*, 69–78.
10. HM Government. (2021). *Approved document volume 2: Buildings other than dwellings—Means of ventilation*. https://assets.publishing.service.gov.uk/media/62a761edd3bf7f03667c667e/ADF2_revised.pdf
11. CEN. (2007). *CSN EN 15251: Indoor environmental input parameters for design and assessment of energy performance of buildings—Addressing indoor air quality, thermal environment, lighting and acoustics*. https://www.en-standard.eu/csn-en-15251-indoor-environmental-input-parameters-for-design-and-assessment-of-energy-performance-of-buildings-addressing-indoor-air-quality-thermal-environment-lighting-and-acoustics/?gclid=EAIaIQobChMI5bPlyLWM2gIV6ZPtCh0_OgA8EAAYASAAEgJ1afD_BwE
12. Brelih, N., & Seppänen, O. (2011). Ventilation rates and IAQ in European standards and national regulations. In *Proceedings of the 32nd AIVC Conference and 1st TightVent Conference in Brussels* (pp. 12–13). AIVC. http://www.rehva.eu/fileadmin/Old_website_content/technology-and-research-committee/aivc-paper%2D%2D-ventilation-rates-and-iaq-in-eu.pdf

13. CIBSE. (2016). *Guide Section B3: Air conditioning and refrigeration*. Chartered Institution of Building Services Engineers.
14. Javed, S., Ørnes, I. R., Dokka, T. H., Myrup, M., & Holøs, S. B. (2021). Evaluating the use of displacement ventilation for providing space heating in unoccupied periods using laboratory experiments, field tests and numerical simulations. *Energies, 14*(4), 952.
15. ASHRAE. (2015). *ASHRAE handbook—HVAC systems and equipment*. ASHRAE.
16. EC. (1989). *Council Directive of 30 November 1989 concerning the minimum safety and health requirements for the workplace (first individual directive within the meaning of Article 16 (1) of Directive 89/391/EEC) (89/654/EEC).*
17. UNEP. (2017). *U4E policy guide series: Accelerating the global adoption of energy-efficient lighting.* http://wedocs.unep.org/bitstream/handle/20.500.11822/20406/Energy_efficient_lighting.pdf
18. CCC. (2018, November). *Biomass in a low-carbon economy*. Committee on Climate Change. https://www.theccc.org.uk/wp-content/uploads/2018/11/Biomass-in-a-low-carbon-economy-CCC-2018.pdf
19. International Electrotechnical Commission. (2016). *IEC 60904-3, Edition 3: Photovoltaic devices—Part 3: Measurement principles for terrestrial photovoltaic (PV) solar devices with reference spectral irradiance data*. IEC.
20. Schneider, A. (2018). *Introduction to OPVs and comparison with traditional PV technologies*. University of Chicago. https://www.renewableinstitute.org/history-and-function-of-polymer-based-organic-photovoltaic-cells-2/
21. Cuce, E. (2016). Toward multi-functional PV glazing technologies in low/zero carbon buildings: Heat insulation solar glass—Latest developments and future prospects. *Renewable and Sustainable Energy Reviews, 60*, 1286–1301.
22. Najera-Trjo, M., Martin-Dominguez, I., & Escobedo-Bretado, J. (2016). Economic feasibility of flat plate vs evacuated tube solar collectors in a combisystem. *Energy Procedia, 92*, 477–485.
23. Stemmie, R., et al. (2025). Current research on aquifer thermal energy storage (ATES) in Germany. *Springer Nature Link, 30*, 107–124. https://doi.org/10.1007/s00767-025-00590-3
24. BS. (2013). *BS EN 15804:2012+A1:2013 Sustainability of construction works—Environmental product declarations—Core rules for the product category of construction products.*
25. BS EN. (2011). *BS EN 15978:2011: Sustainability of construction works—Assessment of environmental performance of buildings—Calculation method.*
26. International Organization for Standardization. (2006a). *ISO 14044:2006: Environmental management—Life cycle assessment—Requirements and guidelines*. ISO.
27. Andrew, R. (2018). Global CO_2 emissions from cement production. *Earth System Science Data, 10*, 195–217.
28. BEIS. (2017). *Fly ash and blast furnace slag for cement manufacturing* (BEIS Research Paper No. 19). https://assets.publishing.service.gov.UK/government/uploads/system/uploads/attachment_data/file/660888/fly-ash-blast-furnace-slag-cement-manufacturing.pdf
29. International Organization for Standardization. (2014). *ISO 14046:2014: Environmental management—Water footprint—Principles, requirements and guidelines*. ISO.
30. Mekonnen, P., Gerbens-Leenes, P., & Hoekstra, A. (2015). The consumptive water footprint of electricity and heat: A global assessment. *Environmental Science: Water Research Technology, 1*(3), 285–297.
31. Gerbens-Leenes, P., Hoekstra, A., & Bosman, R. (2018). The blue and grey water footprint of construction materials: Steel, cement and glass. *Water Resources and Industry, 19*, 1–12.
32. BS. (2014). *BS 8233:2014: Guidance on sound insulation and noise reduction for buildings.*
33. HM Government. (2003) (Incorporating 2004, 2010, 2013, 2015 Amendments). *Approved document E resistance to the passage of sound.* https://assets.publishing.service.gov.uk/government/uploads/system/uploads/attachment_data/file/899279/Single_stitched_together_pdf_of_all_ADs__Jun20_.pdf

34. CIBSE. (2015). Environmental criteria for design. In *Guide section A: Environmental design*. Chartered Institution of Building Services Engineers.
35. HM Government. (2015) (Incorporating 2016 amendments). *Approved Document G: Sanitation, hot water safety and water efficiency*. https://assets.publishing.service.gov.uk/government/uploads/system/uploads/attachment_data/file/899279/Single_stitched_together_pdf_of_all_ADs__Jun20_.pdf
36. Browne, et al. (2013). *Patterns of water: The water related practices of households in Southern England, and their influence on water consumption and demand management*. GB: University of Lancaster.
37. DEFRA. (2000). *Water supply (water fittings) regulations 1999: WC suite performance specifications*. GB: Department for Environment, Food and Rural Affairs. https://webarchive.nationalarchives.gov.uk/20130403142446/http://archive.defra.gov.uk/environment/quality/water/industry/wsregs99/documents/waterregs99-wcspec.pdf
38. Wang, J., Wang, L., Sun, N., Tierney, R., Li, H., Corsetti, M., Williams, L., Wong, P., & Wong, T.-S. (2019). Viscoelastic solid-repellent coatings for extreme water saving and global sanitation. *Nature Sustainability, 2*, 1097–1105.
39. CIBSE. (2014). *Guide Section G: Public health and plumbing engineering*. Chartered Institution of Building Services Engineers.
40. EA. (2011). *Greywater for domestic users: An information guide*. Environment Agency. https://sswm.info/sites/default/files/reference_attachments/ENVIRONMENT%20AGENCY%202011%20Greywater%20for%20Domestic%20Users.pdf
41. EC. (2011). *Communication from the Commission to the European Parliament, the Council, the Economic and Social Committee and the Committee of the Regions: Our life insurance, our natural capital: An EU biodiversity strategy to 2020*. https://eur-lex.europa.eu/legal-content/EN/TXT/PDF/?uri=CELEX:52011DC0244&from=EN
42. House of Commons. (2020). *Environment bill*. House of Commons. https://publications.parliament.uk/pa/bills/cbill/58-01/0009/20009.pdf
43. EC. (2007). *Directive 2007/60/EC of the European Parliament and of the Council of 23 October 2007 on the assessment and management of flood risks*.
44. Brookfield, K. (2017). Residents' preferences for walkable neighbourhoods. *Journal of Urban Design, 22*(1), 1–15. https://doi.org/10.1080/13574809.2016.1234335
45. Charles, P., & Edwards, P. (2015). *Environmental good practice on site guide*. Construction Industry Research and Information Association. https://www.ciria.org/ItemDetail?iProductCode=C741D&Category=DOWNLOAD&WebsiteKey=3f18c87a-d62b-4eca-8ef4-9b09309c1c91
46. EC. (2008). *Directive 2008/98/EC of the European Parliament and of the Council of 19 November 2008 on waste and repealing certain Directives*.
47. ODPM. (2006). *Housing health & safety rating system: Operating guidance*. ODPM.
48. Piddington, J., Nicol, S., Garrett, H., & Custard, M. (2020). *The housing stock of the United Kingdom*. BRE Trust. https://files.bregroup.com/bretrust/The-Housing-Stock-of-the-United-Kingdom_Report_BRE-Trust.pdf

Chapter 8
Sustainable Transport: From Global to Local

Two roads diverged in a yellow wood,
And sorry I could not travel both
And be one traveller, long I stood
And looked down one as far as I could
To where it bent in the undergrowth;
Then took the other, as just as fair,
And having perhaps the better claim
Because it was grassy and wanted wear,
Though as for that the passing there
Had worn them really about the same,
And both that morning equally lay
In leaves no step had trodden black.
Oh, I kept the first for another day!
Yet knowing how way leads on to way
I doubted if I should ever come back.
I shall be telling this with a sigh
Somewhere ages and ages hence:
Two roads diverged in a wood, and I,
I took the one less travelled by,
And that has made all the difference.

A Road Not Taken by Robert Frost (1916)

8.1 Introduction

According to the data published by the International Energy Agency, the global carbon dioxide emissions associated with transport contributed around 23% towards the total in 2019 or just below 30% of total energy demand for the planet [1]. This sector's contribution to greenhouse gas (GHG) emissions, on the other hand, was just 14% in 2013 according to the data produced by the World Resource Institute.[1] The difference between these percentages can be explained by the fact that some

[1] https://www.c2es.org/content/international-emissions/

P. Appleby, *The Overburdened Planet*,
https://doi.org/10.1007/978-3-032-01510-5_8

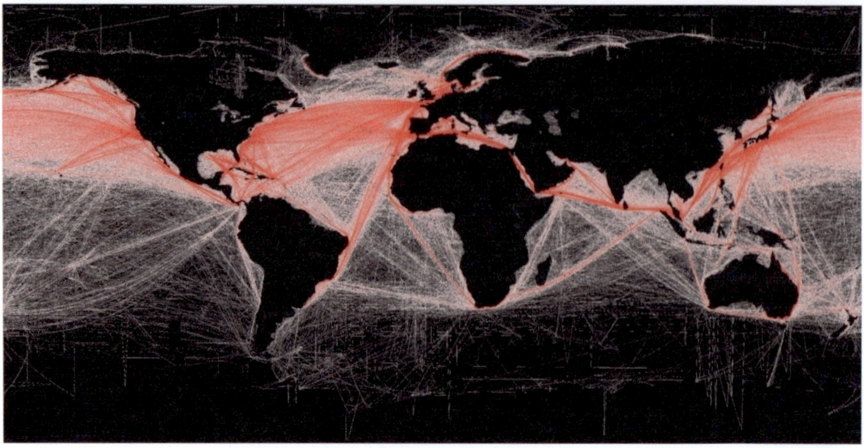

Fig. 8.1 Snapshot of commercial water bound traffic (Image: B.S. Halpern; T. Hengl; D. Groll; From: https://en.m.wikipedia.org/wiki/File:Shipping_routes_red_black.png)

95% of GHG emissions associated with transportation are due to the fossil fuels used in engines, the emissions from which are dominated by CO_2, whilst other more potent greenhouse gases such as methane play a minor role in the total global GHG emissions.

Humankind in the twenty-first century is dependent to a great extent upon motorised transportation: for commuting, for business and leisure travel, for commerce and for motor sports. Car ownership tends to be a sign of prosperity, especially in the developing world, whilst elsewhere for many it is looked upon as essential. At the other end of the scale, global trade depends on millions of movements per annum by road, rail, air and sea. Figure 8.1 shows a snapshot from 2019 of just how congested the waterways can be worldwide with ships carrying freight, cars and passengers to meet consumer demand, keep industry functioning, countries connected and tourism flourishing.

Figure 8.2, on the other hand, shows all global air traffic in 2019: flying more than 31,500 routes between 4041 airports. The thick red lines represents the busiest routes, whilst the other colours represent diminishing frequency with the green lines being the quietest routes. A 2016 report from the World Economic Forum gives the total 'revenue passenger km' (RPK) in 2015 as nine trillion whilst predicting that this could reach 20 trillion by 2040.[2] According to a 2024 report by the International Air Transport Association (IATA) between January and April 2020 RPK fell by about 95% but steadily recovering over the next 4 years to slightly over the January 2020 volume by early 2024.[3] On top of this air freight contributed around 0.26

[2] https://www.weforum.org/agenda/2016/04/the-number-of-cars-worldwide-is-set-to-double-by-2040

[3] https://www.iata.org/en/iata-repository/publications/economic-reports/air-passenger-market-analysis-june-2024/#:~:text=Industry%20total%20Revenue%20Passenger%2DKilometer,for%20air%20travel%20in%20June

Fig. 8.2 A snapshot of global air traffic in 2025 (Image courtesy of http://airnavradar.com/)

trillion 'freight and mail tonne-km' to the air traffic density in 2018, although some of this used the same aircraft as passengers.

These maps demonstrate just how crowded the world's waterways and air space have been, whilst a report from Hedges & Co estimated that there were some 1.475 billion cars on the road globally in 2024,[4] a figure which is forecast to increase to around two billion by 2040.

This chapter looks at what influences this demand, the trajectory of greenhouse gas emissions and other environmental impacts, whilst also investigating what are the infrastructure needs for each mode, discussing the systemic and behavioural changes required to drive down demand and what technological innovations are in the pipeline to reduce impacts.

8.2 Demand: Commuting, Leisure, Business and Goods

With the invention of the wheel, the domestication of animals as beasts of burden, the construction of the first dug-out canoe and the use of logs to transport heavy objects—all between 3000 and 4000 BC—humans have looked for ways to transport themselves and their goods faster and more easily than on foot. The invention of steam engines and subsequently internal combustion, jet and electric motors resulted in the multi-modal means of transport that are available today. The ever increasing number, type and efficiency of trains, boats, cars, commercial vehicles

[4] https://hedgescompany.com/blog/2021/06/how-many-cars-are-there-in-the-world/#:~:text=CLICK%20TO%20ENLARGE%3A%20This%20graphic,are%20in%20the%20United%20States

and planes has influenced almost every aspect of modern life and shaped the world in which we live.

According to *Transport Statistics Great Britain 2017* published by the Department of Transport (DoT), since 1952, the number of passenger-kilometres travelled on land in Britain has quadrupled to over 800 billion in 2016. The DoT's 2023 report showed that the number dropped significantly during the pandemic but recovered to 740 billion by 2022. The majority of the increase since the 50's has been due to car, van and taxi use which rose from 60 bn passenger-km (30% of the total) in 1952 to around 660 billion (83% of the total)[5] in 2016. According to the 2017 report, around 19% of all journeys in 2016 were for commuting, business or shopping, whereas 58% of trips by surface rail were for commuting or business.

According to another report published by the DoT in 2017: 'Between 1995/7 and 2013/14, England's population grew 12% whilst the total number of annual commuting journeys decreased from 8.5 billion to 7.9 billion'.[6] This report, titled *Commuting trends in England 1988–2015,* attributes this to a combination of shorter working weeks for some, an increase in 'trip-chaining' (e.g., dropping off children at school before going on to work), a growth in people not having a fixed place of work or working from home. Interestingly, this decline has not resulted in fewer journeys by car, bus or train but those on foot have dropped from 29% of the total in 1988/92 to 22% in 2013/14. Car journeys dominated, either as driver or passenger, whilst train, tube and bike hovered around the 1–2% mark for each. The picture is very different for inner London, however, where in 2013/14 commuting by car or van represented only 12% of journeys compared with 10% walking, 8% by bike, 26% by tube and 18% by rail. A 5% reduction in car journeys compared with 2002/4 demonstrates the impact of the congestion charge introduced in central London in 2003. During the pandemic commuting into London by public transport reduced to around 10% of pre-pandemic levels, although commuters continued to travel by car, but to a level at around 50% of that before the first lockdown. However, commuting into central London quickly returned to pre-pandemic levels towards the end of 2022.[7]

Unsurprisingly the percentage of commutes by car increases markedly in rural communities, particularly those without easy access to public transport nodes. This is a phenomenon that occurs in all countries where GDP per capita and car ownership is high, although some countries have better developed rail and public transport provision than others, whilst oil-rich countries, such as the USA and Saudi Arabia, have maintained low fuel prices and hence car use has never been deterred.

Business and leisure travel tend be more intermittent than commuting, involving air, road and rail travel, with a greater use of vans and taxis for business use in

[5] https://assets.publishing.service.gov.uk/government/uploads/system/uploads/attachment_data/file/661933/tsgb-2017-report-summaries.pdf

[6] https://assets.publishing.service.gov.uk/government/uploads/system/uploads/attachment_data/file/877039/commuting-in-england-1988-2015.pdf

[7] https://content.tfl.gov.uk/travel-in-london-2023-consolidated-estimates-of-total-travel-and-mode-shares-acc.pdf

particular. In the UK, as with commuting, cars and vans dominate business travel. According to the British National Travel Survey, the average distance travelled per person annually for business purposes in 2018 was 375 miles by van or car compared with 120 miles by train, whilst all other modes were below 10 miles.[8] In the USA, despite the size of the country, road transport still dominates as the favoured mode of transport for business with some 80% of trips compared with 16% by air and 2% by rail. In 2019, Americans took 462 million business trips compared with 1858 million trips for leisure. [9, 10] Globally, however, tourism is dominated by air travel with flights contributing 59% of international arrivals in 2019, with road at 35%, water 5% and rail 1%.

Of course in 2020, as we saw above for London, travel in all modes was heavily impacted by a global pandemic. This resulted in a major increase in people working from home and communicating via the internet and mobile phone networks. Roads were quiet, trains and plane schedules disrupted, whilst the demand for cycles rocketed. Even before COVID-19 had created a major hiatus in the travelling habits of much of the world's population, there had been a growing uptake in video-conferencing with, for example, in the U,K a doubling of people working from home between 2009 and 2019. A 2023 survey of 42,000 people in 34 countries in the developed world by the Institute of Economic Research found that 33.5% of respondents worked at least some of the time away from their workspace for an average of 0.9 days per week with 7.9% working fully from home.[11] The use of video-conferencing platforms has been a major part of this trend and is likely to continue in a post-COVID-19 world.

It can be seen from Figs. 8.1 and 8.2 how crowded the waterways and airspace have been in the early twenty-first century. Despite the pandemic and worldwide financial crises, freight traffic is likely to continue increasing as population, standards of living and associated consumer demand and GDPs grow. The statistics for inland transport paint a more nuanced picture. According to the data published by the Organisation for Economic Cooperation and Development (OECD) between 1998 and 2018 inland freight transport in China increased by a factor of 13 and in India by a factor of 5.7, whilst in the USA, there was only a marginal increase and in the UK and Germany, for example, the volume of internal freight traffic remained pretty stable for the whole period.[12] At 7.125 trillion tonne-km in 2018 Chinese inland freight traffic was more than twice that in the USA, although the per capita quantity was about 56% of that in the USA.

---

[8] https://www.statista.com/statistics/467302/average-distance-travelled-for-business-purposes-by-mode-england/

[9] https://www.statista.com/statistics/305515/international-inbound-tourism-by-mode-of-transport/

[10] https://www.statista.com/statistics/207103/forecasted-number-of-domestic-trips-in-the-us/

[11] https://www.personneltoday.com/hr/home-working-statistics-wfh-global/

[12] https://data.oecd.org/transport/freight-transport.htm

8.3 Greenhouse Gases

As we saw in the introduction to this chapter all of this traffic has a carbon footprint and associated whole life greenhouse gas (GHG) emissions to the atmosphere, thus making a major contribution to climate change. GHG emissions from transport are primarily comprised of CO_2, although other gaseous constituents such as NO and carbon monoxide (CO) can also have an impact. CO destroys hydroxyl (HO) which has a controlling influence over concentrations of methane and NO_2 in the atmosphere, thus also increasing ozone concentrations. Emissions also include black carbon, in particular from diesel engines. Transportation is responsible for around a quarter of global black carbon emissions which, as we saw in Chap. 3 is a significant direct and indirect contributor to global warming.

According to an Indicator Assessment published on-line by the European Environment Agency in December 2020, GHG emissions associated with transport in 27 EU nations (EU_{27}) rose steadily between 1990 and 2007 by around 38%, then fell dramatically until 2013, after which they rose steeply until 2019 when they were 32% above 1990 levels. Predictions indicate a subsequent levelling off so that emissions in 2032 are likely to be similar to 2019. Despite commitments across the EU and elsewhere to phase out road vehicles powered by petrol and diesel engines: any reduction in emissions from road vehicles is likely to be offset by a rise in those from aircraft and shipping.[13]

Globally in 2018, the annual operational CO_2 emissions associated with transport totalled around 8 $GtCO_2$, some three quarters of which were from road vehicles, with 11.6% from aviation, 10.6% from shipping and 1% from rail (excluding electricity generated at power stations).[14] Total operational GHG emissions for transport were approximately 0.5 $GtCO_{2eq}$ greater than CO_2 emissions. Whole life embodied or indirect emissions associated with manufacture, maintenance, etcetera add another 0.1 $GtCO_{2eq}$.[15]

As car ownership in China increased by a factor of 3.4 between 2009 and 2019, it is hardly surprising that carbon dioxide emissions associated with transport also rose dramatically. For example, between 2009 and 2016 transport CO_2 emissions increased by 63%,[16] around 70% of which was due to road vehicles. By comparison, although emissions associated with transportation in 2009 were more than three times higher in the USA than in China the increase thereafter was marginal. The mileage travelled by cars and light trucks in the USA between 1990 and 2018 increased by 46.1%, whilst GHG emissions increased by around 27% in that time, which was was a lower increase than for the EU over the same period.[17]

---

[13] https://www.eea.europa.eu/en/analysis/indicators/greenhouse-gas-emissions-from-transport

[14] https://ourworldindata.org/co2-emissions-from-transport

[15] https://www.ipcc.ch/site/assets/uploads/2018/02/ipcc_wg3_ar5_chapter8.pdf

[16] https://ourworldindata.org/grapher/co-emissions-by-sector?tab=chart&stackMode=absolute&time=earliest..latest&country=~CHN&region=World

[17] https://www.epa.gov/ghgemissions/sources-greenhouse-gas-emissions

8.4 Other Environmental Impacts

Transportation has both direct and indirect impacts on the environment. Direct impacts mostly relate to emissions from engines, whilst indirect impacts are associated with the infrastructure required to convey or support each mode, or the environmental damage enabled by opening new transport routes.

8.4.1 Air Quality

Internal combustion and jet engines both emit a cocktail of pollutants, some of which are potentially injurious to health. Road vehicles are of particular concern as they travel close to areas of human occupation, whilst aircraft and shipping emissions tend to be either concentrated in airports and ports or in remote locations. Diesel locomotives similarly have their greatest impact in stations and goods yards, although, apart from the USA, there has been significant electrification of railways in developed countries across the globe. In 2018, only 1% of trains in the USA were powered solely by electricity, whilst in UK, the proportion was around 71%.

Some 99% of vehicle exhaust emissions comprise non-toxic nitrogen, oxygen and water vapour. Apart from CO_2 and carbon monoxide, the main components of concern that constitute the remaining 1% are oxides of nitrogen (NO_x) and fine particulate matter: in particular those particles smaller than 2.5 μm, which travel deep into the human lung ($PM_{2.5}$). There are also trace quantities of numerous other constituents including polycyclic aromatic hydrocarbons (PAHs) which are carcinogenic at very low concentrations and have been found to attach to fine particles, including the black carbon, that are principally associated with diesel engine emissions. However, an ever increasing number of petrol fuelled vehicles are using gasoline direct injection (GDI) which, although producing less CO_2, generates higher concentrations of $PM_{2.5}$ than the average diesel engine, including black carbon. NO_x emissions occur from both diesel and petrol engines. NO_x comprises both nitric oxide (NO) and nitrogen dioxide (NO_2).

In 2012, the exhaust from diesel engines was classified by the International Agency for Research on Cancer (IARC) as carcinogenic (Group 1) to humans 'based on sufficient evidence that exposure is associated with an increased risk for lung cancer'.[18] The primary constituents implicated were the particulates and $PM_{2.5}$ in particular, including black carbon with PAHs attached. $PM_{2.5}$ is able to penetrate the thoracic region of the lung and has been shown to aggravate asthma, cause respiratory symptoms and in some cases result in lung cancer, cardiovascular or respiratory diseases resulting in premature death.[19]

---

[18] https://www.iarc.who.int/wp-content/uploads/2018/07/pr213_E.pdf

[19] https://unece.org/DAM/env/documents/2013/air/Health-effects-of-particulate-matter-final-Eng.pdf

Similarly, according to a 2015 Statement from the UK Government's independent advisory Committee on the Medical Effects of Air Pollutants (COMEAP) 'NO_2 has been associated with adverse effects on hospital admissions for various diagnoses, decrements in measures of lung function and lung function growth, increases in respiratory symptoms, asthma prevalence and incidence, cancer incidence, adverse birth outcomes and mortality'.[20] However, there is some uncertainty as to what degree NO_2 works synergistically with other pollutants, such as $PM_{2.5}$ and ground level ozone, with which it is correlated in ambient air.

Ozone (O_3) is created by the chemical reaction that occurs when NO is mixed with volatile organic compounds (VOCs) in the presence of sunlight. It can seriously impact the respiratory system: potentially causing shortness of breath, coughing, sore throat, inflamed and damaged airways, whilst also aggravating respiratory conditions such as asthma, emphysema, bronchitis and infections, in some cases leading to premature death.[21]

Although these constituents of engine exhausts tend to form a cocktail in heavily trafficked areas, they also mix with emissions from other sources such as those from industry, power stations and heating plants. In its biannual Air Quality in Europe reports, the European Environment Agency has been able to publish separate estimates of deaths and years of life lost (YLL) for $PM_{2.5}$, NO_2 and O_3, as reported by 41 European countries. For example, in its 2020 report, the premature deaths during 2018 for exposure to $PM_{2.5}$ were estimated as 417,000; 55,000 for NO_2 and 20,600 for O_3, whilst estimates for YLL were given as 4.8 million, 426,000 and 247,000 years, respectively [2]. This report does not identify the sources for the pollutants mentioned since there is no mechanism for linking source with health impact in national statistics. The WHO has stated that 4.2 million premature deaths occur globally every year due to exposure to ambient air pollution, highlighting fine particulates as the primary cause.[22]

Despite these extraordinary statistics, until December 2020, there had not been a single incidence of air pollution having been listed as a cause of death on a death certificate anywhere in in the world. However, on 16th December 2020, the coroner for a case in South London ruled that 9-year-old Ella Kissi-Debrah's death in February 2013 was caused by acute respiratory failure, severe asthma and air pollution exposure and that 'she was exposed to nitrogen dioxide and particulate matter (PM) pollution in excess of World Health Organisation guidelines, the principal source of which were traffic emissions'.[23]

It is important to note that since 2009 under the Euro 5 exhaust emissions legislation, it has been a legal requirement for diesel vehicles sold in the UK and across the whole of the EU to have diesel particulate filters (DPFs) fitted to their exhaust

---

[20] https://assets.publishing.service.gov.uk/government/uploads/system/uploads/attachment_data/file/411756/COMEAP_The_evidence_for_the_effects_of_nitrogen_dioxide.pdf

[21] https://www.epa.gov/ground-level-ozone-pollution/health-effects-ozone-pollution

[22] https://www.who.int/data/gho/data/themes/public-health-and-environment

[23] https://www.theguardian.com/environment/2020/dec/16/girls-death-contributed-to-by-air-pollution-coroner-rules-in-landmark-case

Fig. 8.3 Traffic pollution in China (Photo by Andrew Turner; Source: https://commons.wikimedia.org/wiki/File:Heavy_Traffic,_Heavy_Haze_-_another_day_in_China.jpg)

system, whilst many GDI petrol engine vehicles have had exhaust filters fitted as standard since 2018. Since 2014 it has also been a requirement of the MoT test for diesel vehicles, meaning that all UK registered vehicles, no matter what their age, have had to have diesel particulate filters fitted. The Euro 6 emission standards which came into force in 2015 introduced lower emission requirements for factory tests, followed in 2019 by a requirement for real-world emission tests for diesel cars.

The US Environmental Protection Agency (EPA) introduced legislation that mandated DPFs for heavy duty trucks from 2007 whilst also requiring the processing of diesel to have a much lower sulphur content to facilitate the introduction of DPFs which had been found to be intolerant to high concentration of sulphur in exhaust gases. The EPA system was extended to cover emissions from engines powering everything from motor bikes to aircraft, including locomotives, marine craft, construction machinery and generators (see below).

However, globally there remain millions of vehicles which do not have exhaust filtration. In China, for example, a 2018 report from the Chinese Ministry of Ecology and Environment summarised by Reuters indicated that although only 7.8% of vehicles in China were diesel trucks they have been responsible for 57.3% of NO_x emissions and more than 75% of particulate matter (see Fig. 8.3). This is despite 20 million polluting vehicles having been taken off the road since 2013. However, all new heavy duty diesel vehicles manufactured from July 2021 are required to be fitted with DPFs.[24] The picture has been similar in India which in 2020 introduced legal requirements for emission controls similar to the European system.

Around the same time as the USA was implementing the *EPA Program to Control Pollution from Locomotive and Marine Diesels'* The EU introduced its *2008 Clean Air Non-road Diesel Rule*, both of which required desulphurisation of

[24] https://ccacoalition.org/en/content/china-diesel-emissions-standards

diesel—by 99% in the European case. In parallel with the Euro 6 emission standards, the EU set a requirement for catalytic after-treatment technologies to be installed in diesel locomotives from 2015 onwards. However, this did not by itself tackle NO_2 or $PM_{2.5}$ emissions and so the ENSPIRIT project was launched in 2013 to tackle this problem, focussing on both new and existing locomotives.[25] In 2016, the EU published its *Non-road Mechanical Machinery Regulation* which made exhaust after-treatment systems mandatory, including DPFs for locomotives and other non-road engines [3].

The relationship between exhaust emissions from aircraft and potential exposure to toxins is very different. Airports tend to be more distant from centres of population and once airborne, exhaust emissions have less impact on ground level pollutant concentrations than land-based pollution sources. Those using and working at airports will have the greatest exposure to aircraft emissions, particularly ground crew. There have been a number of studies that have predicted global mortality rates from aircraft, allowing for emissions during take-off, landing and when airborne. These simulation-based studies give widely different results: varying from a few hundred excess mortalities each year to circa 16,000 and with large error ranges [4, 5]. The study by Yim et al. simulates emissions associated with landing, take-off and the cruise phase indicating that the pollution emitted once the plane is airborne could be responsible for three quarters of the premature deaths attributable to aircraft emissions. These studies have mainly looked at exposure to $PM_{2.5}$ and ozone, although occupational exposure of ground staff to carcinogenic polycyclic aromatic hydrocarbons (PAHs) such as benzene from unburnt jet fuel is an additional concern.

8.4.2 Noise

Noise may be defined as unwanted sound and is usually associated with aural discomfort and a range of potential health impacts. Extreme noise can damage hearing of course, but long-term exposure to noise levels that create annoyance or interfere with sleep can have significant health effects.[26] Although individual sensitivity to noise varies, there are statistically derived thresholds for discomfort, annoyance, interference to speech intelligibility and hearing damage that have been published by the likes of the WHO, which issued its *Environmental Noise Guidelines for the European Region* in 2018 [6]. These guidelines refer to a report published by its Joint Research Centre that gave estimated disability adjusted life years (DALYs) lost due to a combination of sleep disturbance, annoyance, tinnitus, cognitive impairment in children and ischaemic heart disease; where DALYs were estimated by adding years of life lost due to early mortality to years lost to disability. It con-

[25] https://cordis.europa.eu/project/id/605019/reporting

[26] https://www.eea.europa.eu/publications/good-practice-guide-on-noise

cluded that 'at least one million healthy years of life are lost every year from traffic-related environmental noise in western Europe. Sleep disturbance and annoyance, mostly related to road traffic noise, constitute the bulk of this burden'.

Hearing loss is generally associated with noise at work, which for much of the world comes under the aegis of health and safety legislation.

Sound occurs across a wide range of frequencies: from low-frequency bass notes to high frequencies which may be inaudible to humans. A healthy auditory system in a young human can detect sounds with frequencies ranging from 20 to 20,000 Hz, although the response is not linear and tends to tail off at the top end, particularly with age. Perceived sound and the mechanisms by which the human brain processes it is complex. The aural environment typically includes numerous sources, some constant, some moving, some with a fixed sound power and spectrum, some variable; and the paths that the pressure waves emanating from these sources follow to reach a human ear will be a combination of direct and reflected from surrounding surfaces. Someone indoors may be exposed to multiple sources located both indoors and outdoors and the perceived noise level will depend on distances between the sources, surfaces and the ear, the volume and absorptive properties of the room, the sound power of the sources, the size and location of openings in the external fabric of the building and the sound attenuating properties of the fabric. There may also be pathways for sound transmission within a building via temporary or permanent openings and internal walls and partitions.

When these multiple sound waves reach the human ear, the resultant sound level will be perceived as the logarithmic addition of the sound pressure levels for each, expressed in decibels (dB). For example, adding 2 dB levels of the same value increases the sound pressure level at the ear by 3 dB, whereas if the two levels are 20 dB apart, then the lower one will not have any impact on the pressure at the ear at all. Noise levels are frequently expressed as a rating that represents a weighted spectrum that corresponds to the sensitivity of the human ear. The most popular of these is the A weighted decibel scale—dB(A), which recognises that the ear is more sensitive to sound between 1000 and 6000 Hz than frequencies below or above these values respectively.

Traffic noise includes noise from road vehicles, trains and aircraft. Globally millions of people are exposed to noise levels that exceed the WHO Guidelines referred to above. These guidelines use the weighted decibel reading at the face of a building averaged over the whole 24 h of day, evening and night (dB L_{den}) and during 8 h of night (dB L_{night}) as defined in ISO 1996-1:2016. Levels are given that, which exceeded for a 24-h period in an outdoor location at an open window, are predicted to create adverse health effects or, if exceeded during the night, would have adverse effects on sleep. For road traffic, these are 53 dB L_{den} and 45 dB $L_{night,}$ respectively; for railway noise 54 dB L_{den} and 44 dB $L_{night,}$ respectively, and for aircraft noise 45 dB L_{den} and 40 dB $L_{night,}$ respectively [6]. Inherent in the definition of L_{den} is that the day is assumed to run from 07.00 to 19.00 h, that the evening level will be 5 dB less and run between 19.00 and 23.00 h and night would be 10 dB lower and run from 23.00 to 07.00 h, although this definition has been loosely applied internationally. Some authorities such as the UK Government in its Aviation Policy

Framework[27] use the alternative measure of L_{AeqT} which is the average A-weighted noise level over a period, T.

The USA has Federal standards for noise from transportation sources that apply to government subsidised residential projects. These have been set by the US Department of Housing and Urban Development (HUD) and are referred to as the 'HUD' Standards. These refer to an 'acceptable' maximum level over 24 h of 65 dB.[28]

Noise generated by motor vehicles is characterised by a combination of engine and tyre noise. The former depends on the age and type of vehicle as well as the presence of junctions and obstructions that require vehicles to accelerate; whilst the latter depends on road surface, tyre design and vehicle speed. The EU introduced a labelling scheme for new tyres in 2012 which requires all tyres to be tested for external sound generation using a standard testing procedure akin to real-world conditions, with decibel levels measured at 7.5 m from source. Intermittent and unexpected noise, such as horns and reversing alarms can be a confounding factor, particularly in sleep disturbance. Ground-borne vibration can also be an issue, particularly where buildings are located near to heavily trafficked roads with a high proportion of heavy vehicles.

Railway noise is in some ways more complex than road traffic noise in that it comprises a mix of engine noise, rolling noise, aerodynamic noise, the occasional squealing of wheels on curves and the change of noise characteristics as trains cross viaducts and bridges. Similar to HGVs, ground-borne vibration can impact nearby buildings whilst train horns tend to be very loud and persistent where there are level crossings or line-side working. Although rolling noise tends to dominate from trains that are not classified as 'high speed', diesel locomotives are much noisier than electric, particularly during acceleration and manoeuvring phases. Rolling noise depends on the surface roughness of rails and wheels, the quality and type of joints between lengths of rail and how thermal movement is catered for. For high-speed trains, aerodynamic noise tends to dominate, with obstructions to air flow such as the pantographs, bogies and coupling areas having a major effect, although the most recent high speed train designs have reduced the noise generated from each of these components considerably.

Most people do not get as close to an aircraft with its engines operating as they do to a road vehicle or train. Airfield operations personnel have to wear hearing defenders because of the high-noise levels they are exposed to on airfields. However, it is those living under the flight paths close to airports that are exposed to fortuitous noise from aircraft landing and taking off. Globally there are thousands of people who are exposed to the noise of aircraft passing directly overhead—in some instances, such as Myrtle Avenue in Hounslow next to London's Heathrow Airport, only metres from the roofs of their homes (see Fig. 8.4).

[27] https://assets.publishing.service.gov.uk/government/uploads/system/uploads/attachment_data/file/153776/aviation-policy-framework.pdf

[28] https://www.hudexchange.info/programs/environmental-review/noise-abatement-and-control/

Fig. 8.4 A Qantas Boeing 747–400 approaching runway 27 L at Heathrow Airport in July 2004 (Source: Arpingstone; Wikimedia Commons: https://commons.wikimedia.org/wiki/File:Qantas_b747_over_houses_arp.jpg)

Although modern jet aircraft are considerably quieter than those in the 1960s, the noise levels experienced at the outside of these houses regularly exceeds 70 dB and, with flights passing over head every few minutes, the L_{den} clearly must exceed WHO Guideline values as well as the UK limit of 57 dBA $\mathrm{L}_{\mathrm{Aeq16}}$ set out in para 3.17 of the 2013 Aviation Policy Framework.

There is no protection in the UK law for those exposed to environmental noise and no legal limits to noise from existing roads and railways for example. However, planning law in England, including the 2019 edition of the National Planning Policy Framework, treats noise as a material consideration: 'Noise needs to be considered when development may create additional noise, or would be sensitive to the prevailing acoustic environment'.

This note from the UK Ministry of Housing, Communities and Local Government includes a hierarchy of noise impacts that defines a significant adverse effect level as when:

> The noise causes a material change in behaviour, attitude or other physiological response, e.g. avoiding certain activities during periods of intrusion; where there is no alternative ventilation, having to keep windows closed most of the time because of the noise. Potential for sleep disturbance resulting in difficulty in getting to sleep, premature awakening and difficulty in getting back to sleep. Quality of life diminished due to change in acoustic character of the area.[29]

[29] https://www.gov.uk/guidance/noise%2D%2D2

If exposure is predicted to be above this level, then 'the planning process should be used to avoid this effect occurring'. At one extreme, this could require a change in site location at the plan-making stage, or some form of mitigation, such as alterations to the design or layout (see the section on *Noise and air quality* in Chap. 7).

Noise from shipping and marine activities such as oil and gas exploration can have major impacts on marine life. Noise from shipping is particularly pervasive with continuous ship movements across the planet as can be seen in Fig. 8.1. Sound transmits effectively through water where there are few obstructions and particularly at the low frequencies that characterise the sound spectrum from marine engines and propellors, which are especially prone to cavitation noise—due to collapse of bubbles formed on propellor blades. Sound level distribution has been modelled by researchers at the National Oceanic and Atmospheric Administration (NOAA—part of the US Department of Commerce) for water bodies that border the North American coastline.[30] Annual average sound levels (L_{eq}) across much of the area are predicted to be in the upper region of the 40–115-dB scale range, even at depths of 1000 m. Long-term monitoring in the North Pacific shows that low-frequency sound levels measured using deep water hydrophones increased from an average of 72 dB in 1950 to 90 dB in 2000.[31] Many marine animals rely on sound to communicate, hear predators and hunt. Background noise at these levels can seriously interfere with these vital aspects of survival for these animals and in some instances creates panic resulting in beaching or rapid diving which can cause multiple fatalities.

Some activities such as low-frequency sonar, used by the military to detect submarines, and seismic air guns, used in oil and gas exploration can have even more damaging impacts. When surveying geological structures under the sea bed typically up to 40 guns are mounted on the sea bed firing simultaneously and continuously every 10 seconds over perhaps many months. In a brief written in 2016 for the UN's Global Sustainable Development Report the authors concluded that:

> These impacts (from seismic air guns) can include permanent damage to an animal's hearing, changes in vocalizations which could affect feeding, mating, or navigation, displacement from habitat, changes in abundance, or lower fisheries catch rates, physiological (stress) effects, and "masking", or obscuring of signals important to an animal, even at long ranges.[32]

[30] https://www.fisheries.noaa.gov/science-data/passive-acoustic-research-southeast-fisheries-science-center#:~:text=The%20Underwater%20Soundscape%20of%20the%20Gulf%20of%20Mexico,-The%20Gulf%20of&text=Low%2Dfrequency%20noise%20from%20seismic,other%20sounds%20from%20the%20environment

[31] https://dosits.org/science/sounds-in-the-sea/how-does-shipping-affect-ocean-sound-levels/

[32] https://sustainabledevelopment.un.org/content/documents/973534_Weilgart_Alternative%20Quieting%20Technology%20to%20Seismic%20Airguns%20for%20Oil%20&%20Gas%20Exploration%20and%20Geophysical%20Research.pdf

The NOAA has plotted sound levels in the Gulf of Mexico,[33] where there are continuous oil exploration activities, showing that the average low-frequency sound levels across much of the Gulf are predicted to be at or approaching 120 dB, which is above the average human pain threshold.

8.4.3 Water Pollution and Ecological Impacts

According to a seminal report from the US Department of Agriculture issued in 2011:

> The development and presence of transportation corridors results in three primary consequences: (1) reduced landscape permeability, (2) habitat loss, and (3) increased habitat fragmentation. These fundamental changes to landscape structure, which can occur during the construction and postconstruction phases of corridor development, can all have profound cascading ecological implications (i.e., direct effects such as increased juvenile mortality, secondary or indirect effects such as reduced reproductive rates, etc.) (Bennett et al in 2011).

One of the most obvious indicators of the damage that transportation has on wildlife is the number of dead animals that are typically found lying in road or railway corridors as well as bird strikes by aircraft. However, there are more subtle impacts at play, such as the barriers presented to migratory and territorial routes and seasonal mating routines, and the consequent mortality and habitat fragmentation. Indirect impacts include those associated with run-off from roads, which has been found to contain up to 300 different pollutants including engine oil, trace metals, hydrocarbons and other organic chemicals, many of which are damaging to the health of rivers and other water bodies, and the flora and fauna that they support. According to a 2018 report from the Environment Agency for England on the *State of the environment: water quality*, transport is responsible for 13% 'of reasons for water bodies not achieving good status',[34] although agriculture, land management and the water industry are the dominant culprits.

As we saw in Chap. 2, run-off from fields can create algal blooms and eutrophication in water bodies with increased turbidity and reduced oxygen content (hypoxia). Run-off from roads can contribute to this also, resulting in regions where little or no life can be supported. In parts of the world that experience snow and ice, salt in run-off water can create additional problems, destroying saline-intolerant species and altering entire ecosystems.

---

[33] https://www.fisheries.noaa.gov/science-data/passive-acoustic-research-southeast-fisheries-science-center#:~:text=The%20Underwater%20Soundscape%20of%20the%20Gulf%20of%20Mexico,-The%20Gulf%20of&text=Low%2Dfrequency%20noise%20from%20seismic,other%20sounds%20from%20the%20environment

[34] https://assets.publishing.service.gov.uk/government/uploads/system/uploads/attachment_data/file/709493/State_of_the_environment_water_quality_report.pdf

Noise and artificial lighting can also have significant impacts on wildlife: for example, bats can have their echolocation communications disrupted whilst also being attracted to moths that are drawn to the overhead lighting. Other animals that are not normally nocturnal become confused by the 24 h availability of light, which occurs in many towns and cities as well as illuminated road corridors.

Where transport corridors and airport curtilages are not accessible by the public, niches may be created where rare and fragile species thrive, away from human impact. Railway verges are particularly prolific as they tend to be subject to less pollution, no salt from de-icing, no artificial lighting and lighter traffic. Because railway and road corridors also provide a route for the distribution of alien species carried on road vehicles and trains, particularly along routes frequented by trans-continental traffic, many of the species that thrive in these corridors are invasive.

Airports also can offer a haven for specific types of flora and fauna. The competition with planes is not good for birds, whilst aircraft noise can stress many animals within and outside of the airport boundary. It is true that most airfields comprise a combination of tarmac and mown grassland, with very little biodiversity. However, with careful management quieter areas can become niches for rare and protected species. For example, at Stansted airport near London, a 2.5 ha site has been set aside, with ponds, rich grassland and even woodland which provide a home for great crested newts and a variety of rare native plants, albeit isolated from sites outside the airport boundary.[35]

Over the past century or more, in both the developed and the developing world, there have been millions of kilometres of roads, railways, pipelines and power lines constructed across open countryside. In the process ancient woodland and ecologically important sites have been destroyed, resulting in widespread and irreversible habitat loss and even the extinction of species of flora and fauna. High speed travel allows for only gradual changes in direction and so connecting major towns and cities frequently results in wide and hostile corridors being driven through anything that happens to lie in their path. New roads and train lines also open up areas for development which previously might have been difficult to access, facilitating green field development and urban sprawl (see section on *Urban regeneration* versus *green field development* in Chap. 7).

Marine transportation on the other hand can have a significant impact on marine ecosystems. According to a 2020 review by Australian authors:

> Such impacts include physical changes to bottom substrate and habitats from sources such as anchoring and mooring and vessel groundings, alterations to the physico-chemical properties of the water column and aquatic biota through the application of antifouling paints, operational and accidental discharges (ballast and bilge water, hydrocarbons, garbage and sewage), fauna collisions, and various other disturbances [7].

Because of the scale of trauma caused, collisions between shipping and whales tend to get the most attention and hence the statistics for these tend to be the most

[35] https://mediacentre.stanstedairport.com/rare-visitor-sets-up-home-at-london-stansted-and-helps-secure-top-international-award/

reliable. Indeed the International Whaling Commission (IWC) maintains a database of the so-called ship strikes which contains reports dating from 1820. Analysis of this data in a conference paper by Winkler et al. [8] indicates that 933 ship strikes were reported between 1820 and 2019, although some 35% of these were either probable or possible. More than 60% of these were recorded as having occurred in the Atlantic. Unsurprisingly with an increase in marine traffic, there was an exponential increase in strikes during the twentieth century, the 10 year total peaking in the decade commencing in the year 2000. Encouragingly, despite increased and more sophisticated reporting systems, the number of ship strikes recorded in the decade from 2010 was approximately 45% lower than the previous decade.

Winkler et al. recognise that there has been a significant underestimation of whale fatalities indicated by these data. The data does not include the other species of marine life that are killed and maimed by collisions with ships and boats, which tend not to be reported.

Global shipping is also responsible for the spread of non-native invasive species and hence multiple threats to marine biodiversity worldwide. A wide variety of species hitch a ride on ships by either attaching themselves to hulls (biofouling) or being drawn into ships with ballast water and subsequently discharged when ballast is no longer required. In its *2023 Guidelines for the control and management of ships' biofouling to minimize the transfer of invasive aquatic species (Biofouling Guidelines),* the International Maritime Organisation considered that 'The spread of invasive species is now recognised as one of the greatest threats to the ecological and the economic well-being of the planet and one of the main drivers of biodiversity loss. These species are causing enormous damage to biodiversity and the valuable natural riches of the earth upon which we depend. Direct and indirect health effects are becoming increasingly serious and the damage to the environment is often irreversible. Moreover, significant economic impact occurs to industries that depend on the coastal and marine environment, such as tourism, aquaculture and fisheries, as well as costly damage to infrastructure'.[36]

Although historically biofouling, primarily from ship's hulls, has been the dominant source of invasive species globally, billions of tonnes of ballast water are transported around the world every year carrying thousands of different species from their native waters to locations in many instances where they are not normally found.

Biofouling on the other hand involves the development of an entire ecosystem on a submerged surface. For example, when a ship's hull is immersed in sea water, a biofilm quickly forms, comprising typically bacteria and algae, which changes the physical characteristics of the surface, making it amenable to colonisation by macro-fouling species such as seaweeds, sponges, sea anemones, corals and other biota, potentially building up to a layer 300 mm thick or greater and weighing perhaps 5 kg/m^2 (see Fig. 8.5). Within this ecosystem, various mobile species can thrive: including crabs, shrimps and fish, as well as whelks, barnacles and molluscs.

---

[36] https://www.imo.org/en/OurWork/Environment/Pages/Biofouling.aspx

Fig. 8.5 An extreme example of hull biofouling (Photo: Lamio; Source: https://commons.wikimedia.org/wiki/File:Antifouling_8212.jpg)

According to a briefing published by the OneOcean Flotilla group of NGOs, 'Invasive species can alter marine ecosystems by:

- disrupting native habitats
- causing the extinction of flora and fauna (by consumption and out-competing for space and resources)
- overwhelming important vulnerable ecosystems such as coral reefs and mangroves
- decreasing water quality
- increasing competition and predation among species
- spreading disease'.[37]

There are examples from around the world where invaders have driven out native species, damaged local ecosystems and had serious economic impacts. For example, the Sea Walnut, a jellyfish-like creature without a sting, which is native to the east coast of North and South America, was discovered in the Black Sea in 1982 and subsequently the Caspian Sea, having been carried in ballast water. In both seas, it formed immense populations and contributed to the collapse of local fisheries through predation of fish eggs, larvae and zooplankton (small sea creatures).

[37] https://www.oceanprotect.org/about-2/

8.5 Alternatives to Fossil Fuel Engines

It is perhaps obvious that if everyone either worked from home or within easy walking or cycling distance from their workplace or school, then transport associated environmental impacts would drop markedly. The reduction would be even greater if agriculture and manufacture of consumer products became more localised. Although some countries in the developing world have communities that function close to this level, the aspiration of most is to get closer to the Western model of car ownership and transport infrastructure.

At the same time Western societies are beginning to transition away from transportation based on fossil fuels and the internal combustion engine. Some countries are phasing out petrol and diesel cars and vans. For example, in 2020, the UK Government proposed that from 2030, all new cars and vans must have zero emission engines: battery or hydrogen powered, for example, or engines that can run for significant periods with zero emissions: such as hybrids, with or without a plug-in capability. Then from 2035, no more hybrids can be sold and all new cars and vans will have to have zero emission engines.[38] This was subsequently dropped then reinstated by a new Labour Government in 2024. It is also putting money into developing an infrastructure of charging points for electric vehicles. With most major car manufacturing companies investing in large-scale battery manufacture, the trend in the UK is towards electric rather than hydrogen drives for motor vehicles and cars in particular.

The European Commission published its *Sustainable and Smart Mobility Strategy* in December 2020 [9] as part of the European Green Deal[39] and supported by European Climate Law. The European Green Deal states that 'to achieve climate neutrality, a 90% reduction in transport emissions is needed by 2050'. The mobility strategy sets out overall targets for EU nations which includes commitments for 30 million zero emission vehicles on the road by 2030, with high speed rail traffic doubling, all mass transit journeys of less than 500 km to be carbon neutral and zero emission shipping to be 'ready for market'; and with a further objective for zero emission aircraft to be ready for market by 2035.

Elsewhere the ambitious targets for installing a network of charging points across the USA has depended on the delicate balance between opposing politicians in the Senate and Congress. However the transition to electric vehicle production has depended largely on the actions taken by manufacturers of vehicles and batteries. By 2021, charging infrastructure had been rolled out across the US by private sector entities, such as the Electrify America Volkswagen subsidiary, '...which was created in 2016 as part of a gigantic package of fees and fines imposed on the

---

[38] https://www.gov.uk/government/news/government-takes-historic-step-towards-net-zero-with-end-of-sale-of-new-petrol-and-diesel-cars-by-2030

[39] https://commission.europa.eu/strategy-and-policy/priorities-2019-2024/story-von-der-leyen-commission/european-green-deal_en

German automaker after the 2016 revelation that Volkswagen diesel engines were bypassing pollution controls'.[40]

However, according to the analysis published by the Yale School of the Environment in March 2021 'China is the most important force affecting EVs today on all fronts: supply, demand, batteries, infrastructure and establishing cultural momentum. China's dominance is due, in part, to its deliberate industrial policies that fostered battery R&D and a battery supply chain in parallel with its push for EV design and production by domestic automakers'.[41] Car ownership in China increased by a factor of 5 between 2009 and 2019 when there were 225 million cars in private ownership. This compares with 276 million in the USA in the same year which is equivalent to 0.84 cars per head of population, whilst the figure for China is 0.16, having a population which was more than four times greater than the USA in 2019.

Electric car sales in China grew exponentially between 2013 and 2019, mostly due to a Government subsidy programme which has been adjusted periodically and from 2021 took the form of tax exemptions. This growth has also been fuelled by steady improvements in battery technology and hence trip range and a Government funded roll-out of charging points which more than doubled in number between 2018 and 2020, representing nearly two thirds of the total number of charging points globally. In October 2020, the State Council of the People's Republic of China released the *New Energy Vehicle Industrial Development Plan for 2021 to 2035*[42] which included a suite of targets for 2025 and 2035. These are all designed to augment the transition to electric, hydrogen, fuel cell and hybrid vehicles (referred to as 'new energy' in China), although the only target that includes a specific metric is for the '100% electrification of the stock of public fleets' by 2035.

8.5.1 Evolution of Battery Technology

Some of the very first road vehicles built in Britain in the 1830s were driven by electric motors. Indeed an electric vehicle (EV) held the world land speed record until 1900. They found application as delivery vans, taxis and city cars towards the end of the nineteenth century. However, they fell out of favour because of limits on battery life and, apart from limited application for utility vehicles such as milk floats and fork lift trucks, were supplanted by road vehicles powered by internal combustion engines fuelled with fossil fuels. In the early days, the only batteries available were of the lead-acid type, which continue to be used due to their relative low cost. However the drive to reduce carbon emissions in the first quarter of the twenty-first

[40] https://www.govtech.com/fs/transportation/sacramento-calif-to-see-massive-electric-vehicle-investment.html

[41] https://e360.yale.edu/features/ev-turning-point-momentum-builds-for-u.s.-electric-vehicle-transition

[42] https://theicct.org/sites/default/files/publications/China-new-vehicle-industrial-dev-plan-jun2021.pdf

century and developments in alternative battery technologies have led to a frantic burst of research activity to improve the performance of EV batteries.

Lithium-ion batteries were developed in the late 1970s initially for use in portable computers and other consumer electronics. Since their commercialisation by Sony in the early 1990s, they have become the most common rechargeable battery for a wide range of consumables, including EVs. Although there have been numerous combinations of materials used in the manufacture of lithium-ion batteries, the essential components remain the same, that is: an anode, a cathode, a separator and electrolyte. The anode is the positive electrode typically made from graphite; the cathode is the negative electrode made from an oxide such as lithium cobalt oxide; the separator is a porous polymer which prevents contact between the anode and cathode, whilst allowing transport of ions between the two and the electrolyte comprises organic carbonates such as an organic solvent containing lithium salt through which lithium ions are transported.

This configuration allows lithium-ion (Li-ion) batteries to provide very high power densities and to be recharged by simply reversing the flow of ions between the anode and cathode. There is not space here to discuss all of the recent and ongoing developments in EV battery technologies but the main objectives of research have been to increase range, reduce weight and size, reduce cost and reduce reliance on materials that have limited stocks or are obtained only from politically sensitive or unstable regions, with the associated potential risks to cost and reliability.

The largest reserves of lithium are to be found in Bolivia, followed by Argentina and Chile, although it is only Chile that has developed a mature lithium mining industry, which in 2019 was second only in output to Australia[43]; whilst China controls 80% of global lithium refining, supplementing its own reserves with imports from the likes of Australia. Like most mining processes, the extraction of lithium has significant environmental impacts. For example, in Chile's Salar de Atacama region, lithium extraction and processing uses 65% of the region's water supply, whilst the hydrochloric acid used in extracting lithium from lithium-rich salt deposits has been known to pollute ground water and rivers.

Figure 8.6 shows one of the many large salt flats in Bolivia. This one lies at an altitude of nearly 4000 m: the lithium-rich brine shows as turquoise areas and lithium evaporation pools can be seen as a grid structure which measures 4.8 km along its longest edge.

A bigger problem in the sourcing of materials for batteries is the extensive use of cobalt in cathode manufacture. Along with nickel, cobalt is used to improve stability, reduce corrosion and hence fire risk, whilst also increasing energy density. However, cobalt is highly toxic, and resources have only been extensively exploited in a small number of locations: in 2019, two-thirds of the world supply came from the Democratic Republic of Congo (DRC) where some 30% of cobalt mining is artisanal, with little oversight in place, and the use of child labour is reported to be

[43] https://www.nsenergybusiness.com/features/six-largest-lithium-reserves-world/

Fig. 8.6 Satellite view of Bolivia's Uyuni salt flats and lithium evaporation pools (Photo: Oton Barros; Source: https://commons.wikimedia.org/wiki/File:Lithium_mine_at_Bolivia's_Uyuni_Salt_Flat,_on_a_CBERS4_MUX_yesterday's_image.jpg)

endemic and fatalities common.[44] This has been recognised by the electronics and battery manufacturing industries resulting in the Responsible Cobalt and Responsible Mineral Initiatives[45] and an intensive research programme to develop cathodes that use alternative materials, such as a derivative of lithium nickelate known as NFA, which combines nickel with iron and aluminium to produce a cathode which it is claimed is sustainable, stable and economic.

Globally there are dozens of projects having the multiple aims of improving battery performance whilst reducing weight and cost, improving safety and tackling the sustainability and reliability of supply chains. For example, historically liquid electrolytes have been used in lithium-ion batteries but these can become unstable under certain voltages and temperatures and catch fire—usually because of a fault or damage. Researchers have therefore been focussing on replacing liquid electrolytes with solid. However, they have found that such materials as germanium and tantalum are required in the manufacture of solid electrolytes which, because they are the by-products of mining coal and tin respectively pose their own supply chain and cost challenges.[46]

Alternatives to lithium are also being explored: for example, with the development of sodium-ion (Na-ion) batteries which, whilst currently providing lower energy densities than Li-ion avoid the problematic exploitation of lithium and cobalt referred to above. According to a 2020 review published by the American Chemical Society:

> The foremost advantage of Na-ion batteries comes from the natural abundance and lower cost of sodium compared with lithium. The abundance of Na to Li in the earth's crust is 23,600 ppm to 20 ppm, and the overall cost of extraction and purification of Na is less than

[44] https://www.sciencedirect.com/science/article/pii/S2214790X21000290

[45] http://www.responsiblemineralsinitiative.org/minerals-due-diligence/cobalt/

[46] https://news.mit.edu/2021/designing-better-batteries-electric-vehicles-0816

> that of Li. Moreover, Na-containing metal oxide and polyanion cathode materials can be fabricated from naturally abundant transition metals such as iron, manganese, vanadium, and titanium, without using cobalt, making Na-ion batteries sustainable and affordable in rich and poor countries alike.[47]

All batteries have a limited life expectancy as their ability to hold a charge decays. Hence the ability to recycle the components of each battery is vital. Each component within a battery has a different lifespan and requires different techniques to recycle them, so the fact that batteries are glued and/or welded together makes separating these components very difficult. One recent British development uses compression technology to hold the components together so that they can be separately serviced, replaced and recycled.[48] Current recycling methods use smelting (pyrometallurgy) or dissolving in a mix of acids and solvents (hydrometallurgy) all of which are energy intensive and result in potential air or water pollution and loss of material.

8.5.2 Charging Infrastructure

The provision of a suitable charging infrastructure for electric vehicles is directly related to the developments in battery technology. The ideal is for EVs to have a range at least equivalent to those powered by internal combustion engines, with charging points at what were previously petrol stations and service station car parks, for example, that can provide a full charge within a few minutes. It is also important that drivers can charge their EVs at home: which is easy for those who can park in their own drive or garage but more challenging for those who live in houses or flats having parking some distance from their nearest power socket. The solution to this is a network of on-street charging stations: which can be stand alone or integrated into lamp posts. A typical on-street array of charging points can be seen in Fig. 8.7. Charging points can be retrofitted to an existing compatible lamp post, with billing facilitated using a dedicated mobile network for example: although the charge rate may be slow, it is suited to trickle charging overnight or whilst at work.

'To complement stationary charging stations and alleviate range anxiety, the Israeli company Electreon has developed a stationary and dynamic wireless charging system embedded beneath road and parking surfaces. The technology uses coil segments installed under the asphalt, which wirelessly transfer energy to specially equipped electric vehicles as they drive over them via magnetic induction (see Fig. 8.8). Electreon's first pilot was launched in Gotland, Sweden, in 2018 and included trials with both an electric bus and a heavy duty truck. The system requires a roadside control unit and a receiver installed on the vehicle's chassis to enable charging whilst in motion'.[49]

---

[47] https://pubs.acs.org/doi/10.1021/acsenergylett.0c02181

[48] https://www.current-news.co.uk/advik-acquires-uk-battery-repurposing-tech-company-aceleron/

[49] Image and text provided by Electreon.

Fig. 8.7 Typical array of on-street EV charging points (Photo: Avda/avda-foto.de; From: https://commons.wikimedia.org/wiki/File:Berlin_-_Potsdamer_Platz_-_E-Mobility-Charging.jpg)

Fig. 8.8 Under-road copper induction system for on-the-go battery charging (Image: Electreon)

8.5.3 *Hydrogen and Fuel Cell Vehicles*

As with the substitution of hydrogen for gas in heating boilers, hydrogen has been offered as an alternative to petrol or diesel in internal combustion engines for many years. Remarkably the first vehicle to use hydrogen as a principal fuel was built in 1806 by Franco-Swiss inventor Franco De Rivaz. In fact this was the first application of an internal combustion engine to power a road vehicle, which had previously used steam engines, although it failed to gain traction commercially.

Although in modern times, there have been a number of attempts to promote hydrogen as an alternative to fossil fuels, these have had limited success because of the low energy density of the fuel, resulting in unrealistically large fuel tanks and limited range. Burning hydrogen in an internal combustion engine (ICE) also releases a certain amount of NO_2, although generally lower than for petrol or diesel engines, it can be minimised by using a much larger air:fuel ratio than a petrol engine. Also, due to the low molecular size of the hydrogen atom, it is more prone to leakage than petrol or gas and is highly explosive when mixed with air.

The adaptation of fuel cells to use hydrogen as a fuel has shown greater promise than for ICEs. Fuel cells are similar to batteries in that they comprise an anode and cathode separated by an electrolyte, but instead of storing electricity a fuel cell generates electricity through catalysing hydrogen at an anode which separates electrons from protons, with the electrons providing power for a load connected to the cathode whilst the protons migrate through the electrolyte to the cathode which is fed with oxygen that combines with the electrons and protons to produce water with zero tailpipe emissions (see Fig. 8.9).

The proton exchange membrane (PEM) fuel cell is the type most commonly used for road vehicles. Comprising platinum-based electrodes and a water-based acidic

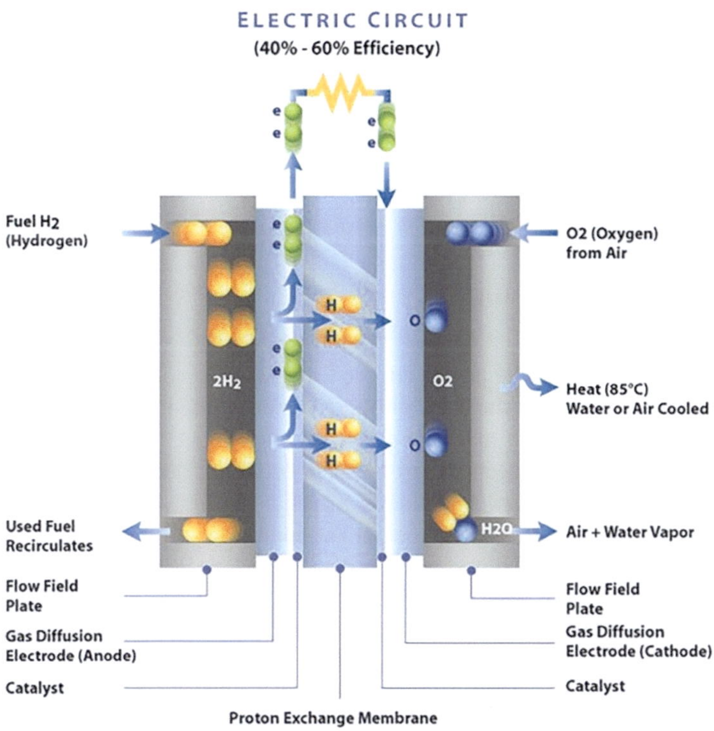

Fig. 8.9 Proton exchange membrane fuel cell (From: https://commons.wikimedia.org/wiki/File:Membrane_Electrode_Assembly_-_Electro-Chemical_Reaction_Diagram.jpg)

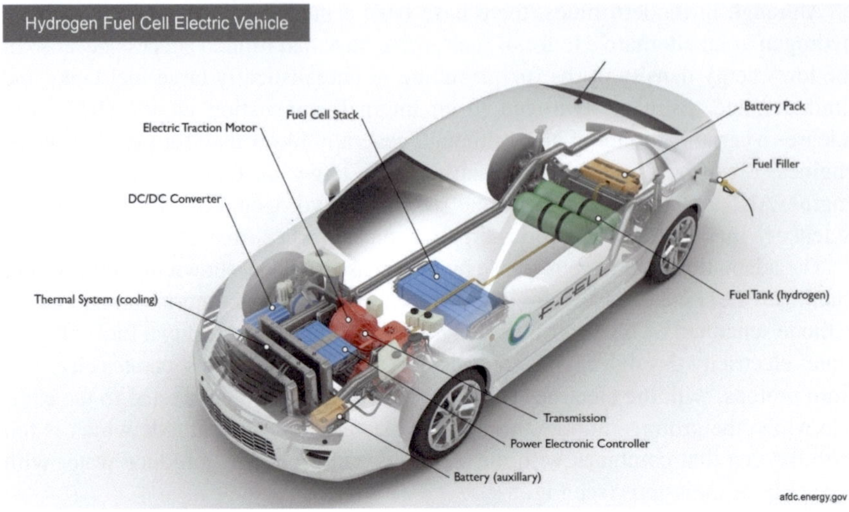

Fig. 8.10 Typical arrangement for a hydrogen fuel cell electric car (US Department of Energy—From: https://afdc.energy.gov/vehicles/how-do-fuel-cell-electric-cars-work)

polymer membrane electrolyte, it is relatively compact and lightweight and can be adapted to operate at a wide range of external conditions. Most modern hydrogen fuel cell vehicles incorporate a battery pack which can be used to supplement the power from the fuel cell and is charged by regenerative braking, as is the case for a conventional hybrid EV (see Fig. 8.10). This reduces hydrogen consumption and smooths out the operation of the fuel cell.

8.6 The Cycle Revolution

There are many reasons why people choose to travel by motor car rather than cycle. Most of these come under the heading of convenience: the journey is too long, the weather unpredictable, too much hassle; or worries about safety. The number of people cycling depends at least in part on socio-economic and demographic factors such as the relationship between income and the cost of buying and running a car, the nature and location of employment and attitudes to cycling and associated cultural influences. A good example of the relationship between socio-economic impacts and cycling has occurred in China. When the People's Republic of China was founded in 1949, the Communist party decreed that the bicycle was the people's vehicle and started a massive bicycle manufacturing programme so that by the 1990s, there were around 670 million bikes in China, falling to about half a billion in the early 2000s as car ownership increased, but still representing some two-thirds of the world's bikes. The local transport infrastructure catered primarily for bicycles, with cities such as Beijing and Shanghai giving right of way to bikes and

Fig. 8.11 Bike parking in Shanghai in the 1980s (Image: H. Grobe; From: https://commons.wikimedia.org/wiki/File:China1982-678.jpg)

experiencing frequent bicycle-related grid locks and high density parking (see Fig. 8.11). However, as car ownership continued to grow cycling decreased and the streets became clogged with cars instead. Cycle ownership dwindled but as pollution in the cities became a major problem and the SARS epidemic drove people away from public transport, cycling experienced a renaissance and from the mid-2010s a massive growth in bike hiring schemes occurred, driven in part by central Government policy.

The uptake of cycling in Europe has also been driven by socio-economic factors, but with less overt influence by government diktat. By far the greatest adoption of cycling overall has been in the Netherlands, where a combination of benign topography and investment in bike-friendly infrastructure has led to cycling becoming embedded in the Dutch lifestyle. In the mid 2010's between 30% and 40% of Dutch respondents to European Cyclists' Federation (ECF) Eurobarometer surveys indicated that they were more likely to cycle on a typical day than adopt other modes of transport.[50] This compares with 2–3% of UK respondents (22nd in a league table of 28 European countries).

According to statistics published by Cycling UK, the mileage cycled in the UK fell steadily by a factor of around 7 between the early 1950s and 1970s, since when it has gone through a number of surges, but dipped in the 1990s.[51] There has been a steady rise since, particularly in London which saw a threefold increase in journeys by bike between 2003 and 2016 following the introduction of the congestion charge in 2003.

[50] https://ecf.com/resources/cycling-facts-and-figures

[51] https://www.cyclinguk.org/statistics

The COVID-19 pandemic, which took hold in 2020 globally, created a surge in cycling in most of the developed world—especially in recreational cycling—with an associated increase in bike ownership. According to a 2021 study published in Transport Review, the USA and Canada experienced an increase in weekend cycling of 29% and 28%, respectively, between 2019 and 2020, whilst the 11 EU countries studied saw an increase of 23%.[52] Many countries experienced a downturn in cycling to work, most likely due to an increase in home working.

According to statistics published for Great Britain by the Department for Transport, the miles cycled in 2020 was 46% higher than in 2019, with a 41% increase in fatalities reported to the police, although serious accidents increased by only 2% and minor accidents actually fell, bearing in mind that many accidents would not have been reported to the police.[53] In 2019, there were 100 reported fatalities whilst across Europe there were just over 4000 with more than half involving collisions with cars.[54] However, despite cyclists being classified as a vulnerable user group and the relatively short distance travelled on bike by the average cyclist, only one death from a cycling accident has been reported in the EU for every 32 million kilometres cycled.

Recreational cycling has potentially significant benefits to physical and mental health but does not generally displace or replace journeys by car or public transport and associated carbon emissions and other environmental impacts. The health benefits from cycling are similar to other forms of exercise taken outdoors but with less detrimental impact on joints.

References

1. International Energy Agency. (2020). *Energy technology perspectives 2020.* https://www.iea.org/reports/energy-technology-perspectives-2020
2. European Environment Agency. (2020). *Air quality in Europe—2020 report.* https://www.eea.europa.eu//publications/air-quality-in-europe-2020-report
3. EU. (2016). *Regulation (EU) 2016/1628 of the European Parliament and of the Council of 14 September 2016 on requirements relating to gaseous and particulate pollutant emission limits and type-approval for internal combustion engines for non-road mobile machinery, amending Regulations (EU) No 1024/2012 and (EU) No 167/2013, and amending and repealing Directive 97/68/EC.* https://eur-lex.europa.eu/legal-content/EN/TXT/PDF/?uri=CELEX:32016R1628&from=EN
4. Morita, H., Yang, S., Unger, N., & Kinney, P. L. (2014). Global health impacts of future aviation emissions under alternative control scenarios. *Environmental Science and Technology, 48*(24), 14659–14667.

[52] https://www.tandfonline.com/doi/abs/10.1080/01441647.2021.1914900?journalCode=ttrv20

[53] https://www.gov.uk/government/statistics/reported-road-casualties-great-britain-pedal-cyclist-factsheet-2020/reported-road-casualties-in-great-britain-pedal-cycle-factsheet-2020

[54] https://ecf.com/news-and-events/news/road-safety-fewer-eu-road-deaths-most-fatal-crashes-still-include-cars

5. Yim, S. H. L., Lee, G. L., Lee, I. H., Allroggen, F., Ashok, A., Caiazzo, F., Eastham, S. D., Malina, R., & Barrett, S. R. H. (2015). Global regional and local health impacts of civil aviation emissions. *Environmental Research Letters, 10*(3), 034001. https://doi.org/10.1088/1748-9326/10/3/034001/meta
6. World Health Organization. (2018). *Environmental Noise Guidelines for the European Region*. WHO. https://www.who.int/europe/publications/i/item/9789289053563.
7. Byrnes, T. A., & Dunn, R. J. (2020). Boating- and shipping-related environmental impacts and example management measures: A review. *Journal of Marine Science and Engineering, 8*, 908.
8. Winkler, C., Murphy, S., Ritter, F., & Panigada, S. (2020). *Global numbers of ship strikes: An assessment of collisions between vessels and cetaceans using available data in the IWC ship strike database*. Report to the International Whaling Commission, IWC/68B/SC HIM09. https://www.researchgate.net/publication/342734400_Global_Numbers_of_Ship_Strikes_An_Assessment_of_Collisions_Between_Vessels_and_Cetaceans_Using_Available_Data_in_the_IWC_Ship_Strike_Database_Report_to_the_International_Whaling_Commission_IWC68BSC_HI
9. European Commission. (2021). *Sustainable and smart mobility strategy—Putting European transport on track for the future*. EC. https://transport.ec.europa.eu/system/files/2021-04/2021-mobility-strategy-and-action-plan.pdf.

Chapter 9
Products, Manufacture and Consumption

Overconsumption and overpopulation underlie every environmental problem we face today.

Jacques Yves Cousteau

9.1 Introduction

And now we return to the dilemma described in Chap. 1: the green growth paradox. The dichotomy between economic growth and environmental impacts with a growing population and increasing standards of living.

Since humans first started exploiting the planet to make things, there has been an inexorable deterioration in the environment and increase in greenhouse gas emissions, gathering momentum during the industrial revolution, which continues to this day. This forms the basis for this chapter: the demand for products driven by consumers associated with a sometimes cynical attitude to product development and advertising resulting in vast quantities of material ending its life in landfill or polluting the natural environment.

This doesn't just apply to motor cars and electronic goods but also to food, fashion, cosmetics and many other consumables: not only in their manufacture, and the raw materials used therein, but also in their packaging, storage, delivery, sale and eventual disposal—from cradle to grave that is, sometimes referred to as the circular economy. A complex relationship has evolved between the developed and developing world whereby western demand for low-cost products has resulted in manufacture migrating to countries with a plentiful supply of cheap labour and with much lower overheads, whilst some industries in the wealthier countries have withered or died, creating local impoverishment and decay.

P. Appleby, *The Overburdened Planet*,
https://doi.org/10.1007/978-3-032-01510-5_9

9.2 Pollution from Industry and Farming

Historically industry has been a major source of pollution to air, water and land, including being an important source of greenhouse gases and hence a contributor to climate change. Agriculture and farming contribute to local air pollution through burning of forests and crops, as well as some intensive farming practices, but is also responsible for greenhouse gas emissions, especially methane and nitrous oxide, and contamination of land and water bodies.

Greenhouse gas emissions associated with industry and farming are spread across a number of sectors: extraction, processing and transportation of raw materials, direct process emissions, process and services energy, transport, construction and waste. Emissions specific to farming and food production include those associated with livestock rearing, fertilisers, peat extraction, soil disturbance, deforestation and crop burning. Figure 3.18 in Chap. 3 shows the haze over Singapore and Southern Malaysia from crop burning in Sumatra. Wood is of course used as a raw material in many products as well as being processed into paper and cardboard. Cultivation of trees for timber production results in CO_2 absorption during the growth period and the carbon is held within the finished product, provided it is not burnt. However, timber obtained from hardwood trees that have taken perhaps hundreds of years to grow results in a resource lost that will take too long to replace and is thus not sustainable.

Fossil fuels, as well as being used in the generation of energy, are used as a raw material in the manufacture of steel, cement, plastics, chemical feedstocks, lubricants, solvents, waxes, cleaning products, personal care products, paint, dies, fertilisers and fabrics. Not all of the processes involved result in greenhouse gas emissions, although in many cases, other forms of pollution are generated, such as plastic waste as litter, in landfill, soils, waterways and oceans. Many of the products, which are not biodegradable, find their way into sewers or waterways and thence into oceans and living organisms.

9.2.1 Environmentally Persistent Waste

According to data published by the United Nations Environment Programme (UNEP), around 9.2 billion tonnes of plastic was manufactured between 1950 and 2017, 57% of which has ended up in landfill or the natural environment, with approximately 300 million tonnes of plastic waste being disposed of annually worldwide.[1] Estimates of how much end up in the oceans vary, but one widely referenced piece of work using data from 2010 made estimates of between 4.8 and 12.7 million tonnes of plastic waste entering the ocean from land sources per annum, with predictions of an order of magnitude increase by 2025 [1]. Non-land sources

[1] https://www.unep.org/interactive/beat-plastic-pollution/

Fig. 9.1 The Citarum river in java, Indonesia (http://www.purewatergazette.net/blog/citarum-river-the-worlds-most-polluted/)

such as fishing boats and commercial shipping probably add another 20% onto these figures. According to research published in 2017, between 88% and 95% of plastic debris entering the oceans is conveyed along ten rivers, eight of which are in Asia, with the worst by far being the Yangste in China which was estimated to deposit nearly 1.5 million tonnes of plastic waste per year into the East China (Yellow) Sea [2]. However, the most polluted river in the world is thought to be the Citarum river in Java, Indonesia which, as can be seen in Fig. 9.1, can become completely blocked with floating debris, most of which is plastic.

9.2.2 *Microplastics*

One particularly insidious form of pollution from plastic results from the dispersal of small particles, or microplastics, which are typically less than 5 mm across. These take various forms defined as primary and secondary microplastics: where primary microplastics are those which start life with small dimensions and include microbeads, microfibres and plastic pellets (also referred to as nurdles). Secondary microplastics start life as large items that are broken down into finer particles over time. The fine particles that are abraded from tyres and painted surfaces may be referred to as secondary microplastics since although they enter the environment as fine particles, they start life as part of a larger object or surface.

Once in the natural environment microplastics are frequently mistaken for food by birds, fish and crustaceans, degrading in time into smaller and smaller particles, eventually becoming nanoparticles which are small enough to cross cellular membranes and affect the functioning of cells in animals. Once in an animal's stomach microplastics cannot be digested and eventually can reach levels that prevent an animal from eating normal food and hence it starves to death.

Microbeads are defined as near-spherical beads typically manufactured from polypropylene, polyethylene or polystyrene with a diameter less than 1 mm. Because of their size, they can escape filtration systems at sewage works and may be mistaken for food by aquatic life forms. Historically they have been used in personal care products to enhance silkiness and assist in exfoliation. From 2017/18, many countries banned the use, sale or import of 'rinse-off' products containing microbeads. This included the UK, USA, Canada, many EU countries and a number of Asian countries.

Synthetic fibres are used primarily in the manufacture of clothing. The aggressive action of washing clothes in a washing machine or drying them in a tumble dryer results in microfibres being shed from the clothing. In a typical load, these will comprise a mix of natural and synthetic fibres, some of which will be trapped in the machine's filter. Research by De Falco et al. [3] found that between 640,000 and 1.5 million synthetic microfibres were shed on average for each load in a washing machine, weighing between 124 and 308 mg/kg of washing. Average fibre dimensions were found to be between 360 and 660 μm long with an average diameter of between 12 and 16 μm: 'indicating dimensions that could pass through wastewater treatment plants and pose a threat for marine organisms' [3]. A study by Gavigan et al. [4] reported on research that had estimated that 167,000 tonnes of microfibres from clothes washing ends up in water bodies each year globally, whilst a further 176,500 tonnes is deposited in terrestrial environments and landfill [4].

An interim stage in the manufacture and recycling of plastic products involves the creation of 'pre-production plastic pellets'—often referred to as nurdles. These pellets are generally smaller than 5 mm across which is a convenient size for processing and shipping—usually in bulk in shipping containers. It has been reported that some 230,000 tonnes of nurdles find their way into waterways and the ocean every year through spillages in factories and in transit (see Fig. 9.2). Whilst there is a steady trickle of nurdles into the environment from factories and lorries there have also been massive spillages directly into the ocean from ships that have sunk, such as the X-Press Pearl container ship that caught fire and sank off the coast of Sri Lanka in May 2021, resulting in the discharge of 1680 tonnes of nurdles into the ocean. Much of these were carried ashore and contaminated hundreds of miles of Sri Lankan shoreline, beaches and coral reefs, resulting in nurdle drifts of up to 2 m deep, the death of numerous marine animals and the catastrophic disruption to fishing and tourism.

Secondary microplastics are fine particles which are derived from larger plastic items, such as bags, tyres and water bottles, which have been discarded and broken down into finer particles or fragments by abrasion, sunlight (photo oxidation) and biological mechanisms. The list of plastic objects that find their way into oceans is extensive. As well as water bottles and plastic bags, the principle culprits include cigarette filters, packaging, cotton buds, tableware and fishing tackle of various kinds.

One little reported source of secondary microplastic pollution of the environment is associated with the stripping of paint through sanding or sand-blasting. Paint also directly enters the environment as it peels away from surfaces due to corrosion or

Fig. 9.2 Nurdles: a common sight around coasts and waterways globally (Photo: hockabilly; Source: https://commons.wikimedia.org/wiki/File:Nurdles.jpg)

wear and tear of painted surfaces: including buildings, structures and road markings, for example. On average paint comprises around 50% plastic polymers, which are used to bind the other ingredients. Some paints have microbeads, hollow microspheres or microfibres added to create a specific property. An important source of microplastic paint particles is from the anti-fouling paints used to protect surfaces in contact with water, such as ship's hulls, piers and jetties. These also tend to contain biocides and heavy metals. Paint may also be considered a source of primary microplastic pollution if poured directly into a drain without application to a surface.

Tyres and brakes are also a source of secondary microplastics. A 2020 study by Evangeliou et al. estimated that approximately 100,000 tonnes of microplastics from tyres end up in the world's oceans each year through airborne transmission, whilst 40,000 tonnes of microplastics shed from brakes follow the same route. They also report on other studies that indicate that some 64,000 tonnes of microplastics from the same sources are carried into oceans via run-off from roads into storm drains and waterways [5]. A typical motor vehicle tyre sheds around 4 kg during its lifetime. Most tyres are manufactured from a mix of natural and synthetic rubbers with additives such as carbon black, silica and plasticisers to create the desired colour, strength and wear characteristics. The particles shed during braking are generally a mix of metals and synthetic polymers which are used in the construction of discs and pads in a suitable combination to ensure maximum strength and friction between contact surfaces under load.

Plastics are a significant contributor to greenhouse gas emissions and global warming: both from the manufacturing processes and disposal. In 2025, 4.5% of global GHG emissions were from plastics manufacture, having doubled since 1995. Plastics are manufactured using a feedstock which is refined and 'cracked', that is, heated, and a catalyst added to form a polymer. This requires the processing of a fossil fuel such as crude oil, coal or natural gas, as well as a wide variety of plant-based materials. The processes initially are different depending on the feedstock: for example, crude oil has to be refined through a fractional distillation process which involves heating the oil which becomes a vapour that is fed into a distillation tower in which a complex series of condensations takes place such that the lightest fluid (petroleum gas) is drawn off at the top of the tower and the heaviest (bitumen) at the bottom. One of the lighter fractions is naptha, which is broken down into long chain hydrocarbons by either steam or catalytic cracking, with further heat and chemicals being added to stimulate the polymerisation process producing polyethylene resin, which forms the building blocks for many plastic products. A similar process uses natural gas as the feedstock, using a similar process to crude oil but tapping a tower for ethane which is 'cracked' at a temperature of around 860 °C to create ethylene and hydrogen. The ethylene gas has a catalyst added and becomes polyethylene syrup under pressure which is used typically to form pellets for the manufacture of a wide spectrum of plastic goods.

Some countries, such as China and South Africa, use coal as a feedstock by converting methanol derived from coal into olefins. This is a much more energy and water intensive process than using oil or gas, resulting in GHG emissions perhaps three times higher and water consumption more than four times higher. It is interesting to note that whilst China is gradually reducing its dependence on coal for electricity generation, there has not been a proportionate fall off in the output from domestic coal mines—the shortfall being largely filled by plastics manufacture.

9.2.3 Other Industrial Pollutants

Historically industry has been the source for a wide variety of substances entering the ground, water bodies and the air. Many of these substances have potential impacts on the health of humans and animals as well as the environment, and some persist in the environment whilst only slowly losing their potency.

The UN *Stockholm Convention on Persistent Organic Pollutants* (POPs) of 2001 introduced restrictions and controls on a list of 12 substances, known colloquially as 'the dirty dozen'. These are substances with well-established negative health impacts such as the related polychlorinated biphenyls (PCBs), polychlorinated dibenzo-p-dioxins (dioxins), polychlorinated dibenzofurans (furans) and the infamous DDT of course. PCBs were used extensively as heat exchange fluids in electrical transformers and capacitors, whilst dioxins and furans are the by-products of incomplete combustion during such processes as incineration and smelting, as well as being a by-product of paper manufacture that uses chlorine as a bleaching agent which combines with organic compounds in wood to produce both dioxins and furans (see below).

Once its insecticidal properties had been discovered in the 1940s, dichlorodiphenyltrichloroethane (DDT) was extensively adopted in agriculture and against disease spreading insects such as mosquitoes, so much so that by 2002 around 1.8 million tonnes had been produced globally, despite being banned in most countries from 1968 onwards. Interestingly the Stockholm Convention does not require the complete elimination of DDT, and indeed it is still manufactured and deployed for mosquito control and agriculture in a number of countries including India, whilst being used in a number of African countries to control the spread of malaria.

Since 2001, the list of POPs has been extended to include a number of other substances which have historically been used as insecticides or flame retardants such as lindane and chlordecone: the former having been used extensively as a pesticide in agriculture for both crops and livestock as well as the treatment of lice and scabies in humans; whilst the latter has been used as an agricultural pesticide and is related to DDT.[2] All of these substances remain in the environment for many years and have potential long-term health impacts, including in many instances cancer.

Although not classified as POPs, polycyclic aromatic hydrocarbons are a class of chemicals which can be persistent in the environment, some of which are known carcinogens or suspected to be carcinogenic. They occur naturally in coal, crude oil and petrol and are emitted widely through incomplete combustion of fossil fuels, wood, garbage, tobacco and certain foods. The most ubiquitous PAH is benzo(a)pyrene which as well as being associated with tobacco smoking is found in relatively low concentrations universally, and in much higher quantities close to roads and both within and external to industrial facilities where fossil fuels are burnt or processed.

Pollutants can enter the environment through ongoing deliberate and controlled emissions from industrial processes, through poorly controlled processes or by accident. The extent to which emissions are controlled depends on many factors and in particularly the effectiveness of national or local environment protection procedures. The worst case scenario occurs when old and poorly maintained plant is handling toxic and unstable chemicals operated by poorly trained and poorly paid staff combined with an ineffective inspection and under-funded environmental protection system.

One of the worst examples of all of these failings combining occurred at the Union Carbide factory located in Bhopal, the capital city of the Indian state of Madya Pradesh, which manufactured the carbanate pesticide Sevin. Although accounts vary, it is clear that during the night of the 2nd to 3rd December 1984, a large volume of water found its way into a tank containing 42 tonnes of methyl isocyanate (MIC). This started an exothermic reaction which resulted in an increase in both temperature and pressure in the vessel. Opportunities to control this reaction were missed by the operatives and none of the safety devices intended to prevent the release of MIC to the atmosphere were functioning at the time resulting in the entire contents of the vessel discharging as a vapour that is heavier than air and transported

[2] http://chm.pops.int/TheConvention/ThePOPs/ListingofPOPs/tabid/2509/Default.aspx

to those living downwind in the local community, with minimal warning and when most were asleep in their homes. Although a siren sounded in the factory which was evacuated fairly quickly, the external siren was sounded only briefly.

According to the US Agency for Toxic Substances and Disease Registry 'Methyl isocyanate is irritating and corrosive to the eyes, respiratory tract and skin. Acute exposure to high-vapour concentrations may cause severe pulmonary oedema and injury to the alveolar walls of the lung, severe corneal damage and death'.[3]

It has been estimated that some 8000 people died in the local community within 2 weeks of the disaster and that there have been a further 8000 fatalities since, with some estimates exceeding 20,000 dying as a result of the disaster, and with many thousands more experiencing severe health impacts including miscarriages and birth defects, through to the present day. The pressure on the local healthcare system at the time of the disaster was unsustainable, there being no strategy in place to deal with MIC poisoning. Litigation following the disaster continued for decades and, despite the death toll and long term health impacts, compensation payments have been both far from generous and slow in their settlement. Far too many publications have attempted to explain what happened on that night in 1984 and detailing the aftermath to list here, however, Bridget Hanna, an independent scientist associated with Harvard University, has written widely about the disaster and her 2007 publication *Bhopal: Unending Disaster, Enduring Resistance* provides a comprehensive account of the events that led to the accident and its aftermath.[4]

9.2.4 Non-energy Sources of Greenhouse Gases

Some industrial processes, such as the manufacture of nitric acid—an important ingredient in fertiliser production—involve the emission of greenhouse gases other than carbon dioxide associated with energy use. Nitric acid is created through the catalytic oxidation of ammonia, releasing nitrous oxide which once in the outer atmosphere has a GWP up to 300 times that of CO_2 with a life expectancy greater than 100 years. On top of this ammonia is manufactured using natural gas as a feedstock, emitting CO_2 in the process. Synthetic fertilisers made up 21.5% of agricultural GHG emissions in 2018 and about 2.4% of total emissions globally.[5] There are a wide variety of industrial processes, such as cement and lime production, that release CO_2 as part of the process. For example, around 60% of CO_2 emissions associated with cement production are emitted during the process of calcination, whilst total CO_2 emissions in 2019 constituted approximately 4.5% of total global GHG emissions (see also Sect. 7.2.3 in Chap. 7).

[3] https://wwwn.cdc.gov/TSP/MMG/MMGDetails.aspx?mmgid=628&toxid=116

[4] https://scholar.harvard.edu/files/bridgethanna/files/Unending_Disaster.pdf

[5] https://grain.org/en/article/6761-new-research-shows-50-year-binge-on-chemical-fertilisers-must-end-to-address-the-climate-crisis

However, it is the steel industry which is the most carbon intensive of all manufacturing globally, contributing in 2019 some 7% of global greenhouse gas emissions. According to the data published by the International Energy Agency, only 11.5% of global CO_2 emissions associated with steel manufacture are from the process, the remainder being energy related.[6] 86% of steel manufacture is through the blast furnace/basic oxygen furnace process which involves the heating of a mixture of coke or natural gas, limestone and iron ore in a furnace to produce pig iron which is fed into the basic oxygen furnace to make steel, which subsequently goes through further refinement depending on the type of steel required. The other most common process employs an electric arc furnace (EAF) which uses electric current to melt scrap/recycled steel or iron ore that has gone through a direct reduction process.

A blast furnace creates a chemical reaction whereby the carbon in coke reacts with the iron oxide in iron ore to produce iron and CO_2. Fossil fuels are used to generate enough heat to melt the iron ore. The mix also includes limestone which carries off the impurities in the form of slag. The iron is then heated in a basic oxygen furnace—typically using a fossil fuel—and supplied with oxygen to produce steel. Coke is produced by baking coal in the absence of oxygen to drive off impurities in a process similar to charcoal production. Only certain types of coal are suitable for use as ‘metallurgical coals’, typically requiring a low sulphur, alkali, phosphorous and chlorine content.

Alternative fuels, such as natural gas and hydrogen, can be used in steel manufacture and there is significant pressure for steel manufacture globally to transition to zero carbon, with an emphasis on the use of ‘green’ hydrogen manufactured through electrolysis using renewable electricity as a prime energy source. Hydrogen is used to convert the iron produced in a blast furnace heated using electricity into ‘sponge iron’ which is turned into steel at significantly lower temperatures than in a conventional basic oxygen furnace, whilst the process emits zero CO_2 provided the electricity is generated renewably. According to analysis commissioned by the European Commission examining the impact of green steel in Germany, electricity demand for the country as a whole would increase by 20% and production costs for steel would increase by a third.[7] A plant in Sweden delivered its first green steel in 2021 and there are other plants in the pipeline in Finland, Germany, Austria and South Korea.

An international team of researchers from universities in Birmingham, England and Beijing, China has developed a method of retrofitting blast furnaces that replaces coke with the thermochemical splitting of CO_2 from the furnace using a double perovskite BCNF1 to produce carbon monoxide for the reduction of iron ore to metallic iron.[8] Perovskites are a versatile class of compounds, of which double perovskites are a particular group, having a crystal structure in numerous forms and

[6] https://www.iea.org/reports/iron-and-steel-technology-roadmap

[7] https://www.europarl.europa.eu/RegData/etudes/BRIE/2020/641552/EPRS_BRI(2020)641552_EN.pdf

[8] https://www.sciencedirect.com/science/article/pii/S095965262300121X?via%3Dihub

having a wide range of properties and applications. The potential for being able to retrofit existing blast furnaces to reduce carbon dioxide emissions to near zero is a game changer and could enhance the feasibility of transitioning to green steel enormously.

Glass manufacture represented around 0.3% of global GHG emissions in 2014. According to a paper on the *Global material flow analysis of glass* published by authors from the University of Cambridge in the Journal of Industrial Ecology in 2021 'There are two main sources of CO_2 emissions in glass making … Energy emissions are released to provide heat for melting the glass, typically from burning natural gas in the melting furnace. Process emissions are released when the limestone and soda ash inputs to glass making decomposes upon heating. In the European Union, combustion emissions account for roughly 62% of the CO_2 emissions burden, other energy uses account for 21% and process emissions for the remaining 16%'.[9]

As we saw in Chap. 4 agriculture, forestry and land use (AFOLU) combined make a considerable contribution to greenhouse gas emissions globally as well as providing a sink for CO_2 and source of oxygen. These factors have a significant impact on the carbon footprint and sustainability of supply chains for food and timber products which we will be examining in Sects. 9.3.1 and 9.3.2.

9.3 Sustainability of Supply Chains

Manufacturing is frequently a global enterprise or at least using materials and components which are derived from elsewhere. Hence, as with construction of buildings, roads and civil engineering structures, the environmental impact, carbon, water and ecological footprints of most products are dependent in part on those of a supply chain that typically commences with the extraction of raw materials, the manufacture of components and the transportation of these through to the delivery to an end user. Note that supply chains are sometimes referred to as 'value chains', such as in the context of European Union terminology.

According to a 2016 report by McKinsey & Company, 'the typical consumer company's supply chain creates much greater social and environmental costs than its own operations. Supply chain impacts account for more than 80% of greenhouse gas emissions and more than 90% of the impact on air, land, water, biodiversity and geological resources'.[10]

When it comes to the manufacturing industries, the importance of supply chains in the overall environmental impact, carbon footprint etcetera varies enormously depending on the product being manufactured—or assembled. For example, a motor

9 https://www.mckinsey.com/capabilities/sustainability/our-insights/starting-at-the-source-sustainability-in-supply-chains

10 https://www.mckinsey.com/capabilities/sustainability/our-insights/starting-at-the-source-sustainability-in-supply-chains

vehicle factory tends to be primarily a production line designed to assemble up to 40,000 components for each vehicle, many of which could be obtained from suppliers all over the world. Typically, the automotive industry relies on three Tiers of suppliers: Tier 1 being purpose-built components, Tier 2 being general-purpose components, such as semiconductors, which normally have application across a number of end products and Tier 3 raw materials, such as plastics and metals. Each one of the Tiers will have its own supply chain which might stretch back to the raw materials being extracted from the earth or the collection and processing of materials for recycling.

Reliance on thousands of suppliers creates an extremely vulnerable manufacturing process, exacerbated in part by the culture of just-in-time ordering and delivery that has been adopted extensively by the automotive industry and elsewhere in order to reduce warehousing costs, increase flexibility and reduce wastage. Hence when there was a shortage of semiconductors due to the COVID-19 pandemic that started in 2019, companies that had long-established just-in-time ordering processes found they had dropped to the bottom of the pecking order compared with companies who had fixed and regular ordering patterns.

However, with so many suppliers, understanding and keeping track of the environmental impact of the supply chain becomes even more difficult when suppliers may change periodically to keep pace with demand. When setting up contracts with suppliers, it is important to carry out due diligence on them and their complete supply chain. This should not only cover quality, financial and capacity issues but also social and environmental factors, applying 'sustainable practices throughout the supply chain' as set out in United Nations Global Compact, which includes 'Ten Principles' covering a range of criteria relating to human rights, labour and anticorruption as well as the environment.[11] These are broad brush principles clearly designed with a global reach in mind.

Key to managing the supply chain is establishing a due diligence process that is thorough and includes a requirement to monitor performance once a supplier has been appointed. The European Commission has proposed a Corporate Sustainability Due Diligence Directive which would cover value chains comprising companies within the European Union, although 80–90% of environmental harm is perpetrated by value chain companies outside the EU. Germany took this one step further by introducing an *Act on Corporate Due Diligence in Supply Chains* in 2021 which came into force in January 2023.[12] At the heart of this legislation is the requirement for supply/value chains to 'Implement a risk management system and perform regular risk analysis, to avoid potential adverse impacts on human rights and the environment' as well as systems for ongoing reporting and dealing with grievances. Critically this legislation applies to global supply chains, regardless of jurisdiction.

[11] https://www.unglobalcompact.org/what-is-gc/mission/principles

[12] https://www.bmas.de/SharedDocs/Downloads/DE/Internationales/act-corporate-due-diligence-obligations-supply-chains.pdf;jsessionid=3646DD86B60AF7B40B62962E9BF4506C.delivery2-master?__blob=publicationFile&v=3

The high percentage of environmental harm in the supply chain referred to above typically occurs in the developing world and artificially improves the environmental performance of developed countries. This applies to impacts across the board, including carbon/greenhouse gas emissions, ecological degradation and pollution to air, water and land. However, the socio-economic impacts are more conflicted since employment is created in these supply chains with potential benefits to living standards and economic growth, although it is not uncommon for standards of health and safety and employment rights to be very different from developed world norms.

9.3.1 Food Miles and Carbon Footprint: Local Versus Imported

The supply chain for food varies in complexity from locally grown seasonal fruit and vegetables to processed meals using multiple ingredients from different countries. The concept of 'food miles' was developed in the UK in the 1990s when concern was growing about the quantity of food being transported globally. However, food miles are only comparable between home and overseas sourced foodstuffs if the growing conditions are comparable. For example, the tomatoes grown in Southern Europe have a lower carbon footprint (greenhouse gas emissions) than in the North because of its shorter and warmer heating season. So despite the fact that 40% of tomatoes grown in Spain, Europe's biggest tomato producer, are grown in greenhouses, on average Spanish tomatoes sold in British supermarkets have a lower carbon footprint than British ones, despite the greater 'food miles' required to get them there. However, the tomatoes with the lowest carbon footprint will be those that are grown organically, outdoors and locally. This is of course true for any fruit or vegetable, wherever outdoor cultivation is possible.

According to research supported by the UN, food systems were responsible for 18 billion tonnes of greenhouse gas emissions in 2015 or about 34% of the global total. Around 39% of these emissions were from production, 38% from land use and 29% from distribution (food miles).[13] Methane emissions from both livestock production and rice cultivation were responsible for a massive 35% of food systems emissions. It is interesting to note that both refrigeration and packaging have been growing steadily as contributors to GHG emissions, both emitting around 5% of the total associated with food systems overall.

It can be seen from Fig. 9.3 that when it comes to greenhouse gas emissions, beef stands out as having by far the highest average global emissions, dominated by deforestation through changes of land use and methane emissions from cattle. At 60 $kgCO_{2eq}/kg$ of product, emissions are more than double those associated with

[13] https://news.un.org/en/story/2021/03/1086822

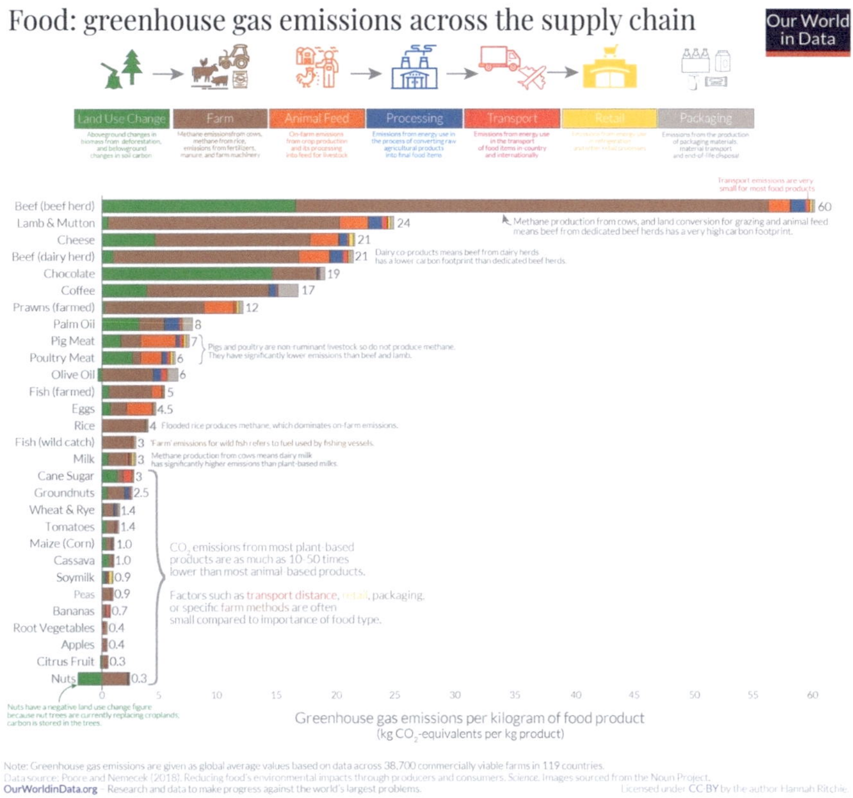

Fig. 9.3 Greenhouse gas emissions associated with food—Hannah Ritchie (From: https://ourworldindata.org/food-choice-vs-eating-local)

lamb production and ten times those from poultry meat production.[14] Some statistics from the Food and Agriculture Organisation stand out:

- Cattle ranching has been responsible for 80% of deforestation in the Amazon.
- Animal raising uses 80% of agricultural land whilst producing 18% of calories.
- 1695 L of water is required to produce one-quarter pound beef burger
- North Americans eat on average three beef burgers per week.
- Global meat consumption is predicted to rise 76% by 2050.

It can also be seen from Fig. 9.3 that for most foodstuffs, it is changes in land use and on-farm emissions that dominate and it is clear that meat, and beef in particular, is associated with very high emissions. According to research carried out by Asian Research and Engagement, a significant contribution to increased meat consumption is due to growing wealth and urbanisation within Asian countries, with consumption predicted to increase by 78% by 2050 compared to that in 2017.[15]

---

[14] https://ourworldindata.org/food-choice-vs-eating-local

[15] https://asiareengage.com/wp-content/uploads/2022/12/ChartingAsiasProteinJourney.pdf

In the early 2020s, a movement towards regenerative agriculture started to grow, popularly referred to as carbon farming, focussing in particular on soil health. In the USA, this was driven by the California-based Carbon Cycle Institute,[16] whilst in Europe, the EC included carbon farming and sustainable food systems in its 'Farm to Fork' strategies as part of the European Green Deal. Farmers are encouraged to increase the carbon uptake of their land and be paid within carbon offsetting systems.[17]

As well as GHG emissions and water consumption, agriculture and food production have a cumulative impact on the natural environment, having shaped the countryside for centuries. Replacing forest with livestock farms, rice paddies or palm oil plantations not only results in higher GHG emissions but also a loss of biodiversity—with potential extinction risk for multiple species, increased soil erosion and flood risk downstream, whilst also polluting waterways and land with fertilisers and animal waste. As we saw in Chaps. 2 and 3, this can lead to algal blooms and eutrophication in rivers and coastal waters, creating a sterile environment in which lifeforms cannot survive.

The intensification of livestock rearing has been an inevitable response to increased demand and resulted in a concentration of harmful impacts. The impact on animal welfare of high density farming methods has been well documented, along with the potential for antibiotic resistance growing with the widespread use of these drugs as a preventative measure in meat production. These issues, along with the impact of pesticides on pollinators, are discussed in detail in Chap. 2 under Sect. 2.5.

9.3.2 Forestry, Timber, Paper and Card Production

The timber industry globally supplies material for furniture, construction, fencing, pallets, biofuels and paper products. The wood pulp industry, supplying products ranging from toilet paper to cardboard packaging, represents a significant user of both virgin and recycled material, consuming between 33% and 40% of wood traded globally. According to the World Wildlife Fund: 'The forest practices associated with some pulp and paper operations have had devastating impacts on some of the world's most ecologically important places and species. Unsustainable pulp and paper operations have contributed to conversion of high conservation value forests, illegal harvesting, human rights and social conflicts and irresponsible plantation development'.[18]

Some 420 million tonnes of paper and card are manufactured per year, although this figure from 2018 is predicted to increase, particularly in China and Asian

[16] https://www.carboncycle.org/what-is-carbon-farming/

[17] https://food.ec.europa.eu/system/files/2020-05/f2f_action-plan_2020_strategy-info_en.pdf

[18] https://www.worldwildlife.org/industries/pulp-and-paper

countries. However, the average per capita usage in the USA and Europe, for example, is between 40 and 50 times higher than in India. The statistics for paper and card manufacture are stark: for every 1 kg of paper 2 or 3 kg of timber is required or around 17 trees per tonne on average. For an A4 sheet of paper up to 20 L of water are required—on average 26,500 L per tonne of paper/card.[19] The GHG emissions associated with the pulp and paper industry have been estimated to be just below 1% of total global emissions and around 83% of these are associated with energy consumption.

Around 55% of the pulp used in the manufacture of paper comes from virgin trees, the remainder is recycled paper, offcuts from timber mills, other fibrous vegetable matter and recycled cloth, depending on the quality required for the end product, whether it be glossy paper, tissue, cardboard or moulded pulp for packaging. The most common method used for breaking down wood fibres into a pulp suitable for paper and card manufacture is the Kraft process, which requires a mix of water, sodium sulphide and sodium hydroxide to form a white liquor mixed with the wood pulp which is heated, breaks down the lignite in the wood fibres and, with the addition of chemicals as binding, strengthening, sizing, bleaching and brightening agents, can be pressed or moulded into a suitable shape and size for the desired end product.

Binding and strengthening agents may include such chemicals as polyacrylamide—a synthetic linear polymer—and additives such as melamine—a type of plastic used in the manufacture of domestic ware which improves the wet strength of paper goods. Up to the 1970s, chlorine was typically used as a bleaching agent and effluent from pulp/paper factories was discharged into rivers and lakes without any treatment and containing high concentrations of dioxins—a highly toxic POP (see above) leading to significant damage to aquatic ecosystems. Ultimately this resulted in the introduction of legislation requiring controls on organochlorine emissions and a trend towards Elemental or Total Chlorine-Free bleaching. Elemental Chlorine-Free bleaching uses chlorine dioxide instead of chlorine gas and prevents the formation of dioxins.

Timber is being used for an ever expanding range of products, the demand for which has been growing steadily since a consensus was formed that wood could be considered to be a sustainable material. For example, with an increasing emphasis on embodied carbon of buildings and construction products, the low embodied carbon of timber compared with concrete, brick, steel and UPVC has fuelled an increase in the use of timber in construction (see Chap. 7). Timber-framed housing has experienced a renaissance, especially in northern climes where historically there has been good access to suitable timber and construction techniques allow greater thicknesses of insulation to be facilitated. For example, some 90% of 1 and 2 storey buildings in Sweden use timber in their construction: a country that has doubled the area of forest since 1920 so that by 2022, 70% of land was covered in

[19] https://environmentalpaper.org/wp-content/uploads/2018/04/StateOfTheGlobalPaperIndustry2018_FullReport-Final-1.pdf

trees. The UK demonstrates an extreme north/south divide with some estimates indicating that up to 85% of homes in Scotland are timber-framed, whilst the figure in England is closer to 25%.

The sustainable management of forests is essential. A number of certification schemes have been established in recent years that aim to monitor this management and provide assurance to the specifier and purchaser that the end product comes from a sustainable source. The best known and most widely used certification schemes are run by the Forest Stewardship Council (FSC) and the Programme for the Endorsement of Forest Certification (PEFC). FSC was established in 1993 following the Rio Earth Summit of the previous year, whilst PEFC was established in 1999. Both schemes certify the management of forests globally by requiring trees to be harvested responsibly with measures taken to preserve biodiversity, a fair wage and work environment for forest workers and with the local community consulted and their rights protected. The main difference between the FSC and PEFC schemes is that whereas FSC is a monolithic international organisation with central standard setting and certification, whilst using a network of accredited certifiers; PEFC is an umbrella organisation for dozens of national certification schemes globally, providing support in the establishment and quality management of national schemes.

By 2021 more than 200 million hectares of forest had FSC certification and more than 50,000 products had ‘chain of custody’ certification which means that the material in the finished product having FSC certification has been traced back to a FSC certified forest.[20] PEFC certification had been achieved for 286 million hectares of productive forest by 2022, with nearly half of this area being in Canada, whilst more than 20,000 companies and organisations had achieved ‘chain of custody’ certification.

According to the 2020 *Global Forest Resources Assessment* published by the UN Food and Agriculture Organisation (FAO), there was approximately 4.06 billion ha of forest globally, covering some 31% of total land area—178 million ha less than 1990: ‘Globally, about 1.15 billion ha of forest is managed primarily for the production of wood and non-wood forest products. In addition, 749 million ha is designated for multiple uses, which often includes production’.[21] Hence just over 25% of forests totally or partially managed for production have FSC or PEFC certification.

9.3.3 Fashion and Textiles

Like most industries, clothing and textiles have a complex global supply chain with a myriad of different environmental impacts from cradle to grave. Unlike food, clothing and textiles use materials from both natural and petrochemical sources. Many different types of plastic are used as synthetic textiles in everything from

[20] https://us.fsc.org/en-us/certification/chain-of-custody-certification

[21] https://www.fao.org/3/CA8753EN/CA8753EN.pdf

clothing and bedding to car seats and sofas. The most common synthetics are polyester, propylene, nylon, elastane and viscose; with polyester being by far the most common, used in the manufacture of around 60% of garments and 70% of household textiles. The dominant natural material used in textile manufacture is cotton, but according to a report from the European Environment Agency: 'The global consumption of synthetic fibres increased from a few thousand tonnes in 1940 to more than 60 million tonnes in 2018, and it continues to rise. Since the late 1990s, polyester has surpassed cotton as the fibre most commonly used in textiles'.[22]

Whilst synthetic materials share the environmental impacts referred to above for all plastics, cotton is not without its share of environmental and social challenges. According to the World Wildlife Fund, it is the most prolific of non-food crops globally, providing employment for 250 million people, employing almost 7% of productive labour in developing countries. Historically cotton plantations have created vast monocultures, associated with a low-wage economy. A continuous cycle of planting and harvesting has degraded soil quality and, with extensive use of chemical pesticides and fertilisers, impacted pollinators and polluted water resources downstream. Also cotton is thought to be the largest consumer of water of all agricultural commodities.[23]

Although produced on quite a small scale compared with cotton, the demand for linen has been growing. Linen is made from the stalks of the flax plant which is grown in cooler climes: mostly in the coastal regions of Northern Europe. Compared with cotton, flax requires very little or zero irrigation and is remarkably pest resistant, requiring no pesticides and far less fertiliser than cotton. Although the harvesting and processing of flax takes longer than cotton, the resulting fabric can be much stronger and last far longer and is one of the most sustainable fabrics available. The Southern hemisphere equivalent to flax is hemp which grows primarily in hot climates such as those of India and Bangladesh. It is processed in a similar manner to flax to produce fibres that are used in the manufacture of hessian, rope, twine, matting and sacks, for which there is a huge market in Southeast Asia in particular, despite the penetration of synthetic alternatives.

Synthetic textiles such as viscose/rayon have also replaced silk in many applications, although demand for silk in China and the Far East remains quite high. Silk is manufactured primarily from the cocoons of moths which feed on mulberry leaves. The larvae are killed by steaming before emerging from their cocoons. Although some silk is made from cocoons that develop in the wild, most are cultivated in factories under relatively strict environmental conditions and demanding large amounts of energy and water. Hence the carbon and water footprints of most commercially available silk are high.

Other natural materials, such as wool, sheepskin, fur, leather and suede, are associated primarily with livestock farming and although, compared with cotton,

[22] https://www.eea.europa.eu/themes/waste/resource-efficiency/plastic-in-textiles-towards-a

[23] https://www.worldwildlife.org/industries/cotton

demand is relatively small, they tend to have disproportionately negative environmental impacts as referred to above for meat production.

Common to the processing of most of the above textile types is dyeing. The vast majority of fabric dyes used globally are synthetic and the most common by far is the azo dye, an organic compound which constitutes between 60% and 70% of dyes used in the textile industry worldwide. They also have much wider application in the printing, food and pharmaceutical industries. These dyes have been developed to be colour fast and hence not fade with repeated washing or exposure to sunlight. This means that they persist in the environment and can have a half-life in the decades.

The global impacts of the textile industry and clothing in particular have been increasing rapidly during the first decades of the twenty-first century. Clothes have become a disposable item for young people in particular as the unit costs have fallen with the growth of the so-called sweat shops in both the developing world and places that have access to cheap labour. This has been due to either the vagaries of fashion or the short lifespan of poorly made products. As a result, the environmental impact of the clothing sector (including footwear) has been increasing accordingly with GHG emissions at around 6.7% of the global total in 2016[24]—forecast to rise to 10% by 2030, whilst water consumption and pollution are also on the rise. Dyeing and finishing are responsible for more than a third of GHG emissions associated with fashion and textiles and are a major contributor to downstream water pollution. GHG emissions are exacerbated because textile manufacture globally is dominated by China (52%) and India (7%) which generate much of their electricity from fossil fuels and coal in particular.

9.4 The Retail Sector: Physical Shops Versus Online Sales

Historically with the evolution of a culture based on shops and factory outlets, there have been those that have provided a delivery service for consumers: ranging from local stores to manufacturers who deliver directly to customers. Although there were some computer-based transactions in the 1960s, such as an American Airlines booking system that linked travel agents to a central main-frame computer, e-commerce and on-line shopping did not become possible until the world wide web was launched in 1991 and subsequently became accessible via browsers in 1993. By 1994, a number of sites had started selling specific items such as records and chocolate, which is also the year that Jeff Bezos launched Amazon as an on-line book shop, followed by ebay, which was established by Pierre Omidyar as AuctionWeb in 1995 as a vehicle for selling second hand items.

Since then both companies have become international juggernauts and been joined by others, such as Alibaba and Jingdon (JD.com) which are both Chinese e-commerce retailers reflecting that China has a massive demand for online transactions, whilst Asia has upwards of 50% of global online sales.

[24] https://www.oneplanetnetwork.org/knowledge-centre/resources/measuring-fashion-insights-environmental-impact-global-apparel-and

Based on a study of the retail environment in the USA in the 2010s, when compared with conventional in-store shopping, the carbon footprint of online shopping was found to be significantly smaller. The dominant carbon emissions for in-store shopping were found to be from private vehicle use, whilst for online shopping, it was packaging.[25] Between 2014 and 2019, the volume of transactions globally through online shopping increased by a factor of 3, growing still further during the COVID-19 pandemic that followed.

The increase in paper, card and plastic packaging associated with deliveries and returns is a major contributor to the increase in both the size of the wood pulp industry and plastic waste going to landfill, as referred to above. In China alone, some 83 billion parcels were shipped in 2020 requiring 1.8 million tonnes of paper and card and 10 million tonnes of plastic packaging, with most of the latter going to landfill due to extremely low recycling rates.[26]

The World Economic Forum has predicted an increase in delivery vehicles on the road in 100 cities globally of 36% by 2030 compared with 2020, with an extra 6 million tonnes of CO_{2eq} being emitted and resulting in increased congestion adding 11 min on average to the daily commute.[27]

Online shopping was a lifeline for many people during the COVID-19 pandemic, enabling them to avoid crowded shops and remain in self-isolation when necessary. However, there are many who succumbed to the nudges that occur extensively within the online environment: such as bots that contain adverts and offers that can prove too tempting to miss. There is evidence from around the world that levels of indebtedness have been increasing due to online purchasing and the increased availability of buy now pay later schemes.

9.5 Overproduction, Obsolescence and Waste

There is a direct link between overproduction and waste. One of the most obvious examples of this is agriculture and food waste. According to a UN report on Sustainable Development Goal 12: *Responsible Consumption and Production,* one-third of all food produced annually is wasted. This represents some 1.3 billion tonnes of food worth around $1 trillion.[28] This is food that is wasted at the farm, during storage, manufacture, at the shop and by consumers. This is food that could be used to feed homeless and impoverished people worldwide, but ends up in landfill or elsewhere, potentially contributing to methane emissions and global warming as it decomposes.

---

[25] https://ctl.mit.edu/pub/thesis/environmental-analysis-us-online-shopping

[26] https://earth.org/online-shopping-and-its-environmental-impact/

[27] https://www.weforum.org/agenda/2020/01/carbon-emissions-online-shopping-solutions/

[28] https://www.un.org/sustainabledevelopment/sustainable-consumption-production/

When production outstrips demand the excess product either has to be stored or disposed of. For products that are not durable, this can result in significant wastage, whilst others may become obsolescent because of fashion or updated versions becoming available.

Waste doesn't only occur due to overproduction but inaccurate estimating and overordering. This doesn't just occur in the construction industry, but in most industries that require assembly of components. Waste can occur due to scaling up from working drawings, allowing contingencies to cater for breakages and inaccuracies and 'rounding up' when ordering packets or bags of products. Items can also languish in stores for eventual disposal to waste due to overoptimistic estimate of demand.

Electronic goods such as cell phones, TVs, laptops and games consoles are key examples of products which have built-in obsolescence, sometimes because consumers are persuaded that they must keep up with technical developments, security updates are no longer supported, battery life has deteriorated, quality or reliability is poor, or maintenance and repair are difficult or impossible.

Electronic waste (e-waste) represents a major challenge because of its longevity and toxicity in the environment. Up to 59 million tonnes was generated in 2021 globally, representing a steady annual rise: with some 80–83% ending up in landfill, where it presents a potential source of contamination and fire risk. A 2019 UN report predicts that global e-waste could reach 150 million tonnes by 2050.[29] 'According to the report, in addition to health and pollution impacts, improper management of e-waste is resulting in a significant loss of scarce and valuable raw materials, such as gold, platinum, cobalt and rare earth elements. As much as 7% of the world's gold may currently be contained in e-waste, with 100 times more gold in a tonne of e-waste than in a tonne of gold ore'. Unfortunately uncontrolled recovery of precious metals from e-waste can have serious impacts on the environment and the health of those involved. 'For example, open-air burning and acid baths being used to recover valuable materials from electronic components release toxic materials leaching into the environment. These practices can also expose workers to high levels of contaminants such as lead, mercury, beryllium, thallium, cadmium and arsenic, and also brominated flame retardants (BFRs) and polychlorinated biphenyls, which can lead to irreversible health effects, including cancers, miscarriages, neurological damage and diminished IQs'.[30]

Much of the above also applies to household appliances and white goods such as cookers, fridges, freezers, washing machines, tumble dryers, dishwashers, vacuum cleaners and so on, many of which incorporate electronic components. Maintenance and repair is generally easier and life expectancies longer than for computers, TVs and cell phones but additional challenges exist, such as the safe disposal, storage and recycling of refrigerants from fridges and freezers, for example. Figure 9.4 is

[29] https://www.unep.org/news-and-stories/press-release/un-report-time-seize-opportunity-tackle-challenge-e-waste

[30] https://www.genevaenvironmentnetwork.org/resources/updates/the-growing-environmental-risks-of-e-waste/

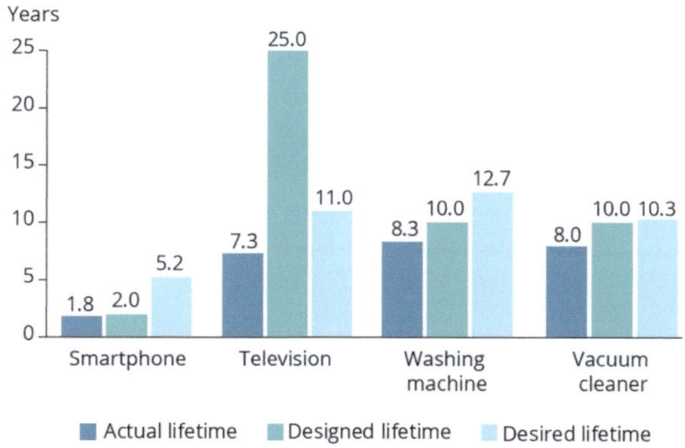

Fig. 9.4 Lifetimes for smartphones, televisions, washing machines and vacuum cleaners (Source: European Environment Agency—From: https://www.eea.europa.eu/publications/europe2019s-consumption-in-a-circular/benefits-of-longer-lasting-electronics)

taken from a European Environment Agency analysis of research that compares actual, design and desired lifetimes for four household products. It can be seen that in each case actual lifetimes are on average shorter than either design lifetimes or what people would expect or desire.[31] Remarkably TVs are expected by manufacturers to last more than three times longer than the typical gap between acquisition and disposal.

References

1. Jambeck, J., Geyer, R., Wilcox, C., Siegler, T., Perryman, M., Andrady, A., Narayan, R., & Lavender Law, K. (2015). Plastic waste inputs from land to the ocean. *Science, 347*(6223), 768–771. https://doi.org/10.1126/science.1260352
2. Schmidt, C., Krauth, T., & Wagner, T. (2017). Export of plastic debris by rivers into the sea. *Environmental Science and Technology, 51*(21), 12246–12253. https://doi.org/10.1021/acs.est.7b02368
3. De Falco, F., Di Pace, E., Cocca, M., & Avella, M. (2019). The contribution of washing processes of synthetic clothes to microplastic pollution. *Scientific Reports, 9*, 6633. https://www.nature.com/articles/s41598-019-43023-x#citeas
4. Gavigan, J., Kefela, T., Macadam-Somer, I., Suh, S., & Geyer, R. (2020). Synthetic microfiber emissions to land rival those to waterbodies and are growing. *PLoS One, 15*(9), e0237839. https://doi.org/10.1371/journal.pone.0237839
5. Evangeliou, N., Grythe, H., Klimont, Z., Heyes, C., Eckhardt, S., Lope-Aparicio, S., & Stohl, A. (2020). Atmospheric transport is a major pathway of microplastics to remote regions. *Nature Communications, 11*, 3381. https://www.nature.com/articles/s41467-020-17201-9.

[31] https://www.eea.europa.eu/publications/europe2019s-consumption-in-a-circular/benefits-of-longer-lasting-electronics

Chapter 10
Conclusion: Towards a Different Future

> *Utopia is the process of making a better world, the name for one path history can take, a dynamic, tumultuous, agonizing process, with no end.*
>
> *Kim Stanley Robinson, Pacific Edge (1990)*

10.1 Introduction

This chapter could be an entire book. Indeed there are many books that have set out manifestos for dealing with the challenges we face on this 'overburdened planet'. I am not going to attempt to write a shopping list of all of the actions required to tackle these challenges. Instead I am going to focus on one key question:

Can decoupling of economic growth from environmental degradation work as a long-term strategy?

Throughout history, humans have exploited natural resources to fuel their need to gain riches, power and territory, whilst also improving their standard of living. Prior to the Industrial Revolution, this generally meant cutting down trees and polluting rivers, but once humans started burning wood and fossil fuels on a massive scale to power the new industries spreading across the planet not only did pollution and environmental degradation become a problem but the concentration of greenhouse gases in the atmosphere grew, setting off the seemingly unstoppable journey to climate change.

P. Appleby, *The Overburdened Planet*,
https://doi.org/10.1007/978-3-032-01510-5_10

10.2 Green Growth or 'Degrowth'?

In Chap. 1, we looked at the arguments around the hunger for economic growth which has been fuelling environmental degradation since growth and industrialisation became inextricably linked. We started exploring the concept of green growth, and it is worth looking at some further definitions from international bodies before exploring how useful the concept is in attempting to develop a strategy for 'saving the planet', or whether other options such as degrowth, post-growth, ecomodernism or Doughnut Economics can be followed or adapted.

The Organisation for Economic Cooperation and Development (OECD) defines green growth as 'fostering economic growth and development whilst ensuring that natural assets continue to provide the resources and environmental services on which our well-being relies'.[1] The UN Environment Programme prefers to moot the potential for a green economy where: 'A green economy is defined as low carbon, resource efficient and socially inclusive' where growth is driven by 'investment into such economic activities, infrastructure and assets that allow reduced carbon emissions and pollution, enhanced energy and resource efficiency and prevention of the loss of biodiversity and ecosystem services'.[2]

So why the obsession with growth? Growth in Gross Domestic Product (GDP) is looked upon as a measure of success, whereas if it shrinks (for two successive quarters in the UK, for example), we are said to be entering a recession, which can be politically damaging and lead governments to take extreme measures—or fall. However, as elucidated in the extract from Robert Kennedy's wonderfully philosophical speech at the University of Kansas in 1968, quoted in Chap. 1, not all growth is 'good growth'. There are a number of factors driving the demand for growth including an increasing population and the desire for full employment, a strong welfare state and a high standard of living. All of these drivers vary enormously between and within individual nations. Confounding factors come into play such as the financing of wars or the defence therefrom, and the horrific impact these can have on civilian populations and infrastructure.

Green growth has a great deal in common with ecomodernism, which is actually a movement comprising primarily academics from universities in the USA and Australia that set out its manifesto in 2015 '… in the belief that both human prosperity and an ecologically vibrant planet are not only possible, but also inseparable. By committing to the real processes, already underway, that have begun to decouple human well-being from environmental destruction, we believe that such a future might be achieved. As such, we embrace an optimistic view towards human capacities and the future'.[3]

[1] https://issuu.com/oecd.publishing/docs/gg_brochure_2019_web

[2] https://www.unep.org/regions/asia-and-pacific/regional-initiatives/supporting-resource-efficiency/green-economy

[3] http://www.ecomodernism.org

Green growth and ecomodernism both require a thriving renewables sector, insulating homes, manufacturing electric vehicles and associated charging infrastructure rather than opening new coal mines, exploring for gas and oil, enlarging airports and manufacturing diesel sports utility vehicles. However, every new bit of kit, whether it be a new ecohome or an electric car, has a carbon footprint, increases the exploitation of irreplaceable natural resources and contributes to the negative impacts of the anthropocene era. Green growth is still growth so is degrowth feasible? Can a world whose population is still expanding entertain the potential for economies to shrink?

As can be seen in Fig. 10.1, based on data produced by the UN, the population of planet earth increased fourfold during the twentieth century, passing 8 billion in 2022 and is projected to reach 10.9 billion by the end of the twenty-first century. The rate of population increase surged dramatically following World War II but has been falling steadily since the 1970s and is expected to drop to below eighteenth century levels by the end of the twenty-first century. However projections for individual countries vary dramatically: for example, UN projections indicate that the population of Africa could rise by 145% between 2022 and 2100, whilst the Chinese population could shrink by 46% over the same period.

There are also huge disparities between the impacts that mostly prosperous nations of the Global North have had on the environment—and climate change in particular—and the threats posed to poorer nations by rising sea levels and extreme

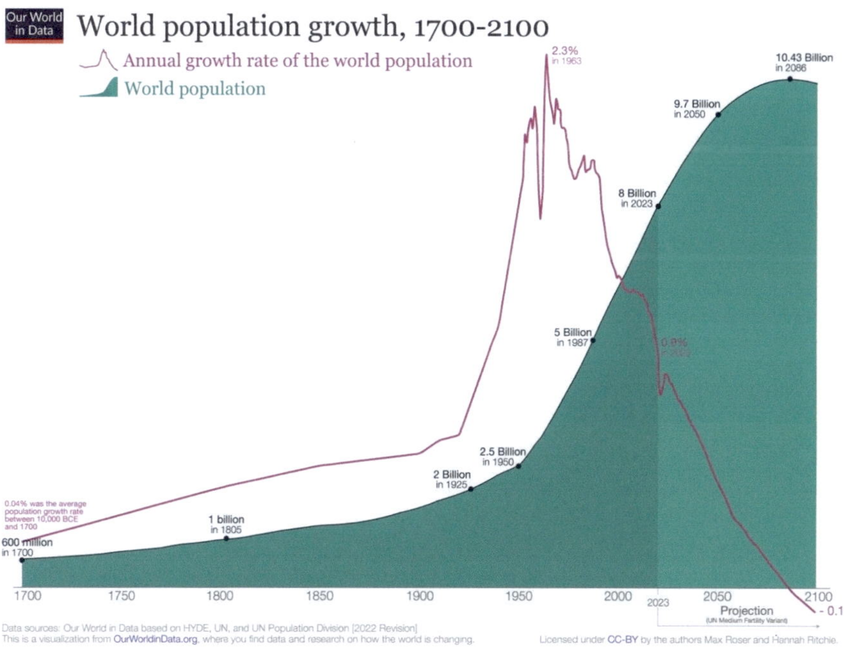

Fig. 10.1 World population growth (Source: Max Roser and Hannah Ritchie. From: https://ourworldindata.org/future-population-growth)

weather events. Many of the poorer countries in the world are low lying, in regions where storms or prolonged drought are becoming more extreme. So the annual 'Conference of the Parties' (CoP) events have not only involved negotiations to reduce carbon/GHG emissions and prepare for the impacts of climate change but also have attempted to wring out commitments from the developed nations to compensate poorer nations for past and future damage caused by climate change: a so-called Loss and Damage fund, which eventually made it onto the agenda at CoP 27 in 2022 after years of resistance from most developed countries.

Whilst the cost of the world reaching net zero carbon emissions by 2050 is predicted to be astronomical, the economic cost of rising sea levels, rising air and water temperatures and extreme weather events would be even higher—with some predicting mass extinctions, migrations and conflict. Currently much of the planet is dependent on fossil fuels for almost every aspect of daily living.

Estimates for the economic cost to the planet of reaching 'net zero' carbon emissions by 2050 range from 105 to 125 trillion dollars according to analysis carried out by McKinsey—quoted by the World Economic Forum—and the UN High-Level Climate Action Champions, respectively.[4] This represents between 3% and 3.5% of global GDP, although the impact varies considerably between individual countries. With regard to the cost of the impacts of climate change on the other hand SwissRe analysis reported in 2021 predicts that global GDP will drop by 18% if average temperatures are allowed to rise to 3.2K above pre-industrial levels, which again would not be evenly distributed globally: with, for example, China losing up to 24%, the United States 10% and Europe 11%.[5] This corresponds to around 19.8 trillion dollars annually at 2024 values.

And so we come to the concept of degrowth, not as a result of climate change, but as a strategy for reducing the emissions that cause it. It is defined on the Degrowth Collective's website as follows:

> Degrowth is an idea that critiques the global capitalist system which pursues growth at all costs, causing human exploitation and environmental destruction. The degrowth movement of activists and researchers advocates for societies that prioritize social and ecological well-being instead of corporate profits, over-production and excess consumption. This requires radical redistribution, reduction in the material size of the global economy, and a shift in common values towards care, solidarity and autonomy. Degrowth means transforming societies to ensure environmental justice and a good life for all within planetary boundaries.[6]

The concept has been around since the early 1970s—about the time of the publication of *The Limits to Growth* by Meadows et al. for the Club of Rome that we discussed in Chap. 1. However, the movement has never gained momentum: probably because no politician that wants to remain in power would run on a platform of shrinking the economy—which is implied by taking a literal view of the definition

---

[4] https://www.weforum.org/agenda/2022/01/net-zero-cost-3-5-trillion-a-year/; https://climatechampions.unfccc.int/whats-the-cost-of-net-zero-2/

[5] https://www.swissre.com/media/press-release/nr-20210422-economics-of-climate-change-risks.html

[6] https://degrowth.info/en/degrowth

of degrowth. Many poorer nations on the other hand have been attempting to exploit their natural resources or pull in investment from external actors, sometimes to line the pockets of the ruling classes and their acolytes or to improve the standard of living for their populations.

The definition of degrowth may sound idealistic—utopian even, but the concept, although relatively simple to define, has to be evaluated within the context of the real world and the scale of the ambition it represents.

In a 2020 paper Jason Hickel of the Department of Anthropology at the University of London posited the following alternative definition: 'Degrowth is a planned reduction of energy and resource use designed to bring the economy back into balance with the living world in a way that reduces inequality and improves human well-being'. He suggested that 'degrowth differs fundamentally from a recession' and that action should be 'primarily focussed on high-income nations', that is, those who caused the problem in the first place.[7] Hickel suggested that GDP should not be considered as the key metric used to judge the effectiveness of a degrowth strategy but those metrics by which harm is measured: greenhouse gas emissions, exploitation of natural resources, air quality, biodiversity, and many more. This brings degrowth closer to the concept of 'green growth' which recognises that there can be some economic growth provided it doesn't result in an increase in damage to the planet. Indeed Hickel prefers to refer to the aim of reducing 'energy and material throughput' and not shrinking GDP.

The relationship between GDP, GHG emissions and 'material footprint' is elegantly illustrated in Fig. 10.2, which has been produced by the European Environment Agency in its Sustainability Transitions series.[8] Here it can be seen that between 1970 and 2018, there was an almost identical growth rate in material footprint and total GHG emissions, with increase in mean global GDP not far behind.

Crucially Hickel's thesis sets out a strategy for degrowth that is the opposite of engineered recessions but focuses on creating low carbon and fairer economies for countries that are already wealthy whilst rethinking the relationship between developed and developing countries. Many poorer countries are sitting on a wealth of natural resources which are frequently being exploited by foreign companies—the new colonisers in fact—and exporting almost their entire output to the developed world, creating some mostly low-paid jobs, whilst making a few wealthy close to government, but without improving living standards for the population as a whole.

Others have argued for 'post-growth' or 'non-growth' strategies, using the same arguments used by the degrowth protagonists, but with slightly different aims. Post-growth foresees the end of growth per se, but followed by a period of non-growth or degrowth, which could of course be sequential.

Fundamental to the success of degrowth, post-growth, non-growth, green growth or ecomodernism strategies is that banking, welfare, public transport, utilities, local government, and so on must remain viable. Governments must be able to balance

[7] https://www.tandfonline.com/doi/full/10.1080/14747731.2020.1812222

[8] https://www.eea.europa.eu/publications/growth-without-economic-growth/file

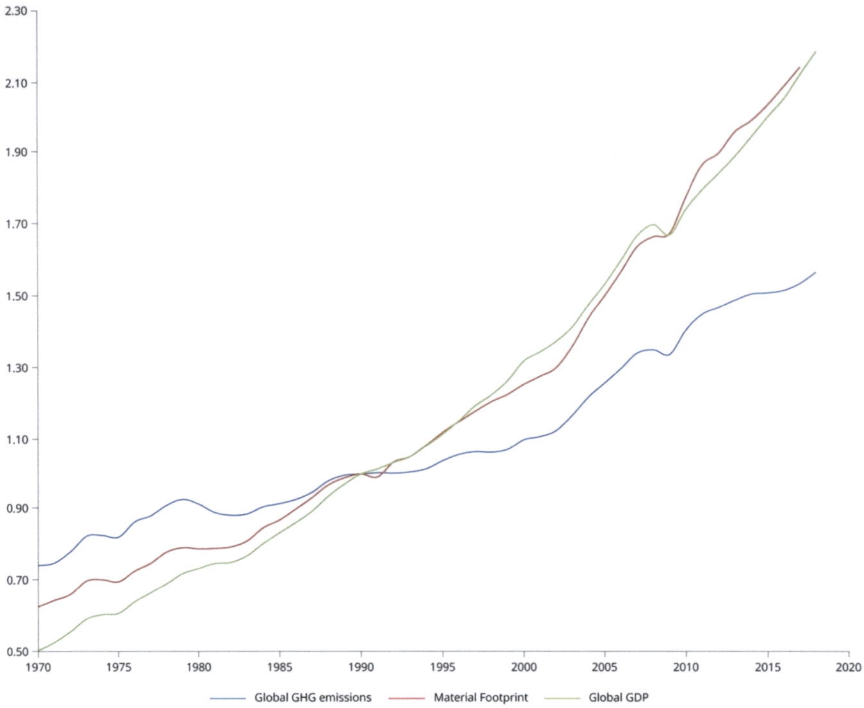

Fig. 10.2 Relative change in main global economic and environmental indicators from 1970 to 2018 (https://www.eea.europa.eu/publications/growth-without-economic-growth/file)

the books without jeopardising their ability to borrow on the financial markets and control inflation without increased unemployment, poverty, homelessness, crime and other indicators of failed governance. These issues are addressed by Kate Raworth in her 'Doughnut economics' concept. This takes the planetary boundaries developed by Rockstrom which we discussed in Chap. 2 (see Fig. 2.1) and proposes a comfort zone ('the safe and just space for humanity') described by the ring of a doughnut between the not to be exceeded 9 planetary boundaries (Ecological Ceiling) and 12 social criteria derived from the UN's Sustainable Development Goals (Social Foundation) (see Fig. 10.3).[9] Although it is worth noting the Sustainable Development Goal is headed 'Decent Work and Economic Growth' to 'promote inclusive and sustainable economic growth, employment and decent work for all', which creates a potential conflict with the post-growth agenda.

Although all of these strategies have been the subject of extensive research, conferences and publications for many years, it is interesting to note that the IPCC has recognised their importance in the armoury of mitigation strategies against climate change set out in the Working Group III contribution to its 6th Assessment report

[9] https://www.kateraworth.com/doughnut/

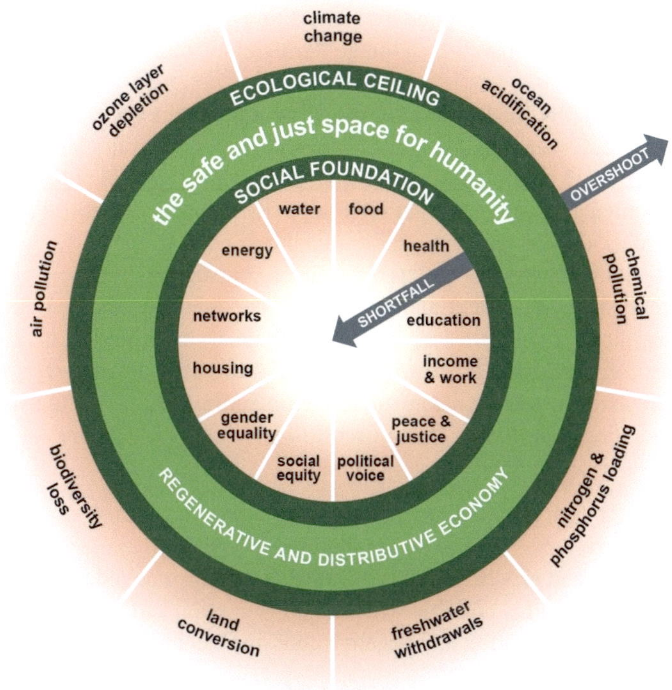

Fig. 10.3 Doughnut economics model (Source: Doughnut Economics. From: https://commons.wikimedia.org/wiki/File:Doughnut_(economic_model).jpg)

published in 2022. Reference is made to degrowth and post-growth throughout the report. For example, the authors state that 'several studies find that only a GDP non-growth/degrowth or post-growth approach allow to reach climate stabilisation below 2°C'.[10] Looking at the relationship between growth and well-being, Chap. 1 of the report refers to 'Literature on degrowth, post-growth and post-development (which) questions the sustainability and imperative of more growth especially in already industrialised countries and argues that prosperity and the "Good Life" are not immutably tied to economic growth'. In what appears to be an endorsement of degrowth as an important component of mitigation strategies against climate change Chap. 5 of this report suggests that 'Consumption reductions, both voluntary and policy-induced, can have positive and double-dividend effects on efficiency as well as reductions in energy and materials use'. The WG III authors refer to 'systems-dynamics models linking strong emissions-reducing policies and strong social equity policies (which) show that a low-carbon transition in conjunction with social sustainability is possible, even without economic growth. Such degrowth pathways

[10] https://www.ipcc.ch/report/ar6/wg3/downloads/report/IPCC_AR6_WGIII_Full_Report.pdf

may be crucial in combining technical feasibility of mitigation with social development goals.' Doughnut economics is referred to as one of the 'development pathways that take sustainability, climate mitigation and environmental limits seriously'.[11]

One 2021 study by Keyßer and Lenzen, referred to in the IPCC report, has noted that all of the 222 scenarios for which integrated assessment models (IAMs) have been run for the 1.5 °C global warming limit agreed by all parties at CoP 21 in Paris in 2015 assumed a steady growth in GDP globally and hence required the development of ambitious and unproven carbon capture and storage (CCS) technologies associated with bioenergy as well as extensive reforestation.[12]

To demonstrate the potential impact of a degrowth/post-growth strategy on global warming, the Heinrich Böll Foundation has produced a *Societal Transformation Scenario for Staying Below 1.5 °C*.[13] This detailed report uses the Global Calculator to predict mean surface temperature, which is a transparent and simple modelling tool that has been used to factor in a whole series of behaviour modifications based on the following premises:

1. The Global North (UNCCC Annex 1 countries) bears primary responsibility for climate change.
2. Consumption and production in the Global North must be reduced.
3. A 'good life' for all is possible with less consumption and production.
4. Nuclear power generation and negative emissions technologies are not required.

The main areas of behaviour change for which metrics are proposed for modelling are transport, housing and food, with different assumptions for Annex 1 (developed) and non-Annex 1 (developing) countries and assuming a steady transformation through to 2050, which is the end date for the Global Calculator climate model. The metrics for transport assume a reduction in road-based journeys and flights per capita, including reductions in 'ground freight transport'. It is also assumed that household densities increase and appliance use reduces whilst measures are introduced to reduce food waste, diets move away from ruminant meats and there will be a trend towards healthier diets.

The parallel body to the IPCC, UN's Intergovernmental Science-Policy Platform on Biodiversity and Ecosystem Services (IPBES) published a study in 2020 on *Biodiversity policy beyond economic growth* based on a collaboration between researchers from 12 countries which concluded that 'economic growth is incompatible with biodiversity conservation'.[14] The authors suggested a series of global strategies 'to ensure prosperity beyond growth and halt the loss of biodiversity'. These include some broad brush proposals such as:

---

[11] https://www.eea.europa.eu/publications/growth-without-economic-growth/file

[12] https://www.nature.com/articles/s41467-021-22884-9

[13] https://eu.boell.org/sites/default/files/2020-12/A%20Societal%20Transformation%20Scenario%20for%20Staying%20Below%201.5C.pdf

[14] https://ipbes.net/node/36960

- The imposition of a cap on land use and resource exploitation allocated on a country by country basis;
- Place limits on and remove subsidies from unsustainable extractive industries;
- Rein in extensive infrastructure expansion;
- Promote reductions in the length of the working week with options such as work sharing;
- Support sustainable agriculture;
- Encourage integrated collective housing and reduce urbanisation;
- Tax unsustainable products and promote sustainable consumption.

The initiatives proposed by the Heinrich Böll Foundation and IPBES would seem to require state intervention on a massive scale globally, focused not only on technical fixes but also on behaviour change requiring political and societal adjustments.

The counter argument to the degrowth and post-growth movements has been that, apart from being difficult to implement globally, degrowth would lead to problems with societal well-being. Indeed it is the case that to facilitate a transition to a post-growth world, it would be essential to gain wide support amongst the populations of those countries most impacted. In a report published in 2022 by Lily Paulson and Milena Büchs of the University of Leeds 60.5% of the nearly 50,000 people interviewed in a European Values Survey covering 34 European countries were broadly supportive of post-growth defined by an improved environment with a smaller economy. However, the nature of support was different between socio-economic groups, demonstrating the need for 'greater emphasis on redistribution and improving opportunities and livelihoods for disadvantaged people in a post-growth economy'… 'However, this conflicts with policy preferences and values such as hierarchy, meritocracy and individualism that tend to be more prominent among people who are well-off'.[15]

During the Post-Growth Conference at the European Parliament in October 2018 Tim Jackson responded to a question which suggested that post-growth had not been taken up because 'no social actor has an interest in it' by stating that 'inequality between rich and poor, the instability that inequality causes, and the further instability of a financial system created to push for growth against fundamental forces such as declining labour productivity, simply end up making the whole system and its politics more unstable. Then you realise every social actor has an interest in post-growth'.[16]

It is important to recognise that the term 'degrowth' is politically loaded and some of the literature exploring its potential is expounding a bottom-up revolution on the assumption that the establishment—politicians, industry, vested interests—will not take the steps required without external pressure. There does appear to be quite an extensive degrowth movement which, although resonating with Green

[15] https://www.sciencedirect.com/science/article/pii/S0016328722001203

[16] https://www.greeneuropeanjournal.eu/the-politics-of-post-growth/

politics across Europe in particular, has not developed beyond the academic and theoretical.

Post-growth on the other hand, although overlapping with degrowth, has a different community and appears to have gained traction with the European Union, for example, through the auspices of the European Research Council which has funded research in post-growth policies[17] and a separate PROSPERA project 'examining the role of innovation in a post-growth era'.

In parallel with this Doughnut Economics has been developing within communities, specifically in individual cities. Some of these—Melbourne, Brussels and Amsterdam, for example—have had short-term funding and local political support, whereas initiatives in Sydney, Berlin and Cambridge are community-based and run by local volunteers. Although there has been a degree of oversight from the Doughnut Economics Action Lab (DEAL) each of these local initiatives has been focussing on local issues, such as Melbourne's drive for a circular economy and zero waste.

All of these strategies have evolved through the promulgation of ideas, generally developed and explored by academics and researchers across a range of disciplines, but with the intention of building sufficient pressure and momentum to influence and persuade those actors that can make a difference.

10.3 The Way Forward

The previous chapters of this book have attempted to paint a picture of the stressors that have led to the situation in which this planet finds itself in the first quarter of the twenty-first century, whilst hinting at some of the measures that are available to help mitigate and adapt to these. Every year that passes makes these more difficult, particularly as the population continues to grow and we approach key tipping points that will have irreversible impacts on climate and life on this planet as we know it.

We have seen that there is a myriad of actions that could reduce or eliminate the causes of environmental damage to this planet; however, the time lag between cause and beneficial effect varies enormously. For example, it has been estimated that due to inertia in the physical climate system that for global mean surface temperatures "time to detection is not shorter than 25 years after emission mitigation starts".[18] On the other hand pollution mitigation at a local level, such as the measures taken in Beijing for the 2008 Olympics, can be very rapid. The temporary closure of factories and construction projects as well as controls of traffic movements in the months leading up to the Games resulted in pollution concentrations as low as 20% of normal.

---

[17] https://www.uab.cat/web/sala-de-premsa-icta-uab/detall-noticia/european-project-to-explore-pathways-towards-post-growth-economics-1345819915004.html?detid=1345872411651

[18] https://www.pnas.org/doi/10.1073/pnas.1300005110#sec-3

To get to a post-growth world, resistance to change would clearly add to this inertia. Indeed without a major change in world and domestic politics, many countries will continue to target growth in GDP as the holy grail. For the rest of this chapter, we will look at some of the implications of a post-growth strategy and how it might be implemented.

It is self-evident that the biggest issue facing humanity is climate change and one of the primary drivers of post-growth economics is to reduce energy demand and greenhouse gas emissions. Although the post-growth strategy may not specifically refer to greenhouse gas or carbon dioxide emissions, each element will have an impact on emissions, whilst also addressing other issues such as air pollution, biodiversity, health, quality of life and socio-economic sustainability.

We saw earlier in this chapter that the European Union is funding research into post-growth policies. However, the European Green Deal, launched in 2020, is essentially advocating economic growth decoupled from resource use—or essentially 'green growth'—whilst targeting carbon neutrality for EU countries by 2050. The associated industrial strategy is primarily concerned with reducing the environmental impact and carbon footprint of industry through green technologies without addressing demand and consumption.

Not addressing demand and consumption is like attempting to reduce water consumption by slowing the flow of water into a leaking bucket without blocking the holes. Demand reduction requires a multi-pronged attack across practically all aspects of our lives: buildings and construction, transport, industry, agriculture, mining. All have associated greenhouse gas emissions and have some relationship with both economic and population growth.

The International Energy Agency (IEA) has used its complex Global Energy & Climate (GEC) Model to build a strategy for achieving Net Zero Emissions by 2050. The strategy they outline in a 2022 report is highly dependent on emerging technologies, acceptance of behaviour change, availability of sustainable bioenergy, international collaboration and an 'orderly transition' away from fossil fuels, which means avoiding volatility in energy markets[19]—a tall order considering ongoing conflicts in Ukraine, the Middle East and elsewhere.

In this context in 2022, the EU launched a *voluntary framework for certification of high quality carbon removals* reinforced by a regulation adopted in 2024[20] including bioenergy with carbon capture and storage (BECCS) and direct air CCS, whilst also encouraging the use of timber in construction and 'carbon farming'—also referred to as regenerative agriculture—as discussed in Chap. 9. Carbon farming involves practices that enhance the carbon uptake of soils, with the potential for an

[19] https://iea.blob.core.windows.net/assets/2db1f4ab-85c0-4dd0-9a57-32e542556a49/GlobalEnergyandClimateModelDocumentation2022.pdf

[20] https://climate.ec.europa.eu/eu-action/carbon-removals-and-carbon-farming_en#:~:text=On%2010%20April%202024%2C%20the,storage%20in%20products%20across%20Europe

income for farmers from carbon offsets, whilst also improving fertility, conserving water and reducing harm to the environment.[21]

Although there may be a disagreement about the role of bioenergy in a decarbonisation strategy, there is clearly some overlap between the post-growth strategy and those postulated by the IEA and EU, although the balance between technological mitigation and changes in behaviour are where the main differences lie. Furthermore, the IEA model assumes an average annual global growth in GDP of 2.8% between 2021 and 2050 with a population growth rate of 0.7% per annum, reaching 9.7 billion by 2050, as per the latest predictions from the UN shown in Fig. 10.1.

One of the problems for those promoting zero or degrowth of GDP is the long established convention of using per capita GDP as an indicator of living standards. It has been argued that the figure for GDP used should be adjusted for inflation, referred to as ‘Real GDP’, giving an indication of average purchasing power. In 1990, the UN published the Human Development Index (HDI) which takes into account life expectancy, amount of schooling (expected and achieved) and per capita Gross National Income (GNI), normalised to provide a rating out of 1.0. Unsurprisingly Northern European countries and the developed world in general had HDI scores between 0.9 and 1.0, whereas sub-Saharan African countries were typically below 0.5.[22] The mean HDI globally rose from 0.6 in 1990 to 0.732 in 2022, with most countries experiencing a similar rate of increase.

It is clear that HDI needs to continue growing for the poorest countries, whilst it could be argued that HDI for the developed world can afford to remain steady, provided that the gap between the wealthy and impoverished in those countries is reduced. It is vital that any increase in living standards is not associated with greater carbon emissions and environmental impact. This can only happen with a massive transition to renewable energy globally. This in turn requires the development of a sustainable supply chain and manufacturing industry for batteries: both for electric vehicles and for storage to cater for periods when there is insufficient electricity generation from wind turbines and solar farms. The mining and processing of the minerals currently required for batteries presents a major environmental, social and geopolitical challenge.

Because nuclear power does not emit greenhouse gases as part of the electricity generation process it is increasingly being considered as an important component of a decarbonisation strategy. It is important to note that as of March 2023, there were 60 nuclear reactors due for completion before 2028, with another 100 ordered or planned and 300 proposed. More than one third of those under construction were in China with 8 in India, 4 in Turkey, 3 in Russia, 2 in the USA and 2 in the UK.[23]

---

[21] https://www.carboncycle.org/what-is-carbon-farming/

[22] https://hdr.undp.org/data-center/human-development-index#/indicies/HDI

[23] https://world-nuclear.org/information-library/current-and-future-generation/plans-for-new-reactors-worldwide.aspx

A nuclear power station typically takes between 5 and 10 years to build, however, prior to construction the design and planning process can take decades, particular if there are major issues with funding, planning and local resistance. Disasters such as that caused by the tsunami that hit the Japanese Fukushima nuclear facility in 2011 have resulted in a decrease in the popularity of nuclear power, and in some cases, the closure of existing plants or the abandonment of construction programmes. Following Fukushima, Japan suspended operations in 46 of its 50 reactors, whilst in Europe, Germany permanently shut down 11 of its 17 reactors, Belgium committed to exit its nuclear facilities by 2025 and Spain, Italy and Switzerland abandoned long planned projects.

However, according to the World Nuclear Association 'Today (in 2023) there are about 440 nuclear power reactors operating in 32 countries plus Taiwan, with a combined capacity of about 390 GWe. In 2021 these provided 2653 TWh, about 10% of the world's electricity.' The IEA's Net Zero Emissions by 2050 strategy referred to above anticipates that, despite the repercussions of Fukushima, nuclear energy generation should increase to 871 GWe by 2050. The construction of each new nuclear reactor represents a massive civil engineering project requiring around 200,000 tonnes of mostly high density concrete and 40,000 tonnes of steel. The impact on the local environment can be enormous since most are built on rural and/or coastal sites and away from centres of population for safety reasons. Along with the issues surrounding the processing and storage of radioactive waste and spent fuel generated by the electricity generation process, the argument for continuing with a programme of constructing new nuclear power stations is controversial.

So is it possible to provide a stable and reliable electricity supply without fossil fuel or nuclear power stations? Recognising that wind and solar resources are variable it has been argued that in the absence of fossil fuels, nuclear power should be used to provide baseline electricity generation to cater for periods of low wind and solar power. However, there are a number of alternative methods of ensuring a supply that matches demand. Hydroelectric schemes are well-established where topography allows, some of which incorporate pumped storage to cater for imbalances between demand and supply. Tidal power installations have been operating for some years in France and South Korea, with some very large projects planned for Russia, and potentially for the Severn estuary in the UK. As of 2021, pumped storage hydropower installations were generating some 160 GWe, whilst tidal power contributed only around 0.5 GWe, although planned projects could add a further 128 GWe.

In 2021, large-scale battery storage installations were able to store 16 GWe globally, whilst the IEA's Net Zero scenario suggests that 680 GWe is required by 2030. The majority of installations to date have used lithium ion batteries but, with the issues surrounding lithium and other materials required in their manufacture, alternatives are being explored such as flow batteries, which are experiencing a renaissance having been originally patented in the US in 1879. The vanadium redox flow battery (VRFB), in particular, is gaining in popularity, with a 100 MWe installation having been commissioned in Dalian, China, in 2022 being the largest to date. These batteries are only suitable for large installations and use vanadium pentoxide as an electrolyte in a sulphuric acid solution, Vanadium is a mineral found in deposits with other minerals, there being large deposits in China, Russia and South Africa.

The problem with variability of output from wind and solar farms can also be alleviated by using another well-established technology—the synchronous condenser, which is essentially an electric motor without a load or an electric generator without an engine or turbine. They are used to stabilise an electricity grid through continuous reactive power control. Reactive power occurs when demand does not match supply, such as when there is significant reliance on weather-dependant renewables.

The synchronous condenser 'provides improved voltage regulation and stability by continuously generating/absorbing adjustable reactive power as well as improved short-circuit strength and frequency stability by providing synchronous inertia'.[24]

Regardless of whether there is wholesale adoption of a post-growth strategy or a continuation of the status quo there will have to be major investment by both governments and private sector investors in mitigation of climate change, biodiversity loss, air pollution and other impacts, as well as adaptation to climate change. In its 2021 Race to Zero report referred to earlier the UN High-Level Climate Action Champions put forward an investment programme for 17 opportunities, covering a range of sectors, regions and actors under four headings:

1. Early technology bets
2. Maturing technologies in emerging regions
3. Market creation opportunities
4. Established investment opportunities

Early technology bets included off-grid power and solar photovoltaics in Africa, and electricity networks in Central and South America. Maturing technologies in emerging regions included green steel, alternative proteins and chemicals electrification in Asia Pacific and green hydrogen globally. Market creation included EV charging and wind power in Europe and North America, solar PV in the Middle East and electricity storage globally. Finally, established investment opportunities included building retrofits in the Middle East and Eurasia, forest, mangrove and peatland restoration in Central and South America and Biomethane globally.[25] It is self-evident that most of these opportunities apply globally, but the list focuses on those regions where the need is most immediate and open. The economic activity required to meet these opportunities will clearly be great and unlikely to be evenly distributed; however, the hope is that there will be a commensurate disinvestment in polluting and carbon intensive activities such as fossil fuel extraction and processing.

[24] https://www.entsoe.eu/Technopedia/techsheets/synchronous-condenser

[25] https://climatechampions.unfccc.int/whats-the-cost-of-net-zero-2/

International bodies such as the United Nations and The International Energy Agency have been attempting to emphasise the immediacy of the challenge faced by the human race as a whole, but it is up to governments and large international corporations, primarily in the developed world, to act and act urgently if the catastrophic events already unfolding are not to become irreversible.

Index

A

Absorption chiller, 178, 242
Acidification, 17, 65, 67, 108, 116, 119–122, 151, 195, 196, 260
Active Network Management (ANM), 182
Afforestation, 89, 90
Agriculture, forestry and land use, 338
Aichi biodiversity targets, 23, 24, 217, 225, 229
Aichi targets, 24, 225, 277
Air conditioning, 141, 160, 190, 234, 235, 239–248264
Air handling, 242, 245, 248
Air handling plant, 245, 248
Air handling system, 245
Air quality, 2, 38, 40, 97, 158, 233, 236, 246, 251, 264–268, 286, 295, 305–308, 312, 355
Air-source heat pump, 257, 294
Air traffic, 300, 301
Albedo, 112, 134
Algal bloom, 67, 120, 152, 313, 342
Antarctic ice sheet, 111–114
Antibiotic resistance, 342
Aquaculture, 149–152, 315
Aquifer, 45, 47, 67, 185, 258
Aquifer thermal energy storage (ATES), 258, 259
Artificial lighting, 236, 248, 249, 314
Atlantic Meridional Overturning Circulation (AMOC), 113, 114
Atmosphere, 8, 20, 42, 43, 48, 50, 51, 62, 66, 81, 91–98, 100–102, 119, 122, 123, 126–128, 134, 145, 156, 163, 167, 186, 190, 191, 194–196, 216, 231–233, 260, 304, 335, 336, 351
Aural environment, 264, 265, 267, 309

B

Ballast, 69, 249, 314–316
Bangladesh, 7, 55, 56, 136, 345
Battery storage, 363
Battery technology, 318–321
Bhopal, 335, 336
Bike parking, 325
Biocapacity, 214–217
Biodiversity, 17–20, 22–24, 48, 65–67, 70, 74, 75, 83, 86–90, 109, 117, 121, 137, 152, 195, 213, 214, 217, 225–229, 274, 276–285, 290, 295, 314, 315, 338, 342, 344, 352, 355, 358, 361, 364
Biodiversity Benchmark, 226
Biodiversity net gain, 277
Biofouling, 315, 316
Biofuel, 48, 228, 342
Biomass, 40, 41, 98, 152, 168, 174, 178, 189, 194, 201, 202, 208, 250, 251, 257, 294
Biome, 122, 126, 131, 217
BioRegional eco-footprint, 219
Biosphere, 17, 24, 86, 126–129, 195, 208, 209, 216
Black carbon, 97, 113, 195, 304, 305
Brownfield site, 234, 278, 295

P. Appleby, *The Overburdened Planet*,
https://doi.org/10.1007/978-3-032-01510-5

Building Research Establishment Environmental Assessment Method (BREEAM), 213, 215, 226, 261, 278, 288, 290
Built-in obsolescence, 78, 348

C
Carbon and energy database, 262
Carbon capture and storage (CCS), 185–189, 358, 361
Carbon cycles, 122, 137, 152
Carbon dioxide, 17, 42, 49, 62, 66–69, 71, 72, 81, 92, 94–100, 103, 106–108, 111, 116, 117, 119–124, 126–128, 136, 148, 155, 163, 164, 171, 173, 176, 185, 186, 188, 191, 193–195, 198, 213, 214, 216, 217, 219, 220, 232, 236, 238, 249, 250, 263, 277, 291, 299, 304, 336, 338, 361
Carbon farming, 342, 361
Carbon footprints, 69, 191, 213–229, 235, 250, 262, 263, 270, 291, 295, 304, 338, 340–342, 347, 353, 361
Carbon sequestration, 72, 194
Carbon sink, 98, 128, 193
Carter doctrine, 26, 27
Ceiling diffuser, 246, 247
Cement, 147, 188, 191, 203, 232, 262–264, 330, 336
Central heating, 240, 258
Charging infrastructure, 317, 321–322
Charging points, 317, 318, 321, 322
Chernobyl, 38
Chilled beams, 241
Circular economy, 329, 360
Climate Action Champions, 354, 364
Climate change, 3, 4, 10, 13–15, 17–22, 24, 25, 48–50, 65–67, 69, 73, 74, 76, 86, 87, 91–160, 164, 172, 190, 195, 206, 228, 229, 234, 278, 291, 304, 330, 351, 353, 354, 356–358, 361, 364
Club of Rome, 2, 3, 13, 15, 17, 198, 354
Coanda effect, 246
Cobalt, 204, 319–321, 348
CO_2 emissions, 5, 16, 98, 114, 191, 193, 201, 202, 217, 233, 234, 238, 249, 263, 304, 336–338
Cogeneration, 177–178, 240, 259
Colony collapse, 73
Combined heat and power (CHP), 85, 177, 178, 180, 240, 250, 251, 259, 260
Community energy, 180, 181, 183, 185
Commuting, 294, 300–303
Condensing boiler, 239, 240
Conference of the Parties (CoP), 19, 23, 25, 26, 174, 202, 203, 258, 354, 358
Congestion charge, 302, 325
Considerate Constructors Scheme (CCS), 289, 290
Contracts for Difference (CfD), 183, 184
Cooling tower, 242–245, 270, 272
Coral bleaching, 116
Coral reef, 24, 67, 116, 118, 119, 150, 168, 316, 332
Corals, 24, 57, 67, 69, 116, 118–120, 150, 168, 315, 316, 332
Coriolis effect, 50
Cradle to grave, 11, 74, 191, 219, 231, 264, 329, 344
Critical raw materials (CRM), 203, 204
CROPWAT, 223
Crystalline PV, 253, 254
Cycling, 68, 137, 175, 196, 218, 233, 317, 324–326

D
Dark snow, 112, 134
Daylighting, 236, 254, 293
Decarbonisation, 181, 182, 234, 239, 257, 362
Decency standard, 291, 292
Decoupling, 5, 10, 11, 351
Deepwater Horizon, 168
Deforestation, 15, 16, 25, 49, 56, 89, 127, 128, 152, 153, 207, 225, 330, 340, 341
Degrowth, 11, 352–360, 362
Delta Works, 58
Democratic Republic of Congo (DRC), 34, 36, 204, 319
Dengue fever, 115, 130
Desalination, 76, 178, 269
Desertification, 20, 22, 65, 89, 128, 129, 142, 209
Detention basin, 282, 286
Diarrhoeal disease, 35, 154, 155
Dichlorodiphenyltrichloroethane (DDT), 6, 64, 65, 72, 73, 334, 335
Diesel, 48, 260, 304–308, 310, 317, 318, 322, 323, 353
Dioxins, 39, 40, 80, 81, 85, 334, 343
Direct expansion refrigeration, 242
Displacement ventilation, 246, 247
Distribution losses, 176
Distribution Network Operator, 182
District heating, 85, 178, 179, 258, 259
Domestic hot water, 178, 239, 240, 256

Drinking water, 2, 35, 37, 45–47, 56, 75, 158, 159, 210, 211, 268, 269
Doughnut economics, 352, 356–358, 360
Drought, 20, 47, 59–60, 67, 75–77, 89, 109, 125, 126, 128, 129, 133–136, 138, 141–145, 152, 153, 209, 210, 218, 224, 268, 354
Drylands, 128, 129, 138, 141–144
Due diligence, 34, 339
Dust bowl, 59–61, 89, 90
Dust storm, 59–61, 89, 138, 142, 143

E
Earthquake, 43, 45, 58, 140, 278
Ecological ceiling, 356
Ecological footprint, 11, 213–219, 224, 225, 338
Ecomodernism, 352, 353, 355
Economic growth, 1–6, 8, 11, 14, 22, 86, 102, 130, 154, 155, 172, 192, 329, 340, 351, 352, 355–358, 361
Ecosystem degradation, 69, 86, 128
Ecosystem services, 22–24, 65, 74, 86, 87, 118, 137, 168, 214, 217, 225, 229, 352
Electric vehicles (EV), 204, 317–319, 321, 322, 324, 353, 362, 364
Electronic waste (e-waste), 80, 82, 348
El Niño, 51, 52, 109, 116, 128, 157, 208
Embodied carbon, 189, 191, 232, 234, 260–264, 270, 292, 343
Embodied impacts, 231, 232, 262
Energy Charter Treaty, 203
Energy cooperative, 181, 184
Energy efficiency, 28, 38, 180, 199, 200, 202, 248, 249, 291, 292
Energy from waste, 38, 39, 77, 84, 85
Energy-intensive industry, 191
Energy security, 28, 181, 199, 200
Enhanced oil recovery (EOR), 186, 188
Environmental degradation, 1, 11, 351, 352
Environmental impacts, 9, 11, 21, 37–48, 70, 73, 74, 79, 160, 168, 169, 191, 200, 212, 216, 223–225, 231, 260–264, 273, 289, 290, 301, 305–317, 319, 326, 329, 338, 339, 344–346, 361, 362
Environmentally persistent waste, 330–331
Environmental Product Declarations (EPD), 261
Eutrophication, 57, 66–68, 70, 120, 152, 195, 260, 313, 342
Exhaust emissions, 305, 306, 308

F
Fabric first, 234
Fashion, 40, 329, 344–346, 348
Fast breeder reactor, 170
Feed-in tariff, 183, 294
Fisheries, 23, 65, 67, 69, 116, 121, 149–152, 216, 312, 315, 316
Fishing, 7, 65–67, 69, 116, 118, 122, 146, 149–151, 214, 331, 332
Flood, 10, 53–58, 132, 135, 136, 138, 140, 278, 279, 282
Flood defences, 58
Flooding, 7, 10, 14, 15, 51–58, 109, 114, 115, 132, 135, 140, 152, 160, 186, 209, 218, 278, 279, 282
Food insecurity, 65, 86, 148
Flood risk, 55, 57, 135, 139–141, 233, 277–282, 342
Flood zone, 278, 279
Food miles, 340–342
Food production, 11, 13, 15, 48, 64–75, 142, 148, 150, 195, 228, 231, 232, 330, 342
Forest fire, 60, 207
Forest Stewardship Council (FSC), 344
Fossil fuels, 8, 10, 11, 14, 17, 26, 43, 48, 97, 100, 102, 120, 163–165, 167, 171–174, 185, 197–204, 233, 251, 257, 260, 300, 317–324, 330, 334, 335, 337, 346, 351, 354, 361, 363, 364
Fracking, 17, 41, 43–48, 169
Freon gas, 100
Freshwater, 9, 15, 17, 21, 56, 58, 115, 116, 121, 125–127, 132, 146, 152, 154, 195, 209, 217, 221, 224–226, 268, 269
Fuel cell, 260, 318, 322–324
Fugitive emission, 42, 163, 220
Future Homes Standard, 234

G
Gas, 8, 10, 11, 17, 25–31, 33, 34, 38–43, 45–48, 81, 92, 94–96, 98, 100, 101, 127, 147, 151, 156, 163, 165, 167–170, 172, 174–178, 185, 187, 189, 191, 194, 196, 198–200, 202, 217, 232, 239, 240, 243, 250, 251, 257, 258, 260, 264, 270, 300, 307, 312, 322, 323, 329, 334, 336, 338, 343, 351, 353
Gas-fired boiler, 250, 260
Genetically modified organisms (GMOs), 74

Geothermal, 178, 180, 187, 250, 258
Geothermal energy, 250, 258
Glass, 239, 253, 254, 256, 338
Glass manufacture, 338
Global Calculator Climate Model, 358
Global Footprint Network (GFN), 214–217, 219
Global overshoot, 216
Global warming, 10, 42, 53, 62, 92–109, 111, 125, 126, 134, 138, 144, 156, 163, 189, 190, 304, 333, 347, 358
Global warming potentials (GWPs), 42, 49, 66, 95–98, 163, 190, 191, 193, 194, 219, 242, 251, 261, 336
The Great Smog, 62
Green belt, 11, 295
Green Building Council, 4
Green Deal, 4, 294, 317, 342
Greenfield site, 82, 234, 287
Green growth, 4, 11, 23, 329, 352–361
Green Guide, 261
Greenhouse effect, 92–109
Greenhouse gas (GHG), 16, 25, 65, 66, 71, 72, 81, 92, 94–103, 106–108, 110, 116, 117, 123, 128, 132, 133, 145, 152, 163–196, 206, 217, 219, 220, 229, 233–264, 299, 300, 304, 329, 330, 333, 336–338, 340, 341, 351, 361, 362
Greenhouse gas (GHG) emissions, 11, 16, 25, 65, 66, 71, 72, 81, 95, 96, 99, 101, 102, 110, 117, 128, 132, 152, 163, 164, 168, 172–174, 184, 189–193, 195, 202, 219–221, 232, 234, 249, 262, 299–301, 304, 329, 330, 333, 334, 336–338, 340–343, 346, 354, 355, 361
Greenhouse Gas (GHG) Protocol, 219, 220
Greenland ice sheet, 111–113
Green roof, 283–285
Grey water, 221, 223, 224, 264, 272–276
Gross domestic product (GDP), 3–5, 14, 27, 130, 139, 140, 148, 154, 171, 302, 303, 352, 354, 355, 357, 358, 361, 362
Ground source heat pump (GSHP), 242, 250, 257, 258

H

Habitat loss, 15, 24, 86, 313, 314
Heating, ventilation and air conditioning (HVAC), 234, 239–248
Heat island, 52, 141, 233
Heat network, 180, 181, 251
Heat pump, 239–242, 248, 250, 251, 257–259, 294
Heat stress, 131, 142, 148, 153
Heat waves, 133, 134, 136, 141–145, 156
Home Quality Mark, 215
Housing Health & Safety Rating System (HHSRS), 292
Human Development Index (HDI), 218, 362
Humidifier, 245, 272
Hurricane, 37, 49, 51, 53, 57, 67, 114, 134, 139, 140, 160
Hydraulic fracturing, 17, 43–48, 169
Hydroelectric, 8, 9, 173, 175, 363
Hydrogen, 98, 170, 269, 317, 318, 322–324, 334, 337, 364
Hypoxia, 67, 68, 120, 122, 313

I

Illegal dumping, 81, 82, 84
Illustrative Mitigation Pathways (IMPs), 102, 103, 105, 106
In-store shopping, 347
Intensive rearing, 72
Intergovernmental Panel on Climate Change (IPCC), 14, 16, 19, 21, 25, 49, 65, 75, 95, 97–99, 101, 102, 104, 106, 108–113, 116, 119–122, 124–129, 131–133, 135–142, 148, 150–153, 155–160, 164–166, 171, 188–190, 192, 193, 195, 356, 358
Invasive species, 24, 67, 69, 75, 86, 87, 127, 229, 315, 316
Investor State Dispute Settlement (ISDS), 202

J

Jet streams, 50–53, 63, 64, 107, 145

K

Kelp, 117, 119
Kyoto Protocol, 19, 25, 174, 219, 220

L

Landfill, 6, 39, 40, 77–80, 83, 85, 146, 329, 330, 332, 347, 348
Landfill site, 78–80, 83, 84, 146, 192
Landfill tax, 83, 84
Landslide, 9, 88, 140, 141, 147, 152

La Niña, 51, 52
Leadership in Energy and Environmental Design (LEED), 213, 288, 290
Life cycle assessment (LCA), 224, 225, 261, 262
Light emitting diode (LED), 249
Light emitting diode (LED) lamps, 248, 249
Light water reactor, 35, 170
Liquefied natural gas, 177, 202
Liquefied petroleum gas, 30
Lithium (Li), 204, 219, 220, 242, 363
Lithium ion (Li-ion) battery, 204, 319, 320, 363
Livestock, 66, 69–72, 92, 129–131, 142, 148, 193, 214, 223, 300, 335, 340, 342, 345
Living Planet Index (LPI), 225–227
Load following, 174, 175
Luminous efficacy, 249

M
Madagascar, 23, 87, 88, 225, 226
Malaria, 21, 72, 130, 154, 155, 335
The Maldives, 78, 79, 115
Malnutrition, 60, 142, 153–155, 209
Mangroves, 57, 116–119, 316, 364
Marine ecosystem, 65, 69, 116, 314, 316
Mass extinction, 354
Masterplanning, 233, 286, 288
Material footprint, 335
Mega Rice Project, 208
Methane (CH_4), 31, 42, 43, 46, 49, 66, 71, 83, 92, 94, 96, 98–100, 103, 106, 117, 120, 123–126, 128, 146, 156, 157, 167, 192–196, 209, 219, 270, 300, 304, 330, 340, 347
Microbeads, 331–333
Microplastics, 331–334
Midway Atoll, 7
Millennium Ecosystem Assessment (MEA), 21, 87, 137, 229
Monocrystalline PV, 253, 254

N
Nanoparticles, 331
National Grid, 174, 177, 183
Natural capital, 2, 4, 22, 23, 88, 137, 216
Natural gas, 28, 41–43, 167–169, 177, 196, 198, 200, 202, 240, 334, 336–338
Natural resources, 1, 3, 8, 10, 11, 15, 16, 20, 23, 27, 34, 91, 128, 132, 137, 142, 191, 198, 231, 260, 351, 353, 355
Natural ventilation, 235, 264, 293
Neonicitinoids, 73, 129
Net primary production, 122, 152
Net zero, 28, 189, 202, 203, 361, 363
Net zero carbon, 234, 262, 354
Nigeria, 27, 32, 79, 80, 160
Night cooling, 236, 239
Night ventilation, 236
Nile, 36
Nile Basin Initiative, 36
Nitrogen, 17, 65, 66, 92, 120, 127, 152, 195, 233, 251, 305
Nitrogen cycle, 18, 66, 165
Nitrogen synthesis, 65
Nitrous oxide (N_2O), 66, 71, 94, 98–100, 103, 104, 106, 192–195, 219, 270, 330, 336
Noise, 38, 43, 232, 233, 235, 236, 238, 241, 251, 264–268, 286, 289, 292, 295, 296, 308–314
Noise criteria, 267, 268
Noise impact, 265, 311
Noise level, 241, 265, 267, 295, 308–311
North Korea, 35, 72
North Pacific Gyre, 7, 8
Nuclear power, 8, 28, 35, 38, 141, 170–171, 174, 175, 202, 205, 206, 358, 362, 363

O
Obsolescence, 78, 347–349
Ocean thermal expansion, 111
Ocean thermocline, 121
Oil, 5, 16, 98, 167, 198, 217, 239, 302, 334, 353
Oil-fired boiler, 240
Oil tar sands, 169
Olf, 244
One Planet Living, 2, 17, 214, 218, 219
On-line shopping, 11, 346, 347
Open waste burning, 193
Organic PV, 254
Organization of the Petroleum Exporting Countries (OPEC), 27, 201
Overproduction, 198, 347–349, 354
Oxygen minimum zone (OMZ), 120, 121, 151
Ozone depleting gases, 94

Ozone depletion potential (ODP), 100, 101, 242
Ozone layer, 17, 92, 190, 242, 260

P
Packaging, 6, 8, 40, 78, 163, 329, 332, 340, 342, 343, 347
Palmer Drought Severity Index (PDSI), 134, 135, 209, 210
Palm oil, 16, 48, 49, 145, 207, 342
Paper, 2, 16, 17, 19, 37, 39, 47, 54, 66, 88, 123, 134, 139, 145, 149, 150, 152, 154, 155, 189, 191, 217, 228, 254, 263, 315, 330, 334, 338, 342–344, 347, 355
Paris Agreement, 25, 172, 182, 203
Particulate matter, 61, 97, 157, 233, 305–307
Passive design, 234–236, 239
Passivhaus, 236, 244
Peak gas, 200
Peak oil, 199, 200
Peat, 49, 98, 123, 128, 165, 166, 168, 195, 197, 206–209, 330
Peatlands, 48, 127, 128, 145, 168, 195, 206, 208, 209, 364
Pelagic zone, 118, 121, 151
Performance gap, 288, 289
Permafrost, 76, 107, 123–126, 146–148, 160, 196
Persistent Organic Pollutants (POPs), 7, 39, 40, 72, 80, 334, 335, 343
Pesticide, 6, 15, 39, 55, 64, 73, 129, 335, 342, 345
Pesticide poisoning, 15
Phenological shift, 121, 122
Phosphorous, 17, 66, 337
Photovoltaic (PV), 136, 174, 180, 183, 250, 251, 253–256, 262, 294, 364
Phytoplankton, 96, 120, 122
Pipeline leaks, 31, 32
Pipelines, 27–33, 146, 167, 185, 188, 189, 301, 314, 337
Planet equivalent, 11, 214, 216
Planetary boundaries, 17, 18, 65, 66, 354, 356
Plant Habitat Indicators, 228
Plastic manufacture, 8, 330, 332–334, 343
Plastics, 6–8, 77, 84, 253, 256, 269, 273, 330–334, 339, 343–345, 347
Plutonium, 170, 171
$PM_{2.5}$, 2
Polar bear, 131
Polar vortex, 53, 63, 64, 134
Pollinators, 73, 129, 149, 342, 345
Polychlorinated biphenyls (PCBs), 7, 39, 334, 348
Polycyclic aromatic hydrocarbons (PAHs), 7, 81, 193, 264, 305, 308, 335
Polyvinyl chloride (PVC), 39, 40
Polymer, 254, 319, 324, 333, 334, 343
Population displacement, 114, 138
Population growth, 1, 3, 11, 13, 15, 37, 74, 87, 99, 102, 103, 114, 125, 130, 136, 138, 142, 159, 190, 295, 353, 361, 362
Post-growth, 11, 352, 355–362, 364
Poverty, 2, 6, 9, 15, 20, 21, 23, 24, 65, 86, 89, 91, 130, 137, 138, 142, 155, 158–159, 210, 218, 356
Power station, 2, 8, 28, 38–42, 61, 98, 127, 157, 166, 170, 173–178, 186, 188, 189, 201, 205, 220, 233, 240, 259, 263, 304, 306, 363
Pulp, 145, 152, 223, 342, 343, 347
PV cell, 253
PV slates, 254, 255

R
Radiative forcing (RF), 95, 97, 98, 101, 102, 106, 164, 190
Radium, 47
Rainwater harvesting, 272–276
Rare earth elements, 11, 203, 348
Reed beds, 274, 275, 282
Refrigerant, 95, 96, 100, 101, 189, 190, 241–243, 248, 348
Refrigeration, 100, 190, 239, 241, 242, 250, 259, 340
Refuse collection, 85, 86
Regenerative agriculture, 342, 361
Renewable energy, 165, 171, 174–175, 181–185, 202, 215, 234, 238, 250, 254, 260, 293, 362
Renewable energy generation, 184, 215, 234, 238
Renewable Heat Incentive (RHI), 294
Representative Concentration Pathway (RCP), 102, 122, 136
Reserves, 8, 10, 16, 17, 29, 41, 88, 147, 164–167, 169–171, 196, 198, 199, 204, 205, 209, 215, 295, 319
Resources, 1, 3, 4, 8, 10, 11, 13, 15–17, 19–24, 26, 27, 34, 35, 37, 41, 43, 45–47, 59, 65, 70, 72–76, 78, 80, 88, 89, 91, 103, 122, 125–126, 128,

132, 137, 138, 142, 145, 146, 149, 150, 164–172, 191, 197–212, 214, 216–218, 221, 224, 231, 251, 260, 270, 290, 316, 319, 330, 338, 344, 345, 351–353, 355, 359, 361, 363
Retention pond, 282, 283
Retrofits, 4, 186, 231–296, 338, 364
Reversible heat pump, 248, 257
Rice, 2, 9, 70, 87, 88, 194, 195, 208, 340, 342
Rio Earth Summit, 19, 23, 344
Run-off from roads, 313, 333

S
Salt marsh, 117, 119
Seagrass, 57, 117–119
Sea ice, 64, 65, 108, 110, 112, 131, 134
Sea-level rise, 15, 101, 108–119, 126, 138, 139, 148
Secured by Design (SBD), 288
Seismic air guns, 312
Shale gas, 28, 30, 41–43, 45–48, 167, 169, 174, 200
Shared Socioeconomic Pathways, 101–105, 107
Ship strikes, 315
Sites of Special Scientific Interest (SSSIs), 295
Site waste management plan (SWMP), 290
Smog, 61–62, 195, 240
Social cost of carbon (SCC), 136, 137, 145
Social foundation, 356
Sodium-ion battery, 320, 321
Soil management, 55
Solar hot water, 250, 256, 259, 294
Solar photovoltaics, 180, 184, 250, 256, 363, 364
Solar radiation, 50, 94, 97, 195, 248, 256
Solar shading, 235, 248
Sound insulation, 265–267
Sound transmission, 266, 309
Species extinctions, 24, 69, 86, 127, 342
Spring Index, 229
Steel, 330, 337, 338, 343
Steel industry, 337
Storm surge, 51, 53, 56–58, 108, 109, 114, 117, 135, 138, 140, 153, 160, 278
Sulphur dioxide (SO_2), 10, 62, 98, 106, 120
Supply chains, 11, 74, 149, 222, 318, 320, 338–346, 362
Surface temperature, 101, 106–108, 110, 111, 123, 134, 147, 171, 198, 202, 240, 358, 360
Sustainability, 1, 3, 10, 11, 17, 22, 65, 74, 77, 150, 213, 214, 218, 221, 222, 224, 261, 279, 320, 338–346, 355, 357, 358, 361
Sustainable communities, 11, 286–288
Sustainable development, 1, 4, 19, 20, 23, 102, 103, 159
Sustainable Development Goals (SDGs), 21, 22, 24, 158, 159, 277, 356
Sustainable drainage, 276–285
Sustainable Drainage System (SuDS), 281, 282, 284
Swale, 282
Synchronous condenser, 364
Synthetic fibres, 332, 345

T
Tantalum, 34, 320
Technically recoverable resources, 43, 200, 201
Textiles, 228, 344–346
Thermal mass, 235, 236, 239
Thermohaline circulation, 67, 114
Thin film PV, 253–255
Three Gorges Dam, 9, 10, 174
Three-North Shelterbelt, 89
Tight gas, 41, 167, 169
Tight oil, 41, 43, 169
Timber, 21, 88, 137, 152, 330, 338, 342–344
Traffic noise, 267, 309, 310
Transmission losses, 176
Trophic cascade effect, 121
Tropical cyclone, 133, 134, 139
Troposphere, 50, 53, 92, 98, 195

U
Ukraine, 28–30, 38, 48, 202, 361
Unconventional gas, 167, 169, 200, 260
Under-road charging, 322
Union Carbide, 335
Unregulated emissions, 234, 236, 238, 289
Upstream emissions, 172, 173, 220
Uranium, 34, 170, 171, 205, 206
Urban regeneration, 11, 294–296
Urban sprawl, 16, 152, 295, 314

V
Vapour compression refrigeration, 241, 242
Variable air volume air conditioning, 241, 248
Variable refrigerant flow (VRF), 248, 363

Vector-borne disease, 15, 130
Ventilation rates, 245, 246
Vibration, 38, 251, 310
Video conferencing, 303
Visual environment, 248–249
Volatile organic compounds (VOCs), 156, 157, 195, 244, 260, 265, 306

W

Walkable neighbourhood, 287
Waste incineration, 40, 77, 84, 85, 260
Waste lagoon, 70
Waste pickers, 79, 80
Waste trafficking, 82
Waste water, 45, 46, 222, 269, 272–276
Waste water recycling, 269, 272–276
Water bound traffic, 300
Water footprints, 11, 212–225, 264, 345
Water Footprint Sustainability Assessment, 224
Water quality, 36, 47, 55, 68, 125, 132, 154, 221, 224, 264, 268, 269, 282, 313, 316
Water resources, 35, 37, 46, 59, 73, 75, 80, 88, 89, 125, 209, 218, 221, 260, 345
Water-source heat pump, 240
Water stress, 35–37, 47, 75, 125–126, 128, 224
Water Stress Index (WSI), 36, 76
Water use, 70, 132, 144, 221, 223, 225, 232, 260, 264, 268–273
Wildfires, 112, 113, 126, 127, 136, 138, 141, 145, 152, 153, 157, 164
Wind turbine, 170, 174, 180, 250, 251, 253, 362
Wood chip boiler, 240
Working from home, 302, 303
Worldwide Governance Indicators, 204
World Wildlife Fund (WWF), 2, 7, 17, 22, 219, 221, 225, 342, 345
WWF Footprint Calculator, 221

MIX
Papier aus verantwortungsvollen Quellen
Paper from responsible sources
FSC® C105338

If you have any concerns about our products,
you can contact us on
ProductSafety@springernature.com

In case Publisher is established outside the EU,
the EU authorized representative is:
Springer Nature Customer Service Center GmbH
Europaplatz 3, 69115 Heidelberg, Germany

Printed by Libri Plureos GmbH
in Hamburg, Germany